Nanophytomedicine

Nanophytomedicine is a branch of medicine that involves the application of nanomedicine-based systems to phytotherapy and phytopharmacology and the use of phytonanoparticles for biomedical applications. Nanophytomedicine covers recent advances in experimental and theoretical studies on various properties of nanoparticles derived from plant sources. This book assesses the recent advancements and applications of plant-based nanoparticles and also highlights emerging concepts of biomimetics.

The book contains 24 chapters encompassing various therapeutic applications of phytochemicals derived from plants, ferns, seaweeds, and so on, mediated through nanotechnology and its allied approaches. A fervent attempt has been made to compile every significant advancement in the field of phytonanomedicine so as to accelerate its momentum in the pharmaceutical sector.

Nanophytomedicine
An Emerging Platform for Drug Delivery

Editors
Parimelazhagan Thangaraj
Lucindo José Quintans Júnior
Nagamony Ponpandian

CRC Press
Taylor & Francis Group
Boca Raton London New York

CRC Press is an imprint of the
Taylor & Francis Group, an **informa** business

First edition published 2023
by CRC Press
6000 Broken Sound Parkway NW, Suite 300, Boca Raton, FL 33487-2742

and by CRC Press
4 Park Square, Milton Park, Abingdon, Oxon, OX14 4RN

CRC Press is an imprint of Taylor & Francis Group, LLC

ISBN: 978-1-032-13728-5 (hbk)
ISBN: 978-1-032-13972-2 (pbk)
ISBN: 978-1-003-23174-5 (ebk)

DOI: 10.1201/9781003231745

Typeset in Times
by KnowledgeWorks Global Ltd.

Contents

Preface

Plants are an emporium of medicinal wealth as they provide abundant resources for combating many present-day ailments. The bioactive compounds that contribute to every medicinal and aromatic plant are the key factors that contribute to their medicinal efficacy. Present-day innovations in the field of phytomedicine involve the incorporation of bioactive compounds into a nontoxic carrier that act as effective drug delivery agents. In-line with the advancements in this field, the conjugation of phytomedicine with nanotechnological approaches has emerged to be one of the most promising technologies for targeted herbal drug delivery. In this book, an attempt has been made to accrue the findings of profound researchers and scientists across the globe in such a manner that a quick glance would give ample information about how to initiate and move forward with nanopharmacology research. This book anthologizes recent trends of scientific information germane to nanotechnology, ethnopharmacology, phytochemistry, biophysics and biotechnology. A researcher would find it helpful for organizing and setting a flow for research that involves extraction and incorporation of herbal bioactives into nanotechnological models such as exemplifying a targeted drug delivery. The review process of the articles has been carried out by experts from various universities and research institutes. We hope that the present compilation will be useful for students, research scholars, academics, industrialists and people associated with nano-based herbal research. The authors would also like to thank the research team of Bioprospecting Laboratory, Bharathiar University, Coimbatore, India, for their contributions in the compilation of the articles in a well-designed manner. We would like to express our special appreciations to the publishers and their team for their sincere efforts for timely publishing of the book.

Dr. Parimelazhagan Thangaraj

Dr. Lucindo José Quintans Júnior

Dr. Nagamony Ponpandian

About the Authors

Dr. Parimelazhagan Thangaraj is Professor and Head of the Department of Botany at Bharathiar University, Coimbatore in India. He received his PhD in botany from Bharathiar University and underwent his postdoctoral fellowship in pharmacology at the university of Hawaii, USA. His field of interest encompasses focused research in bioprospecting of medicinal plants, phytomedicine, pharmacology and nutraceutical development. He and his team work on developing plant-based remedies for diseases in an effort to cure them with fewer side effects. He holds four patents and is the Founder, Director of the Intellectual Property Rights Cell of the Bharathiar University. He is a reviewer of many journals and holds editorship in peer-reviewed journals. Recently, he was conferred as a Fellow of the Royal Society of Biology, London and Fellow of the Linnean Society, London.

Dr. Lucindo José Quintans Júnior is Professor in the Department of Physiology at the Federal University of Sergipe in Brazil. He received his PharmD and PhD (in Pharmacology) degrees from the Federal University of Paraiba in Brazil and performed his postdoctoral fellowship in neuroscience and nanotechnology at the University of Iowa, USA. His research focuses on the study of pharmacological properties of essential oils and their secondary metabolites, especially terpenes, used for the control of chronic pain and inflammatory processes and the molecular or cellular mechanisms of natural products to produce their pharmacological action, especially involving animal models. Moreover, the investigation of novel inclusion complexes containing cyclodextrins with natural products is an additional area of scientific interest.

Dr. Nagamony Ponpandian is Professor and Head of the Department of Nanoscience and Technology, Bharathiar University, Coimbatore in India. He received his PhD in physics-material science from the University of Madras, Chennai. He completed postdoctoral fellowship in Germany and the United States in materials science. His current research interests are nanomaterials and thin films for energy devices, sensors and biomedical applications. His research group is working on biosensors, green energy production by water splitting, environmental remediation and monitoring by preparing and modifying the functional properties of various nanomaterials. He is a reviewer in many peer-reviewed journals and has been on the editorial board member of *Scientific Report*, Nature Publishing Group since 2014. He has been ranked as world's top 2% scientist, based on the Scopus database.

Contributors

Vinoth A.
Department of Botany
St. Xavier's College (Autonomous)
Palayamkottai, Tamil Nadu, India

Heidi Abrahamse
Laser Research Centre
Faculty of Health Sciences
University of Johannesburg
Doornfontein, Johannesburg, South Africa

Jackson Roberto Guedes da Silva Almeida
Universidade Federal do Vale do São Francisco (UNIVASF)
Centro, Petrolina, Pernambuco, Brazil

Izabel Almeida Alves
Departamento de Farmácia
Universidade Federal da Bahia
Salvador, Bahia, Brazil

Tatianny de Araújo Andrade
Department of Pharmacy
Federal University of Sergipe
São Cristóvão, Sergipe, Brazil

Eveline M. Anto
Agroprocessing and Technology Division
CSIR–National Institute for Interdisciplinary Science and Technology
Thiruvananthapuram, Kerala, India

Academy of Scientific and Innovative Research (AcSIR)
Ghaziabad, Uttar Pradesh, India

Adriano A.S. Araujo
Department of Pharmacy
Federal University of Sergipe (UFS)
São Cristóvão, Sergipe, Brazil

Ponnusamy Arulpriya
DRDO–BU Center for Life Sciences
Bharathiar University Campus
Coimbatore, Tamil Nadu, India

Kulanthaiyesu Arunkumar
Seaweed Group, Phycoscience Laboratory
Department of Plant Science
School of Biological Sciences
Central University of Kerala
Kasaragod, Kerala, India

S. Bhagya
Department of Zoology
Vyasa Arts & Science Women's College
Subramaniapuram, Tamil Nadu, India

Sara Elis Bianchi
Departamento de Farmácia
Universidade Federal do Rio Grande do Sul
Porto Alegre, Rio Grande do Sul, Brazil

Rahul Chandran
Laser Research Centre
Faculty of Health Sciences
University of Johannesburg
Doornfontein, Johannesburg, South Africa

Yasmim Maria B.G. de Carvalho
Department of Pharmacy
Federal University of Sergipe
São Cristóvão, Sergipe, Brazil

Noureddine El Aouad
Research Team on Biological Engineering, Agri-Food and Aquaculture
Polydisciplinary Faculty of Larache
University Abdelmalek Essaadi
Morocco

Luiza Abrahão Frank
Department of Pharmacy
Federal University of Rio Grande do Sul

UniRitter, Ciências da Saúde e Bem-Estar
Faculdade de Farmácia Laureate International Universities
Porto Alegre, Brazil

Ren-You Gan
Research Center for Plants and Human Health
Institute of Urban Agriculture
Chinese Academy of Agricultural Sciences (CAAS)

Key Laboratory of Coarse Cereal Processing (Ministry of Agriculture and Rural Affairs)
College of Pharmacy and Biological Engineering
Chengdu University
Chengdu, China

Gopalsamy Rajiv Gandhi
Postgraduate Programme in Health Sciences
Federal University of Sergipe
Aracaju, Sergipe, Brazil

Sathiyabama R. Gandhi
Multiuser Health Center Facility (CMulti-Saúde)
Laboratoty of Neuroscience and Pharmacological Assays (LANEF)
Federal University of Sergipe (UFS)
São Cristóvão, Sergipe, Brazil

Blassan P. George
Laser Research Centre
Faculty of Health Sciences
University of Johannesburg
Doornfontein, Johannesburg, South Africa

Ricardo Queiroz Gurgel
Postgraduate Programme in Health Sciences
Federal University of Sergipe
Aracaju, Sergipe, Brazil

Govindasamy Hariharan
Department of Biochemistry
SrimadAndavan Arts and Science College (Autonomous)
Tiruchirappalli, Tamil Nadu, India

Arif Hussain
School of Life Sciences
Manipal Academy of Higher Education
Dubai, United Arab Emirates

Asha Inbanathan
School of Life Sciences
Manipal Academy of Higher Education
Dubai, United Arab Emirates

Gayathri Jagadeesan
Bioprospecting Laboratory
Department of Botany
Bharathiar University
Coimbatore, Tamil Nadu, India

Nayagham Agnel Arul John
Department of Biochemistry
SrimadAndavan Arts and Science College (Autonomous)
Tiruchirappalli, Tamil Nadu, India

M. Johnson
Centre for Plant Biotechnology
Department of Botany
St. Xavier's College (Autonomous)
Palayamkottai, Tamil Nadu, India

Gnanasekaran Jothi
Department of Biochemistry
SrimadAndavan Arts and Science College (Autonomous)
Tiruchirappalli, Tamil Nadu, India

Lucindo José Quintans Júnior
Department of Physiology
Federal University of Sergipe (UFS)
São Cristóvão, Sergipe, Brazil

Krishna Kadirvelu
DRDO-BU Center for Life Sciences
Bharathiar University Campus
Coimbatore, Tamil Nadu, India

Veera Venkata Satyanarayana Reddy Karri
Department of Pharmaceutics
JSS Academy of Higher Education & Research
JSS College of Pharmacy
Ooty, Tamil Nadu, India

Arjunan Karuppaiah
PSG College of Pharmacy
Coimbatore, Tamil Nadu, India

Lijing Ke
Food Nutrition Sciences Centre
Zhejiang Gongshang University
Hangzhou, China

Kesavan Keerthana
Department of Biochemistry
SrimadAndavan Arts and Science College (Autonomous)
Tiruchirappalli, Tamil Nadu, India

Thangavel Krishnaveni
Department of Chemistry
Bharathiar University
Coimbatore, Tamil Nadu, India

Nandha Kumar
Department of Pharmaceutical Regulatory Affairs
JSS Academy of Higher Education & Research
JSS College of Pharmacy
Ooty, Tamil Nadu, India

Selvaraj Rajesh Kumar
Department of Chemical and Materials Engineering
Chang Gung University
Guishan District, Taoyuan City, Taiwan and
Department of Nanoscience and Technology
Bharathiar University
Coimbatore, Tamil Nadu, India

Gowthamarajan Kuppusamy
Department of Pharmaceutics
JSS Academy of Higher Education & Research
JSS College of Pharmacy
Ooty, Tamil Nadu, India

Krishnasamy Lakshmi
DRDO-BU Center for Life Sciences
Bharathiar University Campus
Coimbatore, Tamil Nadu, India

Hua-Bin Li
Guangdong Provincial Key Laboratory of Food Nutrition and Health
Department of Nutrition
School of Public Health
Sun Yat-Sen University
Guangzhou, China

Vítor Prates Lorenzo
Instituto Federal de Educação
Ciência e Tecnologia do SertãoPernambucano (IF-Sertão)
João de Deus, Petrolina, Pernambuco, Brazil

Ashwini Lydia Manoharan
Bioprospecting Laboratory
Department of Botany
Bharathiar University
Coimbatore, Tamil Nadu, India

Hicham Mechqoq
Laboratory of Biotechnologies and Valorization of Natural Resource
Faculty of Sciences
University Ibn-Zohr
Agadir, Morocco

Irwin R.A. Menezes
Graduate Program of Biological Chemistry
Regional University of Cariri (URCA)
Crato, Ceará, Brazil

Paula dos Passos Menezes
Department of Pharmacy
Federal University of Sergipe
São Cristóvão, Sergipe, Brazil

Dipshika Motwani
Bioprospecting Laboratory
Department of Botany
Bharathiar University
Coimbatore, Tamil Nadu, India

Lavanya Mude
Department of Pharmaceutics
JSS Academy of Higher Education & Research
JSS College of Pharmacy
Ooty, Tamil Nadu, India

Nithyananth Muniswamy
PSG College of Pharmacy
Coimbatore, Tamil Nadu, India

Kasipandi Muniyandi
Bioprospecting Laboratory
Department of Botany
Bharathiar University
Coimbatore, Tamil Nadu, India

Arun Muthukrishnan
Department of Biotechnology
Bharathiar University
Coimbatore, Tamil Nadu, India

S. Muthukrishnan
Department of Botany
University of Kerala Kariavattom
Thiruvananthapuram, Kerala, India

Anaga Nair
Agroprocessing and Technology Division
CSIR–National Institute for Interdisciplinary Science and Technology
Thiruvananthapuram, Kerala, India

Academy of Scientific and Innovative Research (AcSIR)
Ghaziabad, Uttar Pradesh, India

Narendra Narain
Laboratory of Flavor and Chromatographic Analysis
Federal University of Sergipe (UFS)
São Cristóvão, Sergipe, Brazil

José Adão C. Nascimento Júnior
Department of Pharmacy
Federal University of Sergipe
São Cristóvão, Sergipe, Brazil

Gayathri Nataraj
Bioprospecting Laboratory
Department of Botany
Bharathiar University
Coimbatore, Tamil Nadu, India

D. Navadeepthy
Department of Nanoscience and Technology
Bharathiar University
Coimbatore, Tamil Nadu, India

Ana Paula de Oliveira
Instituto Federal de Educação
Ciência e Tecnologia do Sertão Pernambucano (IF-Sertão)
Santa Maria da Boa Vista, Pernambuco, Brasil

Raimundo Gonçalves de Oliveira Junior
Soberana – Faculdade de Saúde de Petrolina
Gercino Coelho, Petrolina, Pernambuco, Brazil

Gurusaravanan Packiaraj
Department of Botany
Bharathiar University
Coimbatore, Tamil Nadu, India

Fabiolla R. Santos Passos
Multiuser Health Center Facility (CMulti-Saúde)
Laboratoty of Neuroscience and Pharmacological Assays (LANEF)
Federal University of Sergipe (UFS)
São Cristóvão, Sergipe, Brazil

Benedict Mathews Paul
Bioprospecting Laboratory
Department of Botany
Bharathiar University
Coimbatore, Tamil Nadu, India

Nagamony Ponpandian
Department of Nanoscience and Technology
Bharathiar University
Coimbatore, Tamil Nadu, India

Jayamurthy Purushothaman
Agroprocessing and Technology Division
CSIR–National Institute for Interdisciplinary Science and Technology
Thiruvananthapuram, Kerala, India

Academy of Scientific and Innovative Research (AcSIR)
Ghaziabad, Uttar Pradesh, India

Jullyana S.S. Quintans
Multiuser Health Center Facility (CMulti-Saúde)
Laboratoty of Neuroscience and Pharmacological Assays (LANEF)
Federal University of Sergipe (UFS)
São Cristóvão, Sergipe, Brazil

Arun R.
Department of Pharmaceutics, JSS Academy of Higher Education & Research
JSS College of Pharmacy
Ooty, Nilgiris, Tamil Nadu, India

Murugan Rajan
Laboratory of Flavor and Chromatographic Analysis
Federal University of Sergipe (UFS)
São Cristóvão, Sergipe, Brazil

Gomathi Rajkumar
Department of Botany
Sri Sarada College for Women (Autonomous)
Salem, Tamil Nadu, India

Muthiah Ramanathan
PSG College of Pharmacy
Coimbatore, Tamil Nadu, India

Andreza G.B. Ramos
Post Graduate Program in Biological Chemistry
Regional University of Cariri (URCA)
Crato, Ceará, Brazil

Priyanka Rathod
Seaweed Group, Phycoscience Laboratory
Department of Plant Science
School of Biological Sciences
Central University of Kerala
Kasaragod, Kerala, India

Marilia Matos Rezende
Multiuser Health Center Facility (CMulti-Saúde)
Laboratoty of Neuroscience and Pharmacological Assays (LANEF)
Federal University of Sergipe (UFS)
São Cristóvão, Sergipe, Brazil

Sandhya Saminathan
Bioprospecting Laboratory
Department of Botany
Bharathiar University
Coimbatore, Tamil Nadu, India

Veintramuthu Sankar
PSG College of Pharmacy
Coimbatore, Tamil Nadu, India

Anamaria Mendonça Santos
Department of Pharmacy
Federal University of Sergipe
São Cristóvão, Sergipe, Brazil

Vinoth Sathasivam
Department of Biotechnology
Sona College of Arts and Science
Salem, Tamil Nadu, India

Saikumar Sathyanarayanan
Department of Botany
PSG College of Arts and Science
Coimbatore, Tamil Nadu, India

Mairim Russo Serafini
Departamento de Farmácia
Universidade Federal de Sergipe
São Cristóvão, Sergipe, Brazil

Saravanan Shanmugam
Department of Pharmacy
Federal University of Sergipe
São Cristóvão, Sergipe, Brazil

B. Shivananthini
Centre for Plant Biotechnology
Department of Botany
St. Xavier's College (Autonomous)
Palayamkottai, Tamil Nadu, India

Bruno A.F. Silva
Multiuser Health Center Facility (CMulti-Saúde)
Laboratoty of Neuroscience and Pharmacological Assays (LANEF)
Federal University of Sergipe (UFS)
São Cristóvão, Sergipe, Brazil

Juliane Cabral Silva
Universidade Estadual de Ciências da Saúde de Alagoas (UNCISAL)
Trapiche da Barra, Maceió, Alagoas, Brazil

Mariana Gama e Silva
Universidade Federal do Vale do São Francisco (UNIVASF)
Centro, Petrolina, Pernambuco, Brazil

E.A. Sirll
Department of Botany
University of Kerala, Kariavattom
Thiruvananthapuram, Kerala, India

R. Sreelekshmi
Department of Botany
University of Kerala, Kariavattom
Thiruvananthapuram, Kerala, India

Gurunagarajan Sridharan
Department of Biochemistry
Srimad Andavan Arts and Science College (Autonomous)
Tiruchirappalli, Tamil Nadu, India

Madhumitha Kedhari Sundaram
School of Life Sciences
Manipal Academy of Higher Education
Dubai, United Arab Emirates

Parimelazhagan Thangaraj
Bioprospecting Laboratory
Department of Botany
Bharathiar University
Coimbatore, Tamil Nadu, India

Muhammad Thanzim
Postgraduate, Department of Pharmaceutical Regulatory Affairs
JSS Academy of Higher Education & Research
JSS College of Pharmacy
Ooty, Tamil Nadu, India

Sakthi S. Vasantha
Centre for Plant Biotechnology
Department of Botany
St. Xavier's College (Autonomous)
Palayamkottai, Tamil Nadu, India

Bhaarathy Velmurugan
Singapore University of Social Sciences
Yung Kuang Road
Singapore

Sahayaraj Jasmin Vinitha
Department of Biochemistry
Srimad Andavan Arts and Science College (Autonomous)
Tiruchirappalli, Tamil Nadu, India

1 Drug Discovery from Natural Products

An Approach Using Recent Patents

S.E. Bianchi, L.A. Frank, I.A. Alves and M.R. Serafini

CONTENTS

1.1 INTRODUCTION

Natural products are used worldwide to prevent and treat different diseases. These products, derived from various plant species, have been the most explored source for developing new medicines (Ahuja et al. 2020). Currently, the most researched molecules are those with anti-cancer and anti-infectious activity. Some drugs are known worldwide: morphine (analgesic activity), erythromycin (antibiotic activity), cyclosporine (immunosuppressive activity) and artemisinin (antimalarial activity) are examples of natural substances with consolidated uses in medicine.

World Health Organization (WHO) estimates that each year 14 million new malignant neoplasms are recorded worldwide, ranking second among the deadliest diseases (Siegel, Miller and Jemal 2020). Both science and the pharmaceutical industry are joining efforts to propose the application of new biomolecules that promise, at the very least, to improve the symptoms of the disease when not cured. Research carried out on the Web of Science (WoS) using keywords "natural product" and "cancer" in the field title between the years 1980 and 2020 showed tremendous and growing interest in applying natural products in the treatment of the most diverse types of cancer, as shown in Figure 1.1.

Innovation in technology represents a fundamental element for developing strategies for growth, differentiation and problem-solving. When creating something new, developing a manufacturing process or improving existing techniques and products, protection occurs through patents (Araújo et al. 2020). Patent is a way to assess the degree of innovation present in the pharmaceutical industry. This chapter presents a review of recent patents on drug discovery from natural products, using the International Classification of Patents (IPC) and a database of patents.

DOI: 10.1201/9781003231745-1

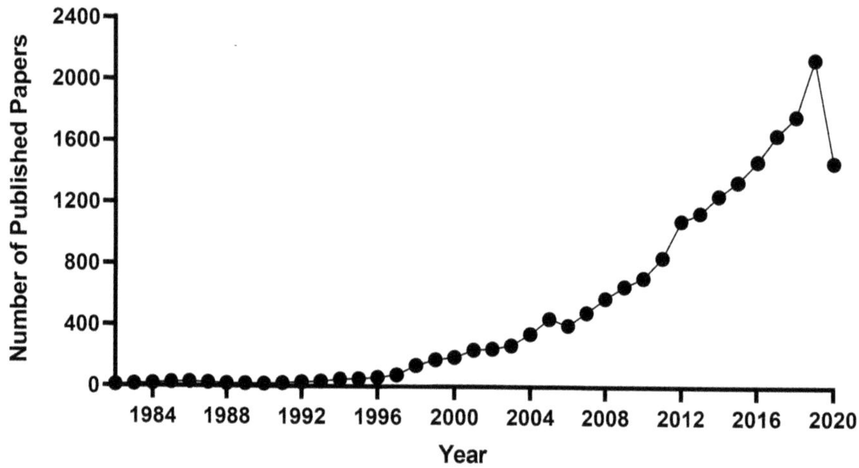

FIGURE 1.1 The number of published papers published per year from 1980 to 2020 including keywords "natural product" and "cancer" in Web of Science.

1.2 SEARCHING STRATEGIES USING INTERNATIONAL PATENT CLASSIFICATION (IPC)

A general approach of patents in new products from natural origin was analysed using the IPC as a source of information. This classification allows indexing patent documents in a specific IPC code and become indispensable since it is used as a filter to find patents in the technological area by authorities in many countries around the world. To establish the novelty and evaluate the inventive step of technical disclosures in patent applications, this system indicates at least one IPC code according to each patent assigned scope (World Intellectual Property Organization 2020).

The complete classification comprises the combined symbols that represent the section, class, subclass and main group or subgroup. An example is shown in Figure 1.2, where A61K 36/00 symbols indicate medicinal preparation of undetermined constitution containing material from algae, lichens, fungi or plants, or derivatives thereof, e.g., traditional herbal medicines and the A61K 36/10 symbols contain a specific combination with Bryophyta (mosses).

To get an approach of patents focusing on drug discovery from natural products, a search was initially performed based on IPC code A61K 36/00, using WoS and European Patent Office (Espacenet) databases. Patents from 2015 to 2020 were mapped and analysed by VOS viewer Software® 1.6.15. In the initial search strategy, 2986 patents were identified. VOS viewer enables each keyword to reflect a specific theme from the text. Thus, when a keyword is related to a particular theme and it has high frequency with the associated keyword, the more important the theme will be.

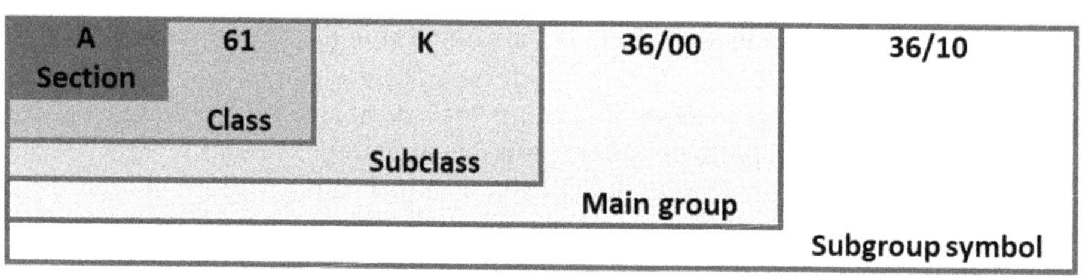

FIGURE 1.2 Informative scheme on the International Patent Classification (IPC) code for a medicinal preparation of undetermined constitution containing material from algae, lichens, fungi or plants, or derivatives thereof, e.g., traditional herbal medicines (A61K 36/00) and containing a specific combination with Bryophyta (mosses) (A61K 36/10).

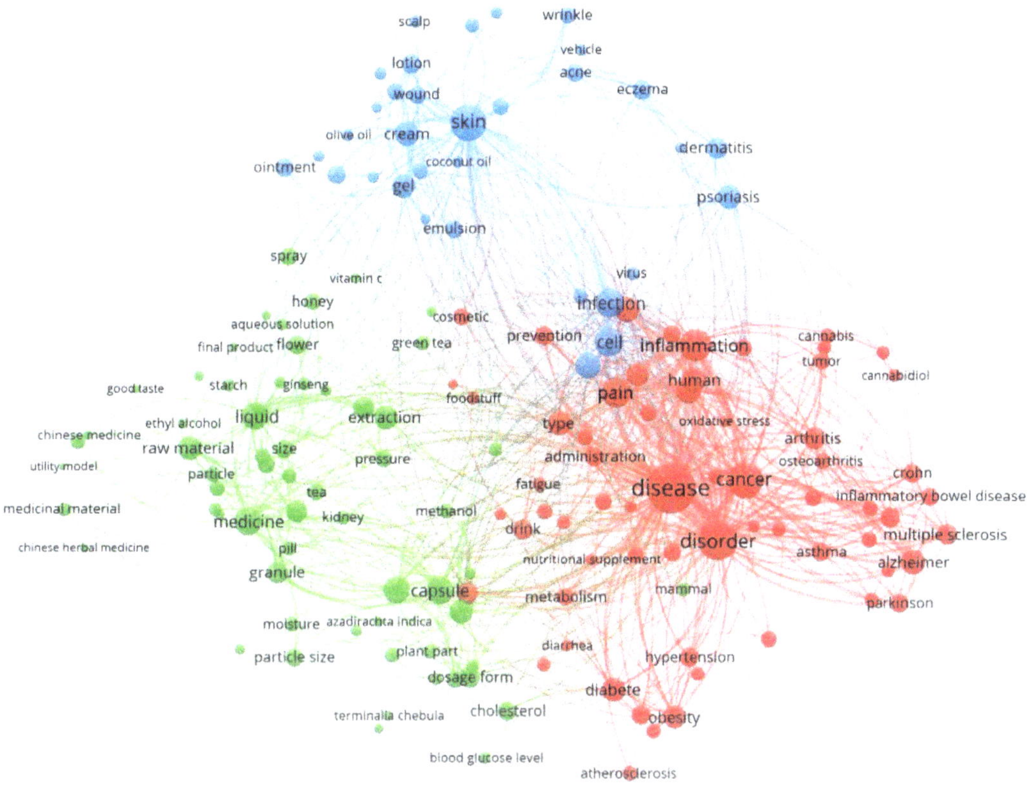

FIGURE 1.3 Map of co-occurrence of keywords related to patents focusing on drug discovery from natural products, based on A61K 36/00 International Patent Classification (IPC) code, through Web of Science and Espacenet databases (between 2015 and 2020).

Figure 1.3 illustrates the most frequent keywords and most associated keyword pairs. The interpretation of the results showed that three clusters were formed from 160 keywords identified algorithmically by the software. Based on the high-frequency keywords in patents and their relevant score to other keywords, cluster 1 (blue points) consists of 33 keywords focusing on pharmaceutical formulations and their application on the skin (to treat eczema, dermatitis, psoriasis, etc.). Cluster 2 (green points), which consists of 63 keywords, shows several patents focused on plant extractions and development of pharmaceutical forms, such as capsules, tablets and granules. Finally, cluster 3 (red points), which consists of 64 keywords, indicates the patenting of products to treat various diseases. This last cluster shows several patents that are especially focused on drug discovery from natural products for cancer treatment.

1.3 NATURAL PRODUCTS AND DISEASE TREATMENT

Products from natural sources have been used in the prophylaxis and treatment of various pathologies. Plants have a wide variety of secondary metabolites that are responsible for several effects of therapeutic interest, creating the potential for the development of new drugs and medications (Katiyar et al. 2012). Despite the growing number of published scientific articles, there is a negative correlation between the number of published scientific articles and the number of innovative products derived from natural products available on the market. The main factor that contributes considerably to the growth in question may be the evolution of scientific studies and patents, particularly chemical and pharmacological studies that prove the effectiveness of natural products (Yuan et al. 2016).

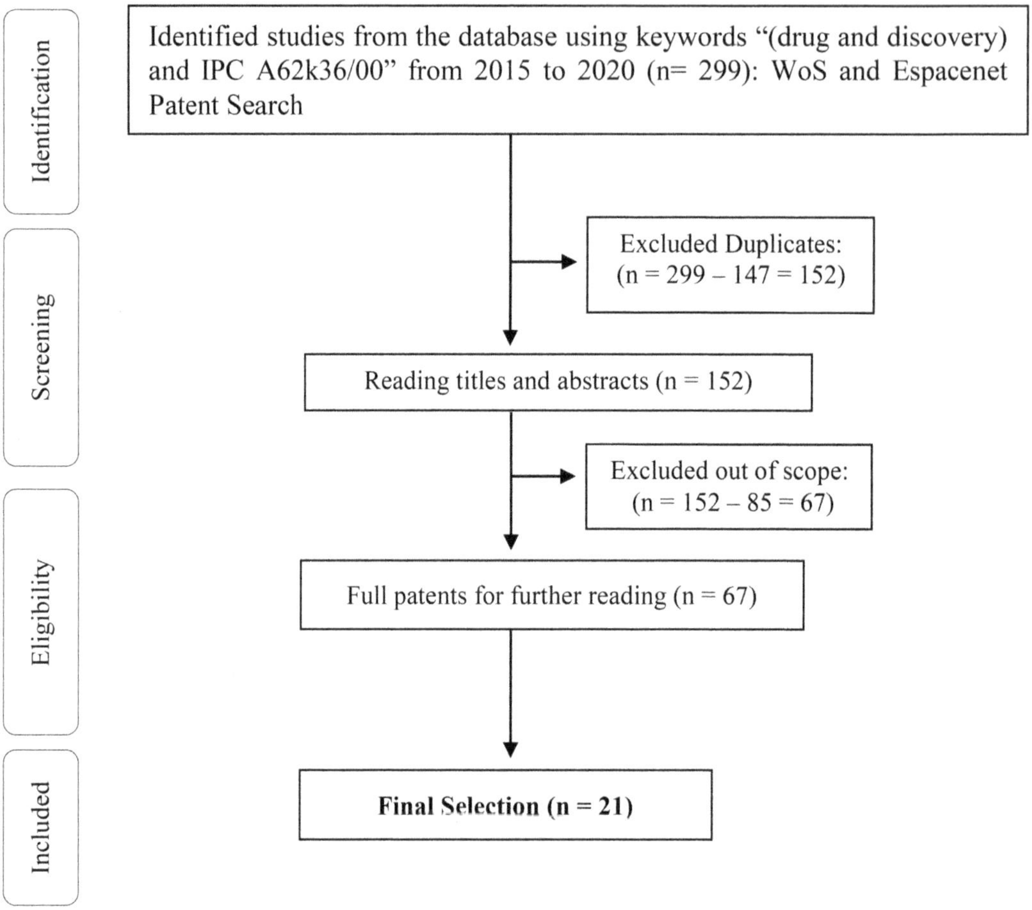

FIGURE 1.4 Flowchart of literature searching and screening.

1.4 RESEARCH METHOD

To acquire more information on technical aspects, after the initial search for IPC code patents (A61K 36/00), patent documents were selected containing a combination of "drug discovery" from "natural products" in the access of title, abstract or full text, that were related to pharmaceutical compositions focused on potential disease treatments. A total of 299 patents were found from the WoS and Espacenet databases (Figure 1.4). Some patents were excluded that were not published in English, duplicated or unavailable. Of these, 21 patents were read, claiming drug discovery from natural products in drug delivery systems. Our search will be presented within the disease sections that the innovation is intended to treat.

1.5 CANCER TREATMENT

Cancer is a global disease that presents high morbidity and mortality around the world. According to the WHO's GCO Program (Global Cancer Observatory), in 2030, there will be more than 20 million people with cancer and 8 million deaths from this disease due to population growth and aging, including effects from lifestyle, unhealthy diet, physical inactivity and abuse of alcohol and tobacco (Siegel, Miller and Jemal 2020). The discovery and development of new drugs from natural products have been the focus of many research works that resulted in natural bioactive products with potent anticancer activity (Yuan et al. 2016). Table 1.1 provides details of new patents including the chemical constituents, formulations and applications in cancer treatment.

TABLE 1.1
Details of New Patents Including the Chemical Constituents, Formulations and Applications in Cancer Treatment

Application Number (Ref)	Country	Year	Disease	Formulation	Application	Compounds	Tests	Other Information
BR102018070770 (A2) (Ribeiro, De Souza and Martins 2020)	Brazil	2020	Cancer	Organic system constituted of self-organized structures of diphenylalanine peptide	Topical use or on electrodes	Coumarin-6 (additive fluorophore) and derivative of peryl alcohol	Formulation studies	The system has the ability, by absorbing UV/Vis light, to convert fundamental oxygen into singlet oxygen
WO2020018759 (A1) (Ciftci 2020)	WO	2020	Inflammatory diseases (cancer, diabetes, cardiovascular and neurodegenerative diseases)	Nanoparticles (biodegradable nanoporous starch aerogels)	Oral	Curcumin	*In vitro* studies	The nanoporous starch aerogels increase curcumin water solubility and bioaccessibility
US10640480 (B2); US2019031635 (A1) (Kato and Hosoda 2019)	United States	2019	Metabolic syndrome, diabetes mellitus, hypertension, gout, cancer, collagen disease, periodontal disease, leukaemia	Medicinal product or a cosmetic	Oral or skin preparation	Melinjo fruit	Extract production	Composition containing at least an emulsifier, a cycrodextrin and an acid into water and/or an organic solvent
SI1991244 (T1) (Giori and Franceschi 2019)	Slovenia	2019	Cancer	Phospholipids complexes	Oral	Curcumin or extracts	*In vitro* and *in vivo* cancer studies	New complexes of curcumin phospholipids or extracts have improved their bioavailability
TR201808733 (T4) (Bror, Saideh and Kefei 2018)	Turkey	2018	Cancer	Nanoparticles	Oral or parenteral routes	Sterol and derived from *Quillaja saponaria* Molina selected from quillaja acid and quillaja saponin	*In vitro* cancer studies	Nanoparticles can be used as adjuvant, especially in vaccines and as agents for treatment of cancer
WO2020058785 (A1) (Kalkundri 2020)	WO	2020	Oral sub-mucosal fibrosis	Amphiphilic complex	Oral cavity	Extracts of neem, fennel, clove and curcuminoids	*In vitro* studies	
KR20190090910 (A) (Choi and Chung 2019)	Republic of Korea	2019	Benign prostatic hyperplasia	Nanoparticles	Oral	Bee pollen	*In vivo* cancer studies	The composition can be administered for long term since the composition does not exhibit toxicity

Over the past five years, interest in naturally occurring compounds remains high, with research being carried out worldwide (Katiyar et al. 2012). Special emphasis has been given to systems involving the improvement in the release of these drugs. The patent application invented by Ciftci (2020) claims a formulation containing curcumin nanoparticles using biodegradable nanoporous starch aerogels for cancer treatment. The authors described the development of low-crystallinity curcumin nanoparticles using controlled supercritical carbon dioxide impregnation into biodegradable nanoporous starch aerogels.

The most common route of application was oral administration as it is well accepted by patient in relation to parenteral route. Giori and Franceschi (2019) invented an oral formulation containing new complexes of curcumin phospholipids or extracts that increased their bioavailability in cancer studies. Alternatively, Bror, Saideh and Kefei (2018) created an oral or parenteral nanoparticle formulation containing sterol derived from *Quillaja saponaria* Molina selected from quillaja acid and quillaja saponin.

The advantages of topical therapy, an alternate to oral administration, are arising as a primary or complementary treatment option for cancer, allowing direct drug delivery to the affected site (Verma et al. 2013). The patent application invented by Ribeiro, De Souza and Martins (2020) is a topical formulation containing coumarin-6 and a derivative of peryl alcohol in an organic system constituted of self-organized structures of diphenylalanine peptide for cancer treatment. The system has the ability, by absorbing UV/VIS light, including the region of the therapeutic window, to convert fundamental oxygen into singlet oxygen, which is the compound that promotes changes in cancer cells. Kato and Hosoda (2020) included oral and topical applications of melinjo fruit composition containing at least an emulsifier, a cyclodextrin and an acid into water and/or an organic solvent.

In another use of curcumin, Kalkundri (2020) reported his invention of a mixture of extracts of neem, fennel, clove and curcuminoids incorporated in phospholipids for the management of oral submucosal fibrosis, a chronic debilitating disease of the oral cavity, which often precedes cancer.

1.6 INFLAMMATORY DISEASES

The inflammatory process is a normal response of the body to the infection and offers protection, through the activation of the innate and adaptive immune system. When there is uncontrolled inflammation, treatment with drugs that reduce inflammation in the body is necessary (Brahmachari 2019). Table 1.2 details five patent applications using natural compounds that were identified in this chapter for treating inflammatory diseases. Of these, two patents – Ciftci (2020) and Kleidon (2019) – also included applications for cancer treatment. Kleidon (2019) reported a micro- and nanoencapsulated cannabinoid effective in inflammatory disease (neurodegenerative and metabolic diseases). The technology claims oral, intranasal or respiratory application for local or systemic effect reduces oxidative stress in cells.

New Zealand's recent patent describes the topical, oral or parenteral application of hydroxytyrosol or a derivative thereof, and a compound having a galloyl group, such as, catechin. In this patent, Ueno, Kominami and Kasajima (2020) demonstrated that the residence time of pharmaceutical formulation (hydroxytyrosol in combination with catechin) was prolonged, which is beneficial for antiageing activity and as an anti-inflammatory agent.

Another patent application, invented by Bharate et al. (2020) from Canada, used an active constituent of *Crocus sativus* extract or fraction, named Crocin-1, for treatment of chronic inflammatory (Alzheimer's disease, type II diabetes or rheumatoid arthritis). The oral administration is considered the preferred and most convenient means of administering drugs to the patient. The formulations described in the patent are for the controlled or prolonged release of bioactive compounds, containing *Crocus sativus* extract or its enriched fraction.

Antony (2019) describes an encapsulated curcumin extract containing saffron essential oil for treatment of inflammation, skin wounds and tumours. The invention has found that when curcumin

TABLE 1.2
Details of New Patents Including the Chemical Constituents, Formulations and Applications in Inflammatory Diseases

Number Application (Ref)	Country	Year	Disease	Formulation	Application	Compounds	Tests	Other Information
NZ716858 (A) (Ueno, Kominami and Kasajima 2020)	New Zealand	2020	Inflammatory and other diseases and for antiageing	Pharmaceutical formulation: tablet, capsule, powder, granule, solution, suspension or syrup	Topical, oral, or parenteral injection	Hydroxytyrosol or a derivative thereof and a compound having a galloyl group (as catechin compound)	Absorption of hydroxytyrosol *in vivo*, oligomeric proanthocyanidins determination in grape seed extract	*In vivo* residence time of hydroxytyrosol in combination with a compound having a galloyl group was prolonged
CA3077335 (A1) (Bharate et al. 2020)	Canada	2019	Chronic inflammatory disease	Pharmaceutical formulation comprising granules of *Crocus sativus* hydroalcoholic extract or fraction; and biodegradable and/or non-biodegradable polymers	Oral	Crocin-1, an active constituent of *Crocus sativus* extract or fraction	*In vitro* dissolution studies; *in vitro* cancer studies and *in vivo* Alzheimer studies	Formulation is useful in treating chronic inflammatory (Alzheimer's disease, gout, type II diabetes, rheumatoid arthritis)
PT1890546 (T) (Antony 2019)	Portugal	2019	Treatment for inflammation, skin wounds and tumours	Encapsulated curcumin extract	Oral	Curcumin with saffron essential oil	Clinical studies	In saffron essential oil, Ar-turmerol is the main constituent
MX2018015583 (A) (Kleidon 2019)	Mexico	2019	Cancer, aging, neurodegenerative and metabolic diseases	Micro and nano-encapsulated cannabinoid	Oral, intranasal or respiratory route for local or systemic effect	Cannabinoid	Ratiometric fluorescence assays and formulation studies	Encapsulated cannabinoid is effective to reduce oxidative stress in cells
WO2020096127 (A1) (Jeong, Jeong and Lim 2020)	WO	2020	Arthralgia or myalgia	Hydrolysed sponge, acetyl glucosamine and dimethyl sulfone	Topical	Includes natural plant extracts and oils as main components	*In vivo* studies	The hydrolysed sponge has the form of a microneedle. Pain relief appears effectively as it can be delivered into the dermal layer

essential oil is added to curcumin, the bioavailability of curcumin increases significantly. The formulation was administered to nine volunteers and blood samples were collected at hourly intervals or at half hour intervals up to eight hours. Maximum absorption was observed three hours after ingestion and the formulation resulted in curcumin levels that were 5–16 times higher compared to curcumin alone.

Jeong, Jeong and Lim (2020) demonstrated their invention of a pharmaceutical and/or cosmetic composition comprising hydrolysed sponge, acetyl glucosamine and dimethyl sulfone to prevent or treat symptoms of chronic inflammatory diseases. The microneedles of the hydrolysed sponge powder allow treatment and pain relief with excellent efficiency, as they are administered in the dermal layer. Hydrolysed sponges can be obtained from sponges of the species *Spongilla lacustris, Spongilla fragilis* Leidy and *Ephydatia fluviatilis*. Through animal testing, patent applicants have demonstrated that this composition provides relief from muscle and joint pain effectively from lower back, shoulders and knees.

1.7 SKIN TREATMENT

The skin is considered an interesting route for the administration of drugs. Through skin administration, many products are offered for beautification every year. In this section, we discuss innovations containing medicines for treating illnesses and products with the cosmetic application (Table 1.3).

Sanders (2020) described an eco-friendly antimicrobial formulation to act against bacterial biofilms in skin, wound and oral care devices, reducing unwanted odours. The formulation is composed of essential oils, such as cinnamon, lemongrass, etc., stabilized into a polymerizable substance. In another invention, Dewar (2020) created a composition for topical application that has, in its base, dispersed elemental silver that acts as a colloidal suspension and plant extracts from at least six plants. The invention might apply to the skin and mucous membranes of human and non-human bodies in the treatment of dermatitis, psoriasis, wounds, blisters and inflammation, etc.

Serebryanskaya (2019) reported a transdermal preparation to treat acne, post-acne and cellulite compounded from a carrier base, which might be made from a plastic biopolymer (such as gelatine, pectin or agar-agar), water, biologically active natural substances such as extracts of animal and plant material or mineral components and an adhesive substance of natural origin. The invention aims to increase the adhesive capacity, ease of use and effectiveness.

Finally, Zhao (2019) developed a skin whitening and sunscreen composition with nanomaterial and Chinese herbal extracts. The nanomaterial is a compound of nanocrystalline cellulose or a mixture of nanocrystalline cellulose-fullerene particles, is a highly effective antioxidant against the appearance of aging in humans and has an anti-UV effect, contributing to the elimination of the harmful effects of UV rays on the skin after a long period of exposure. However, that nanomaterial cannot improve human health.

1.8 APPLICATIONS OF NATURAL PRODUCTS IN OTHER DISEASES

In this patent research, pharmaceutical innovations indicated diseases that are difficult to treat, such as neurodegenerative diseases, coagulation pathology and fungal and viral infections. In this context, this section will present and discuss new treatments and innovations in drug administration or improvements to the management of these diseases (Table 1.4).

Ciesla and Bao (2020) developed nanostructures derived from cell membranes comprising a target membrane protein and one or more magnetic nanoparticles. In these nanostructures, the material derived from the cell membrane encapsulates one or more magnetic nanoparticles.

Regarding infectious diseases, conventional antifungal therapies used for local and systemic candidiasis have shown great disadvantages as well as limitations of use, such as adverse reactions and development of resistance. Thus, Khandelwal and Adepu (2020) invented pharmaceutical compositions and delivery systems for the prevention and treatment of candidiasis. The invention shows

TABLE 1.3
Details of New Patents Including the Chemical Constituents, Formulations and Applications in Skin Diseases

Number Application (Ref)	Country	Year	Disease	Formulation	Application	Compounds	Tests	Other Information
EP3585351 (A1) (Sanders 2020)	United States	2020	Oral and wound care pathogens (bacteria)	Nano and microcrystalline particles	Skin care, wound care or oral care device	Essential oil (cinnamon, lemongrass and peppermint) stabilized into a polymerizable substance	*In vitro studies*	Extended release for up to 30 days
EP3600348 (A1) (Dewar 2020)	United Kingdom	2020	Wounds and other skin conditions	Aqueous suspension formulation	Topical application to treat wounds and other skin conditions	Base, elemental silver and six plants extracts	Not specified	
WO2019088883 (A1) (Serebryanskaya 2019)	Russia	2019	Acne, post-acne and cellulite	masks, creams and patches	Treatment of acne, post-acne and cellulite	Carrier base (plastic biopolymer, base-gelatin or polyethylene oxide) extracts of animal, plant raw materials, mineral components	*In vivo*	Transdermal administration
FR3070862 (A1) (Zhao 2019)	France	2019	Skin whitening and sun protection	Cream	Skin whitening and sun protection	Nanomaterial and a Chinese herbal extract	Not specified	

TABLE 1.4
Details of New Patents Including the Chemical Constituents, Formulations and Applications in Other Diseases

Number Application (Ref)	Country	Year	Disease	Formulation	Application	Compounds	Tests	Other Information
HR2019P002328T (Dandl, Haala and Haack 2020)	Croatia	2020	Cough and cold	Powder, granule, or effervescent tablets	Therapy, management, prevention, or diagnosis of a disease, health condition, or symptom	Water-soluble carrier and an oily liquid	Pharmaceutical form test	
US10556932 (B2); US2019194265 (A1) (Herzog and Daniell 2019)	United States	2019	Hemophilia	Tablets, capsules, granules and powders	Prophylactic protocol; protein replacement therapy	Coagulation factor expressed in chloroplasts, bioencapsulated in plant cells	*In vitro* studies	Oral administration
US10596147 (B1) (Nguyen 2020)	United States	2020	Nausea, vomiting and anorexia associated with chemotherapy; muscle spasticity, pain, anorexia associated with AIDS, epilepsy, glaucoma, Crohn's disease, inflammatory bowel disease, multiple sclerosis, ALS, muscular dystrophy, bronchial asthma, PTSD, mood disorders and migraine	Pressurized metered dose inhaler	Pulmonary	Cannabinoid composition, co-solvent, terpene composition and propellant is disclosed	*In vitro* and *in vivo* studies	

(Continued)

TABLE 1.4 *(Continued)*

Number Application (Ref)	Country	Year	Disease	Formulation	Application	Compounds	Tests	Other Information
WO2020058836 (A1) (Khandelwal and Adepu 2020)	WO	2020	Candidiasis	Essential oils in microcapsules	Oral or vaginal cavity	Preparation of monolithic mononuclear or polynuclear core-shell microcapsules encapsulated with three different oils or active ingredients and the embedding the microcapsules in microbial cellulosic matrix	*In vitro* and *in vivo* studies	The compositions and cellulosic matrix, modified as chewing gum for oral candidiasis, do not require water to swallow and hence are advantageous for patients having difficulty in swallowing
US2020096503 (A1) (Ciesla and Bao 2020)	United States	2020	Alzheimer's disease and other neurodegenerative ailments	The amphiphilic complex formulation comprising a blend of multiple extracts: neem, fennel, clove and curcuminoids		Natural compounds have proven to be a rich source of new biologically active compounds; Magnetic nanoparticles	*In vitro* studies	

delivery systems comprising microcapsules and cellulosic matrix soaked with oil microcapsules, leading to high efficacy in prophylaxis and treatment of candidiasis.

An innovation in cough and cold treatment was described by Dandl, Haala and Haack (2020). The inventors developed a pharmaceutical composition formed by a water-soluble carrier and an oily liquid, which facilitates the attachment of the active ingredient to the carrier particles. The formulation also is a low-dose and small-particle-size drug with a carrier that can act through adhesion or adsorption. The invention is featured as powder, granulate or effervescent tablets with active ingredients for oral use.

Treatment of haemophilia, a genetic disease, is very complex, requiring many medications, constant control of coagulation tests and a lot of patient care in relation to bleeding. Thus, advances in the treatment of this disease represent an improvement in the quality of a patient's life. Herzog and Daniell (2019) invented a coagulation factor expressed in chloroplasts and bioencapsulated in plant cells. The invention provides a lyophilized plant material composed of a coagulation protein or immunogenic fragment, which can retain the immunogenicity in a lyophilized form. The system blocks the formation of inhibitory antibodies in protein replacement therapy; moreover, this invention eliminates anaphylactic reactions and can reverse or reduce undesirable pre-existing inhibitor titers.

Another innovation using natural compounds is the study of Nguyen (2020). This patent developed a pressurized metered dose inhaler device using solvent systems using a composition of cannabinoids, terpenes and propellants. Combinations of terpene compositions can be selected by the user in response to various factors, such as a desired therapeutic benefit, sedative effect, flavour profile or a combination of the above.

1.9 FINAL REMARKS

There are numerous biomolecules and promising compounds of natural origin being studied as alternatives to conventional drugs or as proposals for new pharmacological approaches. With the growing interest in pharmaceutical companies in the chemical diversity of natural products, it is expected that shortly new phytomedicines can be found on the market with greater ease. Our resulting data allowed the identification of recent patents on drug discovery from natural products and their application in innovative technologies in the pharmaceutical industry. These innovative formulations can be considered as a source of new drugs for treating and preventing cancer, diabetes mellitus, cardiovascular diseases, neurodegenerative diseases, gastrointestinal irritation, metabolic syndrome, hypertension and candidiasis.

REFERENCES

Ahuja, Rahul, Neha Panwar, Jairam Meena, Mamta Singh, Debi P. Sarkar, and Amulya K. Panda. 2020. "Natural products and polymeric nanocarriers for cancer treatment: a review." *Environmental Chemistry Letters*, 1–10.

Antony, Benny. 2019. Method for improving the bioavailability of curcumin. PT1890546 (T), issued 23 de janeiro de 2019. https://worldwide.espacenet.com/publicationDetails/biblio?FT=D&date=20190123&DB=EPODOC&locale=en_EP&CC=PT&NR=1890546T&KC=T&ND=1.

Araújo, Cristiany Marinho, Carla Eiras, Moisés das Virgens Santana, and Antonio do Nascimento Cavalcante. 2020. "Technological mapping on the use of palygorskita nano-compositals/silver nanoparticles as a bactericide agent with views in biotechnological applications." *Revista Geintec-Gestao Inovacao E Tecnologias* 10, no. 2, 5456–5470.

Bharate, Sonali Sandip, Vikas Kumar, Rohit Singh, Rani Sarita, Mehak Gupta, Ajay Kumar, Sandip Bibishan Bharate, and Ram Vishwakarma. 2020. "Sustained release formulations of crocus sativus." U.S. Patent Application 16/753,969, filed December 17, 2020.

Brahmachari, Goutam. 2019. *Discovery and Development of Anti-Inflammatory Agents from Natural Products*, pp. 1–6, Netherlands: Elsevier.

Bror, Morein, Berenjian Saideh, and Hu Kefei. 2018. Nanoparticles, process for preparation and use thereof as carrier for amphipatic of hydrophobic molecules in fields of medicine including cancer treatment and food related compounds. TR201808733 (T4), issued 23 de julho de 2018. https://worldwide.espacenet.com/publicationDetails/biblio?FT=D&date=20180723&DB=EPODOC&locale=en_EP&CC=TR&NR=201808733T4&KC=T4&ND=1

Choi, Yun Sik and Il Kyung Chung. 2019. Compositions for improvement benign prostatic hyperplasia comprising bee pollen. KR20190090910 (A), issued 5 de agosto de 2019. https://worldwide.espacenet.com/publicationDetails/biblio?FT=D&date=20190805&DB=EPODOC&locale=en_EP&CC=KR&NR=20190090910A&KC=A&ND=1.

Ciesla, Lukasz, and Yuping Bao. 2020. "Cell membrane coated magnetic nanoparticles and assays for identification of transmembrane protein-binding compounds." U.S. Patent Application 16/523,454, filed March 26, 2020.

Ciftci, Ozan N. 2020. Bioavailable curcumin nanoparticles and methods of making. WO2020018759 (A1), issued 23 de janeiro de 2020. https://worldwide.espacenet.com/publicationDetails/biblio?FT=D&date=20200123&DB=EPODOC&locale=en_EP&CC=WO&NR=2020018759A1&KC=A1&ND=1.

Dandl Klaus, Josef Haala, and Detlev Haack. 2020. "Pharmaceutical compositions incorporating low-dose drugs." U.S. Patent Application 14/766,200, filed December 24, 2020.

Dewar, Laura. 2020. "Colloidal silver combined with plant extracts for use in treating wounds and other skin conditions." U.S. Patent Application 16/499,406, filed February 6, 2020.

Giori, Andrea, and Federico Franceschi. 2019. "Phospholipid complexes of curcumin having improved bioavailability." U.S. Patent Application 16/274,623, filed June 13, 2019.

Herzog, Roland W. and Henry Daniell. 2019. Compositions and methods for suppression of inhibitor formation against coagulation factors in hemophilia patients. US2019194265 (A1), issued 27 de junho de 2019. https://worldwide.espacenet.com/publicationDetails/biblio?FT=D&date=20190627&DB=EPODOC&locale=en_EP&CC=US&NR=2019194265A1&KC=A1&ND=1.

Jeong, Jae Woo, Seong Heon Jeong, and Sang Sook Lim. 2020. Composition comprising hydrolyzed sponge for relieving arthralgia or myalgia. WO2020096127 (A1). Acessado julho 28. https://worldwide.espacenet.com/publicationDetails/biblio?FT=D&date=20200514&DB=EPODOC&locale=en_EP&CC=WO&NR=2020096127A1&KC=A1&ND=1.

Kalkundri, Tejaswini Ashish. 2020. Amphiphilic complex formulation comprising multiple active ingredients and dosage forms thereof. WO2020058785 (A1). Acessado julho 29. https://worldwide.espacenet.com/publicationDetails/biblio?FT=D&date=20200326&DB=EPODOC&locale=en_EP&CC=WO&NR=2020058785A1&KC=A1&ND=1.

Katiyar, Chandrakant, Arun Gupta, Satyajyoti Kanjilal, and Shefali Katiyar. 2012. "Drug discovery from plant sources: an integrated approach." *Ayu* 33, no. 1, 10.

Kato, Eishin, and Shinya Hosoda. 2020. "Gnetin C-rich melinjo extract and production method thereof." U.S. Patent 10,640,480, issued May 5, 2020.

Khandelwal, Mudrika, and Shivakalyani Adepu. 2020. Pharmaceutical compositions and delivery systems for prevention and treatment of candidiasis. WO2020058836 (A1). Acessado julho 29. https://worldwide.espacenet.com/publicationDetails/biblio?FT=D&date=20200326&DB=EPODOC&locale=en_EP&CC=WO&NR=2020058836A1&KC=A1&ND=1.

Kleidon, William. 2019. Methods and compositions for reducing oxidative stress. MX2018015583 (A), issued 16 de maio de 2019. https://worldwide.espacenet.com/publicationDetails/biblio?FT=D&date=20190516&DB=EPODOC&locale=en_EP&CC=MX&NR=2018015583A&KC=A&ND=1.

Nguyen, Jason Ty. 2020. "Metered dose inhaler compositions, systems, and methods." U.S. Patent 10,596,147, issued March 24, 2020.

Ribeiro, Antônio Carlos Chaves, Geovany Albino De Souza, and Tatiana Duque Martins. 2020. sistema orgânico de peptídio, composto luminescente e derivado de álcool perílico, método de preparação do sistema e uso em carreamento de fármaco e terapia fotodinâmica. BR102018070770 (A2), issued 28 de abril de 2020. https://worldwide.espacenet.com/publicationDetails/biblio?FT=D&date=20200428&DB=&locale=en_EP&CC=BR&NR=102018070770A2&KC=A2&ND=4.

Sanders, Mitchell Corey. 2020. "Ecofriendly biofilm-disrupting antimicrobial formulations, their development, and their uses." U.S. Patent Application 16/550,122, filed April 23, 2020.

Serebryanskaya, Anastasiya Vladimirovna. 2019. Transdermal preparation. WO2019088883 (A1), issued 9 de maio de 2019. https://worldwide.espacenet.com/publicationDetails/biblio?FT=D&date=20190509&DB=EPODOC&locale=en_EP&CC=WO&NR=2019088883A1&KC=A1&ND=1.

Siegel, Rebecca L, Kimberly D Miller, and Ahmedin Jemal. 2020. "Cancer statistics, 2020." *CA: A Cancer Journal for Clinicians* 70, no. 1, 7–30.

Ueno, Toshiya, Masaru Kominami, and Naoki Kasajima. 2020. Agent for promoting in vivo absorption of hydroxytyrosol and derivatives thereof and use of same. NZ716858 (A), issued 28 de fevereiro de 2020.

https://worldwide.espacenet.com/publicationDetails/biblio?FT=D&date=20200228&DB=EPODOC&locale=en_EP&CC=NZ&NR=716858A&KC=A&ND=1.

van Eck, Nees Jan and Ludo Waltman. 2014. "Visualizing bibliometric networks." In *Measuring Scholarly Impact*, pp. 285–320. Springer, Cham.

Verma, Ashni, Sukhdev Singh, Rupinder Kaur, and Upendra K. Jain. 2013. "Topical gels as drug delivery systems: a review." *International Journal of Pharmaceutical Sciences Review and Research* 23, no. 2, 374–382.

World Intellectual Property Organization. 2020. Wipo international classifications. https://www.wipo.int/edocs/pubdocs/en/wipo_guide_ipc_2020.pdf.

Yuan, Haidan, Qianqian Ma, Li Ye, and Guangchun Piao. 2016. "The traditional medicine and modern medicine from natural products." *Molecules (Basel, Switzerland)* 21, no. 5, 1–18.

Zhao, Shuihong. 2019. Composition De Blanchiment De La Peau Et De Protection Solaire Comprenant Des Nano Materiaux Et Un Extrait Issu De La Medecine Chinoise. FR3070862 (A1), issued 15 de março de 2019. https://worldwide.espacenet.com/publicationDetails/biblio?FT=D&date=20190315&DB=EPODOC&locale=en_EP&CC=FR&NR=3070862A1&KC=A1&ND=1.

2 Nanophytomedicines as an Emerging Platform for Drug Delivery

Nanoapplications in Drug and Gene Delivery

Arjunan Karuppaiah, Veintramuthu Sankar and Muthiah Ramanathan

CONTENTS

2.1 INTRODUCTION

Phytomedicines have been widely used since ancient times around the world due to their excellent therapeutic effects against many diseases (Sajid et al. 2019). Today, phytomedicines are used as a formulation like *parpam* to deliver the active constituents and a combination of traditional *Siddha* formulations are still used as front-line choices in the management of many diseases.

In *Ayurveda,* natural elements are used to treat the root cause of the disease (Parasuraman et al. 2014). The therapeutic effectiveness of medicinal plants depends on the delivery of active compounds to the site. Phytochemical, phytopharmacological and pharmaceutical sciences have enabled the development of delivering medicinal plant products to improve biological activities.

DOI: 10.1201/9781003231745-2

Though the active constituents of plant extracts (flavonoids, tannins and terpenoids) are highly soluble in water, they have less absorption and less permeability through lipid cell membranes. The plant-derived active constituents have bioavailability and efficacy due to their high molecular size and poor absorption (da Silva et al. 2013). The development of nanotechnology-based formulations might be helpful to deliver the active principle of the herbal extract with the required dose to minimize side effects and potentiate activity of herbal medicines (Sandhiya et al. 2020).

Gene therapy is an innovative technique that modifies the disease-causing gene expression at the cellular level to treat a pathological condition. However, in gene delivery, reaching the target cells with the therapeutic agent is quite difficult as it is challenging to cross the physiological and intracellular barriers.

The carrier systems for delivery of drugs and genes have been developed to date based on lipids (Kunjiappan et al. 2021), polymers (Gagliardi et al. 2021), inorganic materials (Guo et al. 2021), monoclonal antibodies (Lee et al. 2021), proteins (Ganji et al. 2021) and cellular carriers (Lutz et al. 2019). It has been proven that the novel carrier systems could improve the delivery of active compounds effectively through nanocarrier delivery systems versus conventional delivery systems.

One study has shown that the uptake of nanocarrier delivery is 15–250 times greater than that of microparticles. In fact, the delivery systems for drug and gene are important for stability, effective, safe and convenient administration. However, nanocarriers can accommodate and carry multiple active substances like drug, gene and phytocompounds with high therapeutic payload (Kuo et al. 2014).

The nanocarrier systems can be tethered with various targeting moieties for improving both the affinity and specificity for target cells. In drug delivery systems, targeted moieties such as low molecular weight multivalent targeting ligands, antibodies, lectins, peptides, hormones, polysaccharides, aptamers and charged molecules have been widely used (Karuppaiah et al. 2020a).

Nanoparticle conjugation for drug and gene delivery with current technologies can result in reduced toxicity and higher efficacy than the drug without targeting. The combination of drugs loaded into the nanoparticle delivery system can be released simultaneously and serially and thereby bypass multiple drug resistance (Patra et al. 2018).

2.2 TARGETS

The nanoformulations of drug, gene and/or phytomedicines with inert, nontoxic and biocompatible carriers may be coupled with site-specific targeting ligands for *in vivo* delivery would be a promising approach to improving their therapeutic effects with minimized side effects (Puri et al. 2009). The specific drug targeted therapies are more precise to target cancer cells, which could lead to higher efficacy with minimum toxicity (Glasgow et al. 2015).

Researchers have continued for years to try and target the cancer cells to their receptor over expression with suitable ligands. To achieve specific targeting to breast cancer cells with folic acid is a well-known ligand for folate receptors; researchers have developed many nanoformulations with folic acid surface tethering. The targeting moieties such as antibodies, peptides and vitamins, which can bind to the receptors expressed at a tumour site (Zwicke et al. 2012), are also under investigation.

2.3 CARRIERS FOR SPECIFIC TARGETING OF CELLS AND ORGANS AND THEIR APPLICATIONS

The research and development about multifunctional nanoparticles as pharmaceutical drug, gene and phytomedicine carriers have shown exponential growth in medical applications in recent years (Puri et al. 2009). Drug delivery carriers with phyto- and synthetic components including lipids, polymers, inorganic materials, monoclonal antibodies, modified proteins, lipoproteins and cellular carriers, have been widely used. In recent years, a wide range of pharmaceutical nanocarriers have been developed to enhance the physicochemical properties of therapeutic agents such as bioavailability, cellular uptake, sustained delivery and reduced toxicity and macrophage distribution (Chenthamara et al. 2019).

2.3.1 Lipid-Based Nanoparticles

Of the lipid-based formulations, liposomes are the most extensively studied. Lipid nanoparticles primarily consist of phospholipids, which are the major components of biological membranes. The versatility of lipid-based carriers for drug delivery systems makes them a primary choice as carriers of diverse therapeutic agents such as drugs, genes and phytomedicines (Drescher et al. 2020). However, these lipid-based formulations have several drawbacks such as polymorphism and crystalline rearrangements (García-Pinel et al. 2019).

2.3.1.1 Liposomes

Liposomes are small spherical vesicular structured carriers composed of one or more phospholipid bilayers. Due to their amphipathic properties, they readily form concentric bilayers, which can surround hydrophilic drugs in the inner aqueous phase and hydrophobic drugs in the lipid bilayer (Olusanya et al. 2018). Liposomes used in delivery of DNA plays an important role for blood stability, cellular uptake, release of DNA and for enhanced transfection efficiency (Zhao et al. 2020). To enhance the cellular uptake of therapeutic compounds, ligand-bearing liposomes have been developed for active targeting via receptor-mediated endocytosis (Puri et al. 2009).

A number of liposomal formulations combat cancer and infectious diseases, and few are in clinical trials. Doxil®/caylex (Ortho-Biotech), a doxorubicin-loaded liposome formulation for the cancer treatment, was the first formulation approved for application in the clinical patient care (Pillai 2019). Liposomes are an excellent lipid-based nanocarrier for site-specific delivery of drugs and genes.

2.3.1.2 Solid Lipid Nanoparticles (SLNs)

In novel drug delivery, carrier SLNs are rapidly emerging as attractive nanocarriers at the forefront of the drug delivery system. The SLN structure consists of biocompatible and biodegradable solid lipid core such as triglycerides, glyceride mixtures and/or waxes. The size of drug loaded SLN carriers are in nano- to submicron scale at 1–1000 nm. The preparation is easily scaled up at low cost of other lipid-based nanocarrier systems (Ghasemiyeh et al. 2018).

In one study, researchers developed a phytocompound berberine-loaded SLN delivery system with an average particle size of 76 nm. The SLN was used to treat rats and showed higher antidiabetic activity by suppressing gains in body weight, lowering fasting blood glucose levels and promoting islet cell functions (Xue et al. 2013).

2.3.1.3 Nanostructure Lipid Carriers (NLCs)

NLC is a mixture of solid and liquid lipids. It is a better carrier choice for oral delivery of many natural and chemically synthesized compounds. Also, it is considered as a second-generation lipid-based nanoparticle and was originally developed from SLN. Typically, the encapsulation efficiencies of NLC are about 70% (Puri et al. 2009).

In a recent study, authors have discussed the extraction of silymarin from the milk thistle *Silybum marianum* (L.) Gaernt and loading it into NLC to overcome the physicochemical property-related issues (Chauhan et al. 2020).

2.3.1.4 Phytosomes

Phytosomes are small particles developed to incorporate medicinal plant active ingredients and water-soluble phytochemicals into phospholipids. This lipid-compatible complex is used to modify the absorption and bioavailability of plant active ingredients.

Typically, phytosome formulations include the molecular complex and chemical bonding between phosphatidylcholine and plant materials. However, liposomes being macromolecules make thousands of chemical bonding with phosphatidylcholine to enclose water-soluble molecules. In a recent study, researchers have developed and characterized a phytosome-containing ethanolic extract of both *A. indicum* and *P. longum* to enhance to enhance their pharmacological actions as well as safety of the herbal preparation (Sharma et al. 2016).

2.3.2 Polymer-Based Nanocarriers

Polymeric nanoparticles are a solid colloidal dispersion with the size range from 10 to 100 nm. They are made up of biodegradable and biocompatible polymers and consist of a core polymer matrix to encapsulate lipophilic drugs and a hydrophilic corona to ensure the steric stability of nanoparticles (Makadia et al. 2011).

2.3.2.1 Poly Lactic-co-Glycolic Acid (PLGA) Nanoparticles

PLGA nanoparticles have biodegradability, biocompatibility and sustained release properties, making it a suitable polymer for biomedical and pharmaceutical applications. The PLGA polymer has been used in many ways such as imaging, diagnostics and treatment.

The PLGA polymer is used to entrap active materials such as proteins, peptides, genes and drugs (Swider et al. 2018). Recently, quercetin- and curcumin-loaded PLGA nanoparticles were studied to enhance their efficacy in the diabetic model. The antidiabetic phytocompounds were effectively entrapped in PLGA nanoparticles with a size of 100 nm (Elmowafy et al. 2019).

2.3.2.2 Biopolymer-Based Nanocarriers (BBNs)

Biopolymers are derived from proteins such as gelatine, albumin and milk proteins and polysaccharides such as chitosan, hyaluronan, dextran, cyclodextrins, pectin, guar gum, cellulose, sodium alginate and starch. The features exhibited by these biomaterials used in BBN production include biological realization, bioactivity, biodegradability, toxicity reduction and easy modification (Vasile et al. 2020).

Chitosan is a cationic biocompatible and biodegradable polymer that is reported to be the best example of a natural pure biopolymer to deliver plant-derived components such as curcumin, quercetin and trans-resveratrol due to its positive properties such as mucoadhesion, solubility, dissolution rate and specific targeting (Elieh-Ali-Komi et al. 2016).

2.3.2.3 Micelle

Micelle is a polymeric particle or colloidal dispersion measuring about 10–100 nm. Micelle is made up of amphiphilic surfactants called nonpolymeric micelles and amphiphilic copolymers called polymeric micelles. It consists of a single core-shell and di- or tri-block copolymers with both hydrophobic and hydrophilic segments in aqueous media. To enhance the bioavailability and water solubility of artemisinin, researchers synthesized micelles with three series of poly (ε-caprolactone)–poly (ethylene glycol)–poly (ε-caprolactone) (PCL–PEG–PCL) tri-block copolymers (Manjili et al. 2018).

2.3.2.4 Nanosphere

A nanosphere is a colloidal aqueous solution with an amorphous or crystalline nature. It ranges in size from 10 to 200 nm and is composed of a polymeric core encapsulating drug and/or adsorbs them onto the nanoparticles. This system gives improved entrapment of the drug, delayed release, improved plasma drug concentrations, bioavailability and stability in biological fluids. It protects active agents from chemical and enzymatic degradation and reduces toxicity (Rahman et al. 2020).

2.3.3 Inorganic Metal-Based Nanocarriers

Inorganic nanoparticles possess unique properties due to their size, shape, surface structure and aggregation characteristics (Sharma et al. 2016). Metallic nanoparticles have been used widely as a drug delivery carrier due to their physicochemical properties (Rao et al. 2003). Metallic nanoparticles such as gold, silver, iron, selenium and platinum are widely used in cancer research.

Among all the metallic nanoparticles, silver nanoparticles (AgNPs) shows the highest level of commercialization (Xiu et al. 2012) and account for 55.4% of the total nanomaterial-based consumer products available in the market (313 out of 565 products) (Asghari et al. 2012).

A previously reported study has synthesized AgNPs by a chemical reduction method using sodium borohydride as a reducing agent. The authors have further investigated the synergistic cytotoxic effect of a gemcitabine conjugated noncytotoxic dose of AgNP in human triple-negative metastatic breast cancer cells (MDA-MB-453) (Karuppaiah et al. 2020c). The same author has synthesized a biogenically reduced ternary therapeutic nanoconjugate containing andrographolide, silver and doxorubicin to evaluate the cytotoxicity and synergistic effect of nanoconjugates on the MDA-MB-453 breast cancer cell line (Karuppaiah et al. 2020b).

2.4 SILVER NANOPARTICLES (AgNPs)

AgNPs are extensively studied in pharmaceutical sciences because of their antimicrobial activity and cytotoxicity against various cancer cell lines (Yusuf et al. 2018). Therefore, AgNPs have been utilized for a variety of theranostic applications including imaging (e.g., computed tomography, photo acoustic, magnetic resonance and surface-enhanced Raman scattering), delivery (e.g., drugs, proteins, small-interfering RNAs) and therapy (e.g., photothermal therapy, RT).

Nano silver can be easily oxidized in the presence of oxygen leading to the release of silver ions (Ag^+), which are a major source of toxicity. Ag^+ ions induce oxidative stress through the generation of reactive oxygen species and cause damage to cellular components such as cell membrane, protein and DNA. Figure 2.1 shows the mechanism of action of AgNPs. This leads to apoptotic and

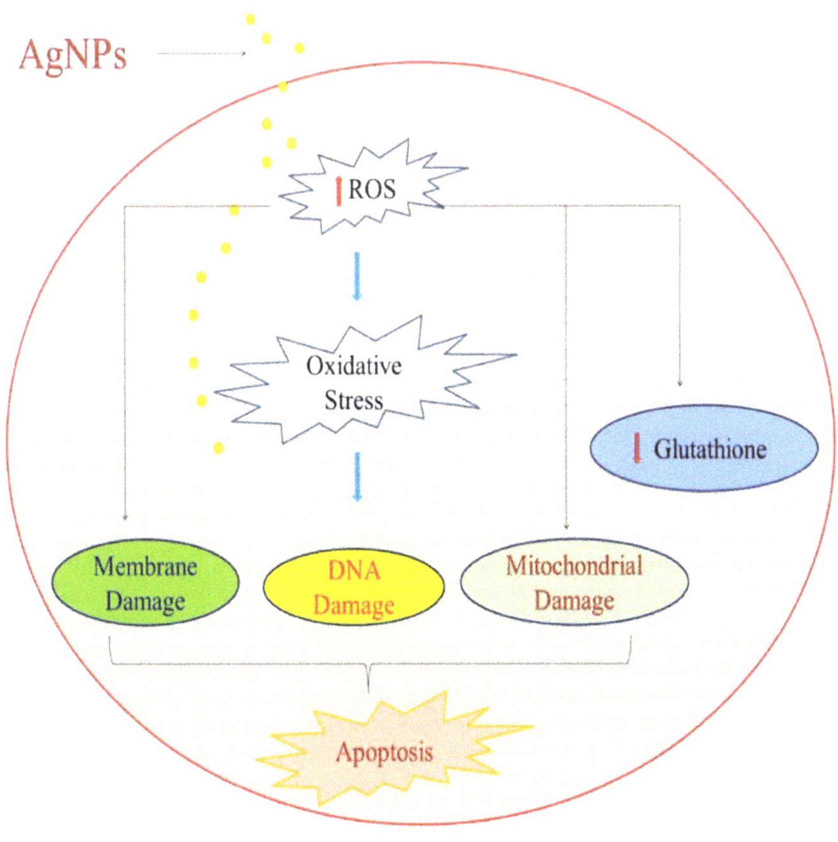

FIGURE 2.1 Mechanism of action of AgNPs.

necrotic cell deaths triggering the release of proinflammatory cytokines that cause further damage to nearby tissues or cells. AgNPs can also deplete antioxidant molecules, like glutathione, and aggravate cytotoxic effects.

2.4.1 Target Based

AgNPs provide the possibility of surface modifications with good stability due their smaller particle size and surface area (Karuppaiah et al. 2020c). A previously reported study has conjugated folic acid with AgNPs for enhancing antibacterial activity. Researchers used triethylamine as a capping as well as reducing agent during the synthesis of AgNPs. Folic acid is conjugated on the surface of functionalized AgNPs through electrostatic interaction. The carbon-dot tagged nanoparticles uptake study reveals that folic acid-conjugated AgNPs show more protective against both Gram-negative and Gram-positive bacteria (Chowdhuri et al. 2015).

2.4.2 Disease Based

AgNPs are extensively studied in pharmaceutical sciences because of their antimicrobial activity and cytotoxicity against various cancer cell lines (Yusuf et al. 2018). Further, researchers have observed that electrostatic attraction (EA) of functionalized drugs on the surface of biogenically synthesized AgNPs could reduce the dose of ternary silver conjugates nanoformulation prepared with plant compounds and anticancer drugs. The biogenic synthesis of AgNPs produced nano particles with expected range (less than 200 nm) and also reduced the utilization of capping agent. The cytotoxicity and synergistic effects of ternary therapeutic nanoconjugates evaluated using MDA-MB-453 breast cancer cells were found to be superior to doxorubicin (Dox) (Karuppaiah et al. 2020b).

2.4.3 Restrictions/Toxicity

The toxicity of AgNPs is mainly due to its aggregation as Ag^+ and accumulation in deep-seated organs because of its smaller particle size and instability in blood circulation. AgNPs are reported to cause genotoxicity, neurotoxicity, pulmonary toxicity, hepatotoxicity and immunotoxicity (McShan et al. 2014). Many research works are in progress aiming to reduce the toxicity of AgNPs related to size, dose, agglomeration rate and excretion rate as well as to improve the therapeutic effect by targeted therapy with reduced dose and adjuvant therapeutic agent.

2.5 CONCLUSION AND FUTURE PERSPECTIVES

Currently, nanotechnology has a wide range of applications in the pharmaceutical industry and research. Typically, nanocarriers for drug delivery have been found to be potentially more beneficial in many aspects than conventional drug delivery systems. Generally, nanocarriers protect drugs and genes from enzymatic degradation, improve their physicochemical properties, overcome biological barriers and enhance systemic bioavailability.

The growing applications of nanotechnology in the medical field spurs researchers to find the impacts of nanoparticles on human organs, tissues and immune system. The surface modifications of the nanocarrier system enable the modulation of the immune system either by evading the immune system to prevent allergic reactions or by enhancing the immunogenic response.

Various nanocarrier systems are currently being developed with the aim of improving drug delivery, especially for brain targeting and to combat cancer and diabetes. Polymeric nanocarriers also have been widely studied and used in targeted drug delivery due to their accessibility, stability, biocompatibility, biodegradability, lack of toxicity, safety and cost effectiveness.

The future perspective of nanocarriers, AgNPs in particular, sees great potential in a broad spectrum of applications as antimicrobial, anticancer and antiviral activities. The AgNPs are also widely used for biomedical device coatings, drug delivery carriers, synergistic therapy, theranostic imaging, biomarkers for diagnostic and optoelectronic properties since it has discrete physical, optical and biochemical functionality. The biological interactions of AgNPs inside cells can be simulated with computer modelling and will be a boon to nano research.

In recent years, different treatment strategies have been adopted, but their therapeutic efficacy are not enough and have shown many side effects. Also, the mechanism of drug delivery by nanotechnology is not still well-explored. Despite various types of nanocarriers in use, the field still requires clearer ideas to deliver the drug safely and conveniently.

In our opinion, future efforts could be focused on several basic concerns to achieve the successful clinical translation of nanoformulations. Like biogenic synthesis of AgNPs, an eco-friendly green approach should be developed for the preparation of nanoformulations and surface coating to overcome aggregation and toxicity. Factors like shape, size, charge and moiety attached to nanomaterials should be well elucidated and evaluated, which is essential for developing ideal drug delivery systems.

REFERENCES

Asghari, Saba, Seyed Ali Johari, Ji Hyun Lee, Yong Seok Kim, Yong Bae Jeon, Hyun Jung Choi, Min Chaul Moon, and Il Je Yu. "Toxicity of various silver nanoparticles compared to silver ions in *Daphnia magna.*" *Journal of Nanobiotechnology* 10, no. 1 (2012): 1–11.

Chauhan, Iti, Mohd Yasir, Madhu Verma, and Alok Pratap Singh. "Nano structured lipid carriers: A groundbreaking approach for transdermal drug delivery." *Advanced Pharmaceutical Bulletin* 10, no. 2 (2020): 150.

Chenthamara, Dhrisya, Sadhasivam Subramaniam, Sankar Ganesh Ramakrishnan, Swaminathan Krishnaswamy, Musthafa Mohamed Essa, Feng-Huei Lin, and M. Walid Qoronfleh. "Therapeutic efficacy of nanoparticles and routes of administration." *Biomaterials Research* 23, no. 1 (2019): 1–29.

Chowdhuri, Angshuman Ray, Satyajit Tripathy, Chanchal Haldar, Soumen Chandra, Balaram Das, Somenath Roy, and Sumanta Kumar Sahu. "Theoretical and experimental study of folic acid conjugated silver nanoparticles through electrostatic interaction for enhance antibacterial activity." *RSC Advances* 5, no. 28 (2015): 21515–21524.

Drescher, Simon, and Peter van Hoogevest. "The phospholipid research center: Current research in phospholipids and their use in drug delivery." *Pharmaceutics* 12, no. 12 (2020): 1235.

Elieh-Ali-Komi, Daniel, and Michael R. Hamblin. "Chitin and chitosan: Production and application of versatile biomedical nanomaterials." *International Journal of Advanced Research* 4, no. 3 (2016): 411. http://www.ncbi.nlm.nih.gov/pubmed/27819009.

Elmowafy, Enas M., Mattia Tiboni, and Mahmoud E. Soliman. "Biocompatibility, biodegradation and biomedical applications of poly (lactic acid)/poly (lactic-co-glycolic acid) micro and nanoparticles." *Journal of Pharmaceutical Investigation* 49, no. 4 (2019): 347–380.

Gagliardi, Agnese, Elena Giuliano, Venkateswara Rao Eeda, Massimo Fresta, Stefania Bulotta, Vibhudutta Awasthi, and Donato Cosco. "Biodegradable polymeric nanoparticles for drug delivery to solid tumors." *Frontiers in Pharmacology* 12 (2021): 17.

Ganji, Nasim, and Geoffrey D. Bothun. "Albumin protein coronas render nanoparticles surface active: consonant interactions at air water and at lipid monolayer interfaces." *Environmental Science: Nano* 8, no. 1 (2021): 160–173.

García-Pinel, Beatriz, Cristina Porras Alcalá, Alicia Ortega Rodríguez, Francisco Sarabia, Jose Prados, Consolación Melguizo, and Juan M. López Romero. "Lipid-based nanoparticles: Application and recent advances in cancer treatment." *Nanomaterials* 9, no. 4 (2019): 638.

Ghasemiyeh, Parisa, and Soliman Mohammadi Samani. "Solid lipid nanoparticles and nanostructured lipid carriers as novel drug delivery systems: Applications, advantages and disadvantages." *Research in Pharmaceutical Sciences* 13, no. 4 (2018): 288.

Glasgow, Micah DK, and Mahavir B. Chougule. "Recent developments in active tumor targeted multifunctional nanoparticles for combination chemotherapy in cancer treatment and imaging." *Journal of Biomedical Nanotechnology* 11, no. 11 (2015): 1859–1898.

Guo, Jiahua, Yuqi Zhang, Shouwen Jin, Xiaoyun Ye, Kaikai Hu, Weiqiang Xu, and Daqi Wang. "Syntheses and structural characterization of eight inorganic organic hybrid solids based on N-Brønsted bases and halometallates of Zn, Cu, Sb and Bi." *Inorganic and Nano-Metal Chemistry* (2021): 1–13.

Karuppaiah, Arjunan, Ravikumar Rajan, Sivaram Hariharan, Dinesh K. Balasubramaniam, Marslin Gregory, and Veintramuthu Sankar. "Synthesis and characterization of folic acid conjugated gemcitabine tethered silver nanoparticles (FA-GEM-AgNPs) for targeted delivery." *Current Pharmaceutical Design* 26, no. 26 (2020a): 3141–3146.

Karuppaiah, Arjunan, Ravikumar Rajan, Muthiah Ramanathan, and Arumugam Nagarajan. "Cytotoxicity and synergistic effect of biogenically synthesized ternary therapeutic nano conjugates comprising plant active principle, silver and anticancer drug on MDA-MB-453 breast cancer cell line." *Asian Pacific Journal of Cancer Prevention: APJCP* 21, no. 1 (2020b): 195.

Karuppaiah, Arjunan, Karthik Siram, Divakar Selvaraj, Mohan Kandhasamy Ramasamy, Dinesh Babu, and Veintramuthu Sankar. "Synergistic and enhanced anticancer effect of a facile surface modified non-cytotoxic silver nanoparticle conjugated with gemcitabine in metastatic breast cancer cells." *Materials Today Communications* 23 (2020c): 100884.

Kunjiappan, Selvaraj, Murugesan Sankaranarayanan, Banoth Karan Kumar, Parasuraman Pavadai, Ewa Babkiewicz, Piotr Maszczyk, Eliza Glodkowska Mrowka et al. "Capsaicin-loaded solid lipid nanoparticles: Design, bio distribution, *in silico* modeling and *in vitro* cytotoxicity evaluation." *Nanotechnology* 32, no. 9 (2021): 095101.

Kuo, Yung-Chih, and Wang Cheng-Chin. "Colloidal drug delivery system for brain targeting therapy." in *Colloid and Interface Science in Pharmaceutical Research and Development* Edited by Hiroyuki Ohshima, Kimiko Makino, Elsevier Publication Chapter 19 (2014): 389–410.

Lee, Hyuk Mi, and Hwan Goo Kang. "Rapid and simultaneous purification of aflatoxin B1, zearalenone and deoxynivalenol using their monoclonal antibodies and magnetic nanoparticles." *Toxicological Research* (2021): 1–7.

Lutz, Halle, Shiqi Hu, Phuong Uyen Dinh, and Ke Cheng. "Cells and cell derivatives as drug carriers for targeted delivery." *Medicine in Drug Discovery* 3 (2019): 100014.

McShan, Danielle, Paresh C. Ray, and Hongtao Yu. "Molecular toxicity mechanism of nanosilver." *Journal of Food and Drug Analysis* 22, no. 1 (2014): 116–27.

Makadia, Hirenkumar K., and Steven J. Siegel. "Poly lactic-co-glycolic acid (PLGA) as bio degradable controlled drug delivery carrier." *Polymers* 3, no. 3 (2011): 1377–1397.

Manjili, Hamidreza Kheiri, Hojjat Malvandi, Mir Sajjad Mousavi, Elahe Attari, and Hossein Danafar. "In vitro and in vivo delivery of artemisinin loaded PCL–PEG–PCL micelles and its pharmacokinetic study." *Artificial Cells, Nanomedicine, and Biotechnology* 46, no. 5 (2018): 926–936.

Olusanya, Temidayo, Rita Haj Ahmad, Daniel Ibegbu, James Smith, and Amal Elkordy. "Liposomal drug delivery systems and anticancer drugs." *Molecules* 23, no. 4 (2018): 907.

Parasuraman, Subramani, Gan Siaw Thing, and Sokkalingam Arumugam Dhanaraj. "Polyherbal formulation: Concept of *Ayurveda*." *Pharmacognosy Reviews* 8, no. 16 (2014): 73.

Patra, Jayanta Kumar, Gitishree Das, Leonardo Fernandes Fraceto, Estefania Vangelie Ramos Campos, Maria del Pilar Rodriguez Torres, Laura Susana Acosta Torres, Luis Armando Diaz Torres et al. "Nano-based drug delivery systems: Recent developments and future prospects." *Journal of Nanobiotechnology* 16, no. 1 (2018): 1–33.

Pillai, Gopalakrishna. "Nanotechnology toward treating cancer." *Applications of Targeted Nano Drugs and Delivery Systems* 9 (2019): 221–256.

Puri, Anu, Kristin Loomis, Brandon Smith, Jae Ho Lee, Amichai Yavlovich, Eliahu Heldman, and Robert Blumenthal. "Lipid-based nanoparticles as pharmaceutical drug carriers: From concepts to clinic." *Critical Reviews™ in Therapeutic Drug Carrier Systems* 26, no. 6 (2009): 523–580.

Rahman, Heshu Sulaiman, Hemn Hassan Othman, Nahidah Ibrahim Hammadi, Swee Keong Yeap, Kawa Mohammad Amin, Nozlena Abdul Samad, and Noorjahan Banu Alitheen. "Novel drug delivery systems for loading of natural plant extracts and their biomedical applications." *International Journal of Nanomedicine* 15 (2020): 2439.

Rao, C. N. R., G. U. Kulkarni, P. John Thomas, and Peter P. Edwards. "Size-dependent chemistry: Properties of nanocrystals." In *Advances in Chemistry: A Selection of CNR Rao's Publications (1994–2003)*, World Scientific Publishing Co Pte Ltd pp. 227–233. 2003.

Sajid, Mohammad, Swaranjit Singh Cameotra, Mohd Sajjad Ahmad Khan, and Iqbal Ahmad. "Nanoparticle based delivery of phytomedicines: Challenges and opportunities." In *New Look to Phytomedicine*, pp. 597–623. Academic Press, 2019.

Sandhiya, V., and U. Ubaidulla. "A review on herbal drug loaded into pharmaceutical carrier techniques and its evaluation process." *Future Journal of Pharmaceutical Sciences* 6, no. 1 (2020): 1–16.

Sharma, Sonam, and Alakh Niranjan Sahu. "Development, characterization, and evaluation of hepatoprotective effect of *Abutilon indicum* and *Piper longum* phytosomes." *Pharmacognosy Research* 8, no. 1 (2016): 29.

da Silva, Patricia Bento, Bonifácio, Bruna Vidal, Matheus Aparecido dos Santos Ramos, Kamila Maria Silveira Negri, Taís Maria Bauab, and Marlus Chorilli. "Nanotechnology based drug delivery systems and herbal medicines: A review." *International Journal of Nanomedicine* 9 (2013): 1.

Swider, Edyta, Olga Koshkina, Jurjen Tel, Luis J. Cruz, I. Jolanda M. de Vries, and Mangala Srinivas. "Customizing poly (lactic-co-glycolic acid) particles for biomedical applications." *Acta Biomaterialia* 73 (2018): 38–51.

Vasile, Cornelia, Daniela Pamfil, Elena Stoleru, and Mihaela Baican. "New developments in medical applications of hybrid hydrogels containing natural polymers." *Molecules* 25, no. 7 (2020): 1539.

Xiu, Zong ming, Qing bo Zhang, Hema L. Puppala, Vicki L. Colvin, and Pedro JJ Alvarez. "Negligible particle specific antibacterial activity of silver nanoparticles." *Nano Letters* 12, no. 8 (2012): 4271–4275.

Xue, Mei, Ming xing Yang, Wei Zhang, Xiumin Li, De hong Gao, Zhimin Ou, Zhipeng Li, Su huan Liu, Xue jun Li, and Shu yu Yang. "Characterization, pharmacokinetics, and hypoglycemic effect of berberine loaded solid lipid nanoparticles." *International Journal of Nanomedicine* 8 (2013): 4677.

Yusuf, Azeez, A. Brophy, Brian Gorey, and Alan Casey. "Liposomal encapsulation of silver nanoparticles enhances cytotoxicity and causes induction of reactive oxygen species independent apoptosis." *Journal of Applied Toxicology* 38, no. 5 (2018): 616–627.

Zhao, Yinan, Tianyi Zhao, Yanyan Du, Yingnan Cao, Yang Xuan, Huiying Chen, Defu Zhi, Shutao Guo, Fangli Zhong, and Shubiao Zhang. "Interaction kinetics of peptide lipids mediated gene delivery." *Journal of Nanobiotechnology* 18, no. 1 (2020): 1–14.

Zwicke, Grant L, G. Ali Mansoori, and Constance J. Jeffery. "Utilizing the folate receptor for active targeting of cancer nanotherapeutics." *Nano Reviews* 3, no. 1 (2012): 18496.

3 The Modulation of IL-6 Levels by Natural Products in Arthritis-Like Animal Models

A Systematic Review of Preclinical Evidence

Fabiolla R.S. Passos, Marilia M. Rezende, Sathiyabama R. Gandhi, Bruno A.F. Silva, Andreza G.B. Ramos, Irwin R.A. Menezes, Parimelazhagan Thangaraj, Lucindo José Quintans Júnior and Jullyana S.S. Quintans

CONTENTS

3.1 INTRODUCTION

Osteoarthritis (OA) is the most common form of arthritis and is one of the leading causes of pain and disability worldwide (WHO, 2021). It affects about 10–15% of the world population with an incidence of approximately 60–70% in the elderly population (WHO, 2003), making it the most common joint disease in the world and the main cause of chronic pain. In Brazil, osteoarthritis affects about 33% of the population over age 25, which corresponds to approximately 40 million people (Coimbra, Plapler, and de Campos, 2019). In the case of rheumatoid arthritis (AR), the worldwide prevalence is about 0.46% (Almutairi et al., 2021). The chronic pain and loss of functionality that occur in arthritis interfere with the patient's quality of life and entails enormous costs for the health system and for the family itself (Sicras-Mainar et al., 2020).

Currently, patients diagnosed with arthritis are treated with lifestyle change, surgery and medicines that include steroidal and nonsteroidal anti-inflammatory drugs (NSAIDs), analgesics, opioids and modifiers of cartilage degradation and bone remodelling, some of which are limited by side effects (Khanna et al., 2007). NSAIDs are effective in reducing pain related to arthritis, but their use is related to the development of gastric and cardiovascular problems (Losina et al., 2013).

The use of disease-modifying drugs has shown promising efficacy against inflammatory parameters linked to arthritis during controlled and randomized studies, although there was still significant structural progression of the disease. Moreover, the chronic administration of disease-modifying

DOI: 10.1201/9781003231745-3

drugs can lead to unwanted side effects and discomfort (Emery, 2006). In a study by Losina et al. (2013), disease-modifying drugs showed about 20% pain relief, which is lower than standard treatment. This indicates that there is a need for new therapeutic targets that can reduce disease progression and pain.

Approximately one-third of the most commonly used medicines in the world are obtained from natural products (NPs) (Calixto, 2019). Of the drugs approved by the FDA between 1981 and 2014, 21% were derived from NPs (Newman and Cragg, 2016). The increase in the study of natural molecules derived from essential oils of medicinal plants can help to identify potential sources of future novel antiarthritic agents (Li and Vederas, 2009).

Secondary metabolites present in medicinal plants can modulate the production or expression of anti-inflammatory substances, such as cytokines (Khanna et al., 2007). The cytokine interleukin-6 (IL-6) plays a key role in the regulation of arthritic inflammation including in the differentiation and proliferation of T and B cells, inflammatory pannus formation (Paleolog, 2002), increasing the number of neutrophils (Lally et al., 2005) and osteoclast activation (McInnes and Schett, 2007). Choy and Calabrese (2018) hypothesized that IL-6 is a primary mediator of joint pain through direct regulation in the nociceptive nervous system. These findings led to the notion that targeting IL-6 might constitute a novel treatment strategy for various diseases. This resulted in the development of tocilizumab, a humanized anti-IL-6R monoclonal antibody of the IgG1 class (Sato et al., 1993) that blocks IL-6-mediated signal transduction by inhibiting IL-6 binding to transmembrane and sIL-6R. This systematic review was carried out to summarize the findings of studies using arthritis-like animal models to examine the potential of NPs as modulators of IL-6 to support the development of effective new treatment options for the disease.

3.2 SEARCH STRATEGY

The systematic review was performed according to the guidelines of the Preferred Reporting Items for Systematic Reviews and Meta-Analyses (PRISMA) (Moher et al., 2009). We conducted searches of the databases PubMed, Scopus, Embase and Web of Science using the descriptors "natural products" and "interleukin6". The search was performed in April 2021. The main objective of this systematic review was to find articles with animal models of arthritis that had investigated the role of NPs in the modulation of cytokine IL-6.

3.3 STUDY SELECTION

Two independent reseachers undertook the search for the studies (F.R.S.P. and M.M.R.) using the Rayyan app (Ouzzani et al., 2016). Articles were selected if they contained the search keywords in the title, abstract or full text of the article. The abstracts of the articles were then examined to see if they met the following inclusion criteria: preclinical studies using animal models that investigated the modulation of the cytokine IL-6 using NPs or substances isolated from them as treatment in experimental models of arthritis. The exclusion criteria were: studies that used associations of substances or extracts, NPs used only as a diet, clinical trials, review and meta-analysis studies, case reports and articles without the full text available.

3.4 SELECTED STUDIES

The primary search found 3321 articles: 896 from Embase, 1684 from Scopus, 536 from Web of Science and 205 from PubMed. After checking the inclusion and exclusion criteria, and removing any duplicates, 42 articles were selected for inclusion in the review. A flowchart of the selection process is shown in Figure 3.1.

Tables 3.1 and 3.2 show the detailed characteristics of the *in vivo* studies. The main models found were collagen arthritis induction and complete Freund's adjuvant (CFA)-induced arthritis. Other

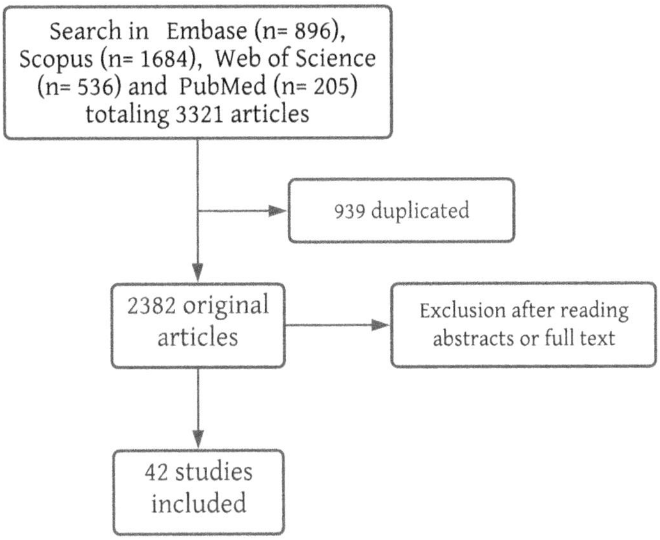

FIGURE 3.1 Flowchart of the included studies.

models used included a surgical procedure (osteoarthritis model) (Wang et al., 2020), the zymosan model, tumour necrosis factor (TNF) overexpression, the Lyme model, and the monosodium iodoacetate-induced osteoarthritis model. The studies evaluated the anti-inflammatory effect of different NPs and whether they were efficient in reducing the production of IL-6 in different types of samples (fluid and synovial tissue, bone marrow, serum, knee joint, plasma, cell cultures, quadriceps muscle and peritoneal lavage).

The routes of administration of treatment used in the selected articles varied between the oral, subcutaneous and intraperitoneal routes, the former being the most used. The duration of treatment varied between one and six weeks according to the protocol developed by each study, both before and after induction.

3.5 OUTCOMES

Arthritis is an autoimmune inflammatory disease that leads to damage of synovial joints (Correa et al., 2016). The search for successful management options for arthritis is still on-going, despite the availability of many synthetic anti-arthritic medications. The existing therapies include NSAIDs, corticosteroids and disease-modifying anti-rheumatic drugs (DMARDs); however, many of the current treatments have dramatic adverse effects or are very expensive. Therefore, new drug candidates from NPs, with potentially reduced and fewer side effects, are being scientifically evaluated in studies worldwide as part of the search for an alternative approach to treat arthritic inflammation (Kumar et al., 2016).

Many researchers are currently focusing on the IL-6 receptor, which has been proven to be a significant target in treating arthritis and its related complications. IL-6 is rapidly produced following infection or tissue damage as part of the acute phase of the immune response, which acts in defence of the host against these stresses. Therefore, IL-6 belongs to the pro-inflammatory cytokine family, which plays a major role in immune regulation and modulation in arthritic inflammatory reactions (Kim et al., 2015). Elevated levels of IL-6 in the sera and synovial fluid indicate arthritic inflammation and generate symptoms of disease activity such as joint pain, swelling, bone erosion and cartilage destruction (Ogata et al., 2019). In this study, we summarized the evidence of the effectiveness of medicinal preparations from plants and their natural molecules as regulators of IL-6 in different animal models of arthritis.

TABLE 3.1
Studies Involving Extracts, Suspensions or Essential Oils from Natural Products That Modulate IL-6 Levels in an Arthritis Model

Article	Animal	Extract, Essential Oil or Fraction	Experimental Model	IL-6 Mensuration	
Ali et al., 2015, Egypt	Wistar rats, M	*Spirulina platensis* suspension 200 or 400 mg/sg p.o.	Adjuvant-induced arthritis	Serum	ELISA
Ananthi et al., 2017, India	Sprague-Dawley rats	Aqueous extract of *Turbinaria ornate* 30, 100, 300 mg/kg p.o.	Adjuvant-induced arthritis	Plasma	ELISA
Cai et al., 2014, China	Sprague-Dawley rats, M	N-Butanol extract from *Ipomoea stolonifera* 100, 200, and 400 mg/kg p.o.	CFA-induced chronic arthritis	RAW 264.7 cells	ELISA
Chakraborty et al., 2010, India	Wistar rats, M	*Lamellidens marginalis* aqueous extract 500 and 1000 mg/kg p.o.	Adjuvant-induced arthritis	Serum	ELISA
Endale et al., 2014, Republic of Korea	DBA/1J mice, M	Korean red ginseng saponin fraction 150 mg/kg p.o.	Collagen-induced arthritis	Serum	ELISA
Hanh et al., 2018, Vietnam	Male albino BALB/c mice	Ethanol extract of *E. bulbosa* 500 and 1000 mg/kg p.o.	Collagen antibody-induced arthritis	Serum	ELISA
Hsu et al., 2016, Taiwan	Sprague-Dawley rats, M	Sesame oil 4 ml/kg p.o.	Osteoarthritis model	Quadriceps muscle	ELISA
Jin et al., 2010, Republic of Korea	Wistar rats, M	Alkaloid-free prenylated flavonoid-enriched fraction of rhizomes of *Sophora flavescens* 100 mg/kg p.o.	Adjuvant-induced arthritis	RAW 264.7 cells	ELISA
Jung et al., 2012, Republic of Korea	DBA/1 mice, M	Root extract of *Dipsacus asperoides* 50 and 100mg/kg p.o.	Collagen-induced arthritis	Serum	ELISA
Kang et al., 2020, Republic of Korea	DBA/1 mice, M	*Humulus japonicus* extract 300 mg/kg p.o.	Collagen-induced arthritis	Paw tissue RAW 264.7 cell	PCR ELISA/PCR
Kumar et al., 2010, India	Wistar rats, F	*Spirulina platensis* suspension 200 and 400 mg kg p.o.	Collagen-induced arthritis	Serum	ELISA
Kumar et al., 2016, India	Wistar rats, F/M	*Melastoma malabathricum* extract 100, 250 and 500 mg/kg p.o.	Complete Freund's adjuvant-induced arthritis	Serum	ELISA

(Continued)

TABLE 3.1 (Continued)

Article	Animal	Extract, Essential Oil or Fraction	Experimental Model	IL-6 Mensuration	
Li et al., 2014, China	Sprague-Dawley rats, M	Lipid extract from *Mytilus coruscus* 100 mg/kg p.o.	Adjuvant-induced arthritis Collagen-induced arthritis	Synovial fluid	ELISA
Park et al., 2017, Republic of Korea	LEW/SSNHSDrats, F	*Tribulus terrestris* extract 200 and 400 mg/kg p.o.	Monosodium iodoacetate-induced osteoarthritis	Cartilage of rat knee joints	Western blot
Pereira et al., 2018, Brazil	Male DBA1/J mice	Hydroalcoholic crude extract from *Chenopodium ambrosioides* leaves 5 and 10 mg/kg p.o.	Type II collagen-induced arthritis	Serum	Cytometric bead array
Rosas et al., 2015, Brazil	C57BL/6 mice, M	*Schinus terebinthifolius* hydroalcoholic extract 100 mg/kg p.o.	Zymosan-induced arthritis	Synovial fluid	ELISA
Sannegowda et al., 2015, USA	Lewis rats, M	*T. cordifolia* extract 1 g/kg p.o.	Adjuvant-induced arthritis	Spleen adherent cells and lymph node cells harvested from arthritic rats Spleen adherent cells and sera	PCR ELISA
Thanh et al., 2020, Vietnam	BALB/c mice, M	Ethanol extract of *P. kingianum* rhizomes 500 and 1000 mg/kg p.o.	Collagen antibody-induced arthritic mice	Serum	ELISA
Zhang et al., 2020, China	Rat, M	Essential oils from *Alpinia oxyphylla, Kaempferia galangal, Amomum kravanh* and *Alpinia galanga* 100 mg/kg p.o.	Adjuvant-induced arthritis	Articular tissue	Immunohistochemical staining

TABLE 3.2
Studies Involving Substances Obtained from Natural Products That Modulate IL-6 Levels in an Arthritis Model

Article	Animal	Substance	Experimental Model	IL-6 Mensuration	
Ananthi et al., 2017, India	Sprague-Dawley rats	*Turbinaria ornata* sulphated polysaccharide 2.5, 5 and 10 mg/kg p.o.	Adjuvant-induced arthritis	Plasma	ELISA
Carvalho et al., 2020, Brazil	Mice, Swiss, F	Phytol 50 mg/kg p.o.	CFA-induced arthritis	Synovial fluid and medulla	ELISA
Chen et al., 2020, China	Rat	Leonurine hydrochloride 10 mm intra-articulary	Osteoarthritis model	Rat chondrocytes	PCR/Western blot
Correa et al., 2016, Brazil	C57BL/6 mice, M	Methyl gallate 7 mg/kg p.o.	Zymosan-induced arthritis	Knee joint macrophage cell supernatant	ELISA
Jain et al., 2014, India	DBA/1J mice, F	Bergenin 3, 10, 30 mg/kg p.o.	Collagen-induced arthritis	Serum	ELISA
Khan et al., 2020, India	Wistar albino rats, M	Stigmasterol 200 mg/kg p.o.	Collagen-induced arthritis	Joint homogenate	ELISA PCR
Lee et al., 2005, Republic of Korea	DBA/1J mice, M	Chebulagic acid 10 and 20 mg/kg i.p.	Collagen-induced arthritis	Serum	ELISA
Liu et al., 2015, China	DBA/1 mice	Tamarixinin A 20 mg/kg c.s.	Collagen-induced arthritis	Cell culture supernatant	ELISA
Ma et al., 2019, China	Wistar rat, F	Pedunculoside 15 mg/kg p.o.	Collagen-induced arthritis	Arthritic FLS cells	PCR
Martin et al., 2009, New Zealand	C57BL/J6 mice, M	6-hydroxypaxidal 6.925 μmol/kg i.p.	Monosodium urate crystal-induced inflammation in gouty arthritis	Peritoneal wash supernatant	ELISA
Munroe et al., 2007, USA	C57BL/6 mice, F	Honokiol 3 mg/kg i.p.	Collagen-induced arthritis	B-cell culture	ELISA
Ren et al., 2013, China	Sprague-Dawley rats, M	5,7,30-Triacetyl hesperetin 66 and 132 mg/kg p.o.	Adjuvant-induced arthritis	Serum synovial tissue	ELISA PCR
Staurengo-Ferrari et al., 2018, Brazil	Mice, Swiss, M	Trans-chalcone 30 mg/kg, p.o	Monosodium urate crystals-induced gout arthritis	Knee joint	ELISA
Tong et al., 2014, USA	Lewis rats, M	Pristimerin 1 mg/kg i.p.	Adjuvant-induced arthritis	Spleen adherent cells	PCR
Wang et al., 2014, China	Sprague-Dawley rats, F	Berberine 200 mg/kg p.o.	Bovine Type II collagen-induced arthritis	Sera	ELISA

(Continued)

TABLE 3.2 *(Continued)*

Article	Animal	Substance	Experimental Model	IL-6 Mensuration	
Wang et al., 2020, China	C57/BL6 mice, F/M	Notopterol 10 mg/kg i.p.	Type II collagen-induced arthritis	Serum and paw tissue homogenate	ELISA
Wang et al., 2020, China	Sprague-Dawley rats	Ursolic acid 1 mM in joint	Anterior cruciate ligament transection (ACLT)-induced osteoarthritis	Chondrocytes	PCR
Weng et al., 2021, China	Sprague-Dawley rats, M	Corynoline 10, 20 and 30 mg/kg i.p.	CFA-induced arthritis	Serum	ELISA
Xiao et al., 2016, China	DBA/1J mice, M	Piperlongumine 40 mg/kg i.p.	Collagen-induced arthritis	Supernatants of dendritic and CD4 + T cell suspension	Cytometric bead array
Xu et al., 2018, China	Sprague-Dawley rats, M	Osthole 20 and 40 mg/kg p.o.	Collagen-induced arthritis	Serum	ELISA
Xue et al., 2019, China	C57BL/6 wild-type mice, M	Troxerutin 30 mg/kg p.o.	Surgical destabilization of the medial meniscus (osteoarthritis model)	Human chondrocytes	ELISA
Yang et al., 2017, China	DBA/1 mice, F	Periplocoside A 20 mg/kg p.o.	Collagen-induced arthritis	Serum	ELISA
Zhai et al., 2017, China	DBA/1 mice, M	Paeonol 10 mg/kg p.o.	Collagen-induced arthritis associated with complete Freund's adjuvant	Human fibroblast-like synoviocytes rheumatoid arthritis	ELISA
Zhao et al., 2017, China	Kunming mice, F	Isoforskolin 1 mg/kg i.p.	Lyme arthritis model	Murine macrophages	ELISA

In general, CFA-induced arthritis is the most used animal model to study and analyse inflammatory polyarthritis. This experimental model is characterized by infiltration of the synovial membrane and associated with joint destruction — the condition that most closely matches the clinical features of human arthritis. A CFA-induced arthritis model was used to evaluate the following NPs: phytol (Carvalho et al., 2020); the aqueous extract and the sulphated polysaccharide isolated from *T. ornate* (Ananthi et al., 2017); *M. malabathricum* extract (Kumar et al., 2016); tamarixinin A (Liu et al., 2015); pristimerin (Tong et al., 2014); triacetyl hesperetin (Ren et al., 2013); alkaloid-free prenylated flavonoid-enriched fraction of rhizomes of *Sophora flavescens* (Jin et al., 2010); essential oils from *Alpinia oxyphylla,* paeonol (Zhai et al., 2017); n-butanol extract from *Ipomoea stolonifera* (Cai et al., 2014); aqueous extract of *Lamellidens marginalis* (Chakraborty et al., 2010); corynoline (Weng et al., 2021); and *Spirulina platensis* suspension (Ali, Barakat, and Hassan, 2015). The NPs tested significantly reduced the joint swelling, decreased the polyarthritis index in CFA animals and down-regulated the production of IL-6. Thus, plant-derived substances and compounds may be developed as IL-6 inhibitors for the treatment of arthritis and other degenerative/inflammatory diseases.

In a collagen-induced arthritis (CIA) animal model pedunculoside, hydroalcoholic crude extract from *Chenopodium ambrosioides* (Pereira et al., 2018), red ginseng saponin fraction (Endale et al., 2014), berberine (Wang et al., 2014), bergenin (Jain et al., 2014), chebulagic acid (Lee et al., 2005), osthole (Xu et al., 2018), stigmasterol (Ahmad Khan et al., 2020) and *Humulus japonicus* extract (Kang et al., 2020) decreased serum IL-6 levels. Polyphenols in various food products, with particularly high levels in fruits, vegetables and red wine, have been shown to have anti-inflammatory, antioxidant, anticancer and immunomodulatory properties. The authors from the studies mentioned here used a wide range of plant-derived polyphenols in a dose-dependent manner to cure CIA specifically by targeting IL-6.

Methyl gallate is the primary phenol compound derived from many medicinal plant species. Oral administration of methyl gallate attenuated the experimental arthritis induced by zymosan via the inhibition of several proinflammatory mediators, particularly IL-1β, IL-6 and TNF-α. The molecular analysis of the study showed that methyl gallate inhibited oedema production, leukocyte migration and had a profound influence on signalling cascades related to inflammatory mediators (Correa et al., 2016). Hydroalcoholic extract *Schinus terebinthifolius* (Brazilian pepper) containing main chemical constituents like gallic acid, methyl gallate and pentagalloylglucose has been traditionally used in the treatment of arthritis, gout, tumours, diarrhoea and respiratory problems in Brazil. Hydroalchoholic extract derived from Brazilian peppers suppressed the secretion of IL-6, IL-1β and TNF-α in synovial tissue of zymosan-induced arthritis (Rosas et al., 2015), therefore showing potential to be further developed as a safe therapeutic agent for treating joint inflammation.

Trans-Chalcone is a kind of flavonoid compound that represents a major set of polyphenolic compounds widely distributed in fruits, vegetables and herbs. *Trans*-Chalcone reduced the arthritis conditions induced by monosodium urate crystals, down-regulating IL-1β, TNF-α and IL-6 in experimental gout arthritis in mice. Moreover, *trans*-Chalcone decreased the activation of NF-κB, indicating its potential as an anti-inflammatory flavonoid for the treatment of gout arthritis (Staurengo-Ferrari et al., 2018).

Isoforskolin is a bioactive compound found in *Coleus forskohlii.* Isoforskolin administration was shown to inhibit expression of the proinflammatory mediators, IL-6 and TNF-α in murine macrophages in a Lyme model of arthritis in mice. Treatment with isoforskolin resulted in a decreased arthritic index, with histological examinations showing improvements in joints (Zhao et al., 2017). Troxerutin suppressed the proinflammatory response in chondrocytes such as cyclooxygenase-2 (COX-2), inducible nitric oxide synthase (iNOS), nitric oxide (NO), prostaglandin E2 (PGE2), TNF-α and IL-6 in a mice osteoarthritis model. In addition, troxerutin reduced the narrowing of the joint space and the loss of proteoglycans in the mice osteoarthritis model (Xue et al., 2019). Sesame oil decreased osteoarthritis-linked joint pain in rats. It decreased muscular IL-6, lipid peroxidation, Nrf2 protein expression and reactive oxygen species generation (Hsu, Chu, and Jou, 2016).

Polygodial, 6-hydroxypaxidal and other sesquiterpene dialdehyde analogs showed significant inhibition of neutrophil infiltration and superoxide production by infiltrating neutrophils in *in vivo* models of monosodium urate (MSU) crystal-induced inflammation in gouty arthritis. It also showed decreased levels of IL-1β, IL-6 and the neutrophil chemokine KC (Martin et al., 2009). *Tribulus terrestris* showed significant reductions in levels of IL-6, IL-10, TNF-α, NO and COX-2 and improved inflammation in monosodium iodoacetate-induced osteoarthritis pain in rats. The expression of matrix metalloproteinases-2 and -9 in the osteoarthritis condition was significantly regulated by the *T. terrestris* treatment (Park et al., 2017). Leonurine hydrochloride administration to the osteoarthritis model in rats resulted in the significant suppression of MMP-1, 3, 13, IL-6 and ADAMTS-5 and attenuated the progress of the inflammation via the NF-κB dependent signalling pathway (Chen et al., 2020).

Terpenes are chemical entities that modulate the production of cytokines and are found in chronic inflammatory conditions such as AR and similar diseases (Carvalho et al., 2020). This immunomodulatory profile of terpenes is significant because they can act (depending on the type and class of the terpene involved) by improving the production of anti-inflammatory cytokines (such as IL-10) and at the same time mitigating the storm of proinflammatory cytokines (IL-6, TNF-α and IL-1β) induced in acute and chronic inflammatory processes (Lima et al., 2013).

A weakness of the articles identified in this review is the general lack of pharmacokinetic studies of the chemical compounds and/or that they are often related to the chosen animal model. The challenges involved in mimicking the complex clinical aspects of arthritis found in humans in animal models are something that limits, at least in part, the findings of the studies in relation to the disease itself. However, the finding that the substances modulate cytokines is more reliable. Despite their limitations, animal models are very important experimental tools and have already produced significant findings that have contributed to the better management of arthritis (Kollias, 2011). The key role of IL-6 in controlling the molecular triggers that combine to produce the painful and debilitating processes associated with arthritis is well established, so substances identified in animal studies that mitigate the effects of IL-6 are potential candidates for clinical application against arthritis or other chronic inflammatory processes. These clinical studies of efficacy and safety, as well as pharmacokinetic studies are already overdue.

This chapter used a systematic review approach to summarize the evidence produced in preclinical animal studies in relation to the use of NPs to target IL-6 in arthritis. However, the topic needs to be further investigated and studies by different scientific groups around the world must seek to deepen the understanding of this subject to shed more light on the both the bottlenecks (in terms of developing clinical treatments) and the potential related to these NPs.

DECLARATION OF INTERESTS

The authors have no relevant affiliations or financial involvement with any organization or entity with a financial interest in, or financial conflict with, the subject matter or materials discussed in the book chapter.

ACKNOWLEDGEMENTS

This work was supported by grants from FAPITEC-SE, CNPq and CAPES, all agencies from Brazil.

REFERENCES

Ahmad Khan, Mahmood, Abu Hasnath Md Golam Sarwar, Rani Rahat, Rafat Sultana Ahmed, and Sadiq Umar. "Stigmasterol protects rats from collagen induced arthritis by inhibiting proinflammatory cytokines." *International Immunopharmacology* 85 (2020): 106642.

Ali, Eman A. I., Bassant M. Barakat, and Ranya Hassan. "Antioxidant and angiostatic effect of *Spirulina platensis* suspension in complete Freund's adjuvant-induced arthritis in rats." *PLoS One* 10, no. 4 (2015): e0121523.

Almutairi, Khalid, Johannes Nossent, David Preen, Helen Keen, and Charles Inderjeeth. "The global prevalence of rheumatoid arthritis: a meta-analysis based on a systematic review." *Rheumatology International* 41, no. 5 (2021): 863–877.

Ananthi, S., V. Gayathri, R. Malarvizhi, Meenakshi Bhardwaj, and Hannah R. Vasanthi. "Anti-arthritic potential of marine macroalgae *Turbinaria ornata* in complete Freund's adjuvant-induced rats." *Experimental and Toxicologic Pathology* 69, no. 8 (2017): 672–680.

Cai, Congyi, Yicun Chen, Shuping Zhong, Bin Ji, Jinzhi Wang, Xueting Bai, and Ganggang Shi. "Anti-inflammatory activity of N-butanol extract from *Ipomoea stolonifera* in vivo and in vitro." *PLoS One* 9, no. 4 (2014): e95931.

Calixto, João B. "The role of natural products in modern drug discovery." *Anais da Academia Brasileira de Ciências* 91, Suppl no. 3 (2019): e20190105.

Carvalho, Alexandra M. S., Luana Heimfarth, Erik Willyame Menezes Pereira, Fabrício Santana Oliveira, Irwin R. A. Menezes, Henrique D. M. Coutinho, Laurent Picot, Angelo Roberto Antoniolli, Jullyana S. S. Quintans, and Lucindo J. Quintans-Júnior. "Phytol, a chlorophyll component, produces antihyperalgesic, anti-inflammatory, and antiarthritic effects: possible NFκB pathway involvement and reduced levels of the proinflammatory cytokines TNF-α and IL-6." *Journal of Natural Products* 83, no. 4 (2020): 1107–1117.

Chakraborty, Mousumi, Sourav Bhattacharya, Pushpak Bhattacharjee, Rinku Das, and Roshnara Mishra. "Prevention of the progression of adjuvant induced arthritis by oral supplementation of Indian fresh water mussel *(Lamellidens marginalis)* aqueous extract in experimental rats." *Journal of Ethnopharmacology* 132, no. 1 (2010): 316–320.

Chen, Cheng, Zhenglin Zhu, Ning Hu, Xi Liang, and Wei Huang. "Leonurine hydrochloride suppresses inflammatory responses and ameliorates cartilage degradation in osteoarthritis via NF-κB signaling pathway." *Inflammation* 43, no. 1 (2020): 146–154.

Choy, Ernest H. S., and Leonard H. Calabrese. "Neuroendocrine and neurophysiological effects of interleukin 6 in rheumatoid arthritis." *Rheumatology* 57, no. 11 (2018): 1885–1895.

Coimbra, Ibsen Bellini, Pérola Grinberg Plapler, and Gustavo Constantino de Campos. "Generating evidence and understanding the treatment of osteoarthritis in Brazil: a study through Delphi methodology." *Clinics* 74 (2019): e722.

Correa, Luana Barbosa, Tatiana Almeida Pádua, Leonardo Noboru Seito, Thadeu Estevam Moreira Maramaldo Costa, Magaiver Andrade Silva, André Luis Peixoto Candéa, Elaine Cruz Rosas, and Maria G. Henriques. "Anti-inflammatory effect of methyl gallate on experimental arthritis: inhibition of neutrophil recruitment, production of inflammatory mediators, and activation of macrophages." *Journal of Natural Products* 79, no. 6 (2016): 1554–1566.

Emery, Paul. "Treatment of rheumatoid arthritis." *British Medical Journal* 332, no. 7534 (2006): 152–155.

Endale, Mehari, Eun Ju Im, Joo Young Lee, Sung Dae Kim, Taddesse Yayeh, Yong-Bum Song, Yi-Seong Kwak et al. "Korean red ginseng saponin fraction rich in ginsenoside-Rb1, Rc and Rb2 attenuates the severity of mouse collagen-induced arthritis." *Mediators of Inflammation* 2014 (2014): 1–14.

Hanh, P. T. B., D. T. Thao, N. T. Nga, et al. "Toxicity and anti-inflammatory activities of an extract of the *Eleutherine bulbosa* rhizome on collagen antibody-induced arthritis in a mouse model." *Natural Product Communications* 13, no. 7(July 2018): 883–886.

Hsu, Dur-Zong, Pei-Yi Chu, and I-Ming Jou. "Enteral sesame oil therapeutically relieves disease severity in rat experimental osteoarthritis." *Food & Nutrition Research* 60, no. 1 (2016): 29807.

Jain, Shreyans K., Surjeet Singh, Anamika Khajuria, Santosh K. Guru, Prashant Joshi, Samdarshi Meena, Janhavi R. Nadkarni et al. "Pyrano-isochromanones as IL-6 inhibitors: Synthesis, in vitro and in vivo antiarthritic activity." *Journal of Medicinal Chemistry* 57, no. 16 (2014): 7085–7097.

Jin, Jeong Ho, Ju Sun Kim, Sam Sik Kang, Kun Ho Son, Hyun Wook Chang, and Hyun Pyo Kim. "Anti-inflammatory and anti-arthritic activity of total flavonoids of the roots of *Sophora flavescens*." *Journal of Ethnopharmacology* 127, no. 3 (2010): 589–595.

Jung, H. W., J. K. Jung, K. H. Son, D. H. Lee, T. M. Kang, Y. S. Kim, and Y. K. Park. "Inhibitory effects of the root extract of *Dipsacus asperoides* C.Y. Cheng et al T.M.Ai on collagen-induced arthritis in mice." *Journal of Ethnopharmacology* 139, no. 1 (2012): 98–103.

Kang, Eun-Jung, Hyun Jong Kim, Ji Hyun Choi, Jung-Ran Noh, Jae-Hoon Kim, In Bok Lee, Young-Keun Choi et al. "*Humulus japonicus* extract ameliorates collagen-induced arthritis in mice through regulation of overall articular inflammation." *International Journal of Molecular Medicine* 45, no. 2 (2020): 417–428.

Khan, Mahmood A., Sarwar, Abu H. M. G., Rani Rahat, Ahmed, Rafat S., Sadiq Umar. "Stigmasterol protects rats from collagen induced arthritis by inhibiting proinflammatory cytokines." *International Immunopharmacology* 85 (2020): 106642.

Khanna, Dinesh, Gautam Sethi, Kwang Seok Ahn, Manoj K. Pandey, Ajaikumar B. Kunnumakkara, Bokyung Sung, Amita Aggarwal, and Bharat B. Aggarwal. "Natural products as a gold mine for arthritis treatment." *Current Opinion in Pharmacology* 7, no. 3 (2007): 344–351.

Kim, G. W., N. R. Lee, R. H. Pi, Y. S. Lim, Y. M. Lee, J. M. Lee, H. S. Jeong, and S. H. Chung. "IL-6 inhibitors for treatment of rheumatoid arthritis: past, present, and future." *Archives of Pharmacal Research* 38 (2015): 575–584.

Kollias, G., P. Papadaki, F. Apparailly, M. J. Vervoordeldonk, R. Holmdahl, V. Baumans, C. Desaintes, J. Di Santo, J. Distler, P. Garside, M. Hegen, T. W. Huizinga, A. Jüngel, L. Klareskog, I. McInnes, I. Ragoussis, G. Schett, B. Hart, P. P. Tak, R. Toes, W. van den Berg, W. Wurst, and S. Gay. "Animal models for arthritis: innovative tools for prevention and treatment." *Annals of the Rheumatic Diseases* 70, no. 8 (2011):1357–1362.

Kumar, Vikas, Prakash Chandra Bhatt, Kalicharan Sharma, Mahfoozur Rahman, Dinesh Kumar Patel, Nikunj Sethi, Atul Kumar et al. "*Melastoma malabathricum* Linn attenuates complete Freund's adjuvant-induced chronic inflammation in Wistar rats via inflammation response." *BMC Complementary and Alternative Medicine* 16, no. 1 (2016): 1–16.

Kumar, N., S. Singh, N. Patro, and I. Patro. "Evaluation of protective efficacy of *Spirulina platensis* against collagen-induced arthritis in rats." *Inflammopharmacology* 17, no. 3 (2009): 181–190.

Lally, Frank, Emily Smith, Andrew Filer, Michael A. Stone, John S. Shaw, Gerard B. Nash, Christopher D. Buckley, and G. Ed. Rainger. "A novel mechanism of neutrophil recruitment in a coculture model of the rheumatoid synovium." *Arthritis & Rheumatism* 52, no. 11 (2005): 3460–3469.

Lee, Sang-Ik, Pung-Mi Hyun, Seung-Hyung Kim, Kyoung-Shin Kim, Sang-Keun Lee, Byoung-Soo Kim, Pil Jae Maeng, and Jong-Soon Lim. "Suppression of the onset and progression of collagen-induced arthritis by chebulagic acid screened from a natural product library." *Arthritis & Rheumatism: Official Journal of the American College of Rheumatology* 52, no. 1 (2005): 345–353.

Li, G., Y. Fu, J. Zheng, and D. Li. "Anti-inflammatory activity and mechanism of a lipid extract from hard-shelled mussel *(Mytilus coruscus)* on chronic arthritis in rats." *Marine Drugs* 12, no. 2 (2014): 568–588.

Li, J. W.-H., and John C. Vederas. "Drug discovery and natural products: End of an era or an endless frontier?" *Science* 325, no. 5937 (2009): 161–165.

Lima, M. da S., L. J. Quintans-Júnior, W. A. de Santana, C. Martins Kaneto, M. B. Pereira Soares, and C. F. Villarreal. "Anti-inflammatory effects of carvacrol: evidence for a key role of interleukin-10." *European Journal of Pharmacology* 699 (2013): 112–117.

Liu, Jia Bao, Ya Si Ding, Ying Zhang, Jia Bao Chen, Bao Song Cui, Jin Ye Bai, Ming Bao Lin, Qi Hou, Pei Cheng Zhang, and Shuai Li. "Anti-inflammatory hydrolyzable tannins from *Myricaria bracteata*." *Journal of Natural Products* 78, no. 5 (2015): 1015–1025. doi:10.1021/np500953e.

Losina, Elena, Meghan E. Daigle, L. G. Suter, D. J. Hunter, D. H. Solomon, R. P. Walensky, J. M. Jordan, Sara A. Burbine, A. David Paltiel, and Jeffrey N. Katz. "Disease-modifying drugs for knee osteoarthritis: can they be cost-effective?" *Osteoarthritis and Cartilage* 21, no. 5 (2013): 655–667.

Ma, X., G. Chen, J. Wang, J. Xu, F. Zhao, M. Hu, Z. Xu, B. Yang, J. Guo, S. Sun, and M. Liu. "Pedunculoside attenuates pathological phenotypes of fibroblast-like synoviocytes and protects against collagen-induced arthritis." *Scandinavian Journal of Rheumatology* 48, no. 5 (2019): 383–392.

Martin, William John, Patries M. Herst, Elizabeth W. Chia, and Jacquie L. Harper. "Sesquiterpene dialdehydes inhibit MSU crystal-induced superoxide production by infiltrating neutrophils in an in vivo model of gouty inflammation." *Free Radical Biology and Medicine* 47, no. 5 (2009): 616–621.

McInnes, Iain B., and Georg Schett. "Cytokines in the pathogenesis of rheumatoid arthritis." *Nature Reviews Immunology* 7, no. 6 (2007): 429–442.

Moher, D., A. Liberati, J. Tetzlaff, and D. G. Altman. "Preferred reporting items for systematic reviews and meta-analyses: the PRISMA statement." *British Medical Journal* 339 (2009): b2535.

Munroe, M. E., J. L. Arbiser, and G. A. Bishop. "Honokiol, a natural plant product, inhibits inflammatory signals and alleviates inflammatory arthritis." *Journal of Immunology* 179, no. 2 (2007): 753–763.

Newman, David J., and Gordon M. Cragg. "Natural products as sources of new drugs from 1981 to 2014." *Journal of Natural Products* 79, no. 3 (2016): 629–661.

Ogata, A., Y. Kato, S. Higa, and K. Yoshizaki. "IL-6 inhibitor for the treatment of rheumatoid arthritis: a comprehensive review." *Modern Rheumatology* 29, no. 2 (2019): 258–267.

Ouzzani, Mourad, Hossam Hammady, Zbys Fedorowicz, and Ahmed Elmagarmid. "Rayyan—A web and mobile app for systematic reviews." *Systematic Reviews* 5, no. 1 (2016): 1–10.

Paleolog, Ewa M. "Angiogenesis in rheumatoid arthritis." *Arthritis Research & Therapy* 4, no. 3 (2002): 1–10.

Park, Young Jin, Young-Rak Cho, Joa Sub Oh, and Eun-Kyung Ahn. "Effects of *Tribulus terrestris* on monosodium iodoacetate-induced osteoarthritis pain in rats." *Molecular Medicine Reports* 16, no. 4 (2017): 5303–5311.

Pereira, Wanderson S., Girlaine P. da Silva, Mariana V. Vigliano, Nathalia R. F. Leal, Fabiana A. Pinto, Daniele C. Fernandes, Shirley V. M. Santos et al. "Anti-arthritic properties of crude extract from *Chenopodium ambrosioides* L. leaves." *Journal of Pharmacy and Pharmacology* 70, no. 8 (2018): 1078–1091.

Ren, Dan Yang, Tao Xu, Rong Li, Cheng Huang, Yan Huang, Ren Qiu Li, Hui Ying Li, and Jun Li. "5, 7, 3′-Triacetyl hesperetin suppresses adjuvant-induced arthritis in rats through modulating JAK2/STAT3 pathway." *American Journal of Chinese Medicine* 41, no. 03 (2013): 601–614.

Rosas, Elaine Cruz, Luana Barbosa Correa, Tatiana de Almeida Pádua, Thadeu Estevam Moreira Maramaldo Costa, José Luiz Mazzei, Alan Patrick Heringer, Carlos Alberto Bizarro, Maria Auxiliadora Coelho Kaplan, Maria Raquel Figueiredo, and Maria G. Henriques. "Anti-inflammatory effect of *Schinus terebinthifolius* Raddi hydroalcoholic extract on neutrophil migration in zymosan-induced arthritis." *Journal of Ethnopharmacology* 175 (2015): 490–498.

Sannegowda, K. M., S. H. Venkatesha, and K. D. Moudgil. "*Tinospora cordifolia* inhibits autoimmune arthritis by regulating key immune mediators of inflammation and bone damage." *International Journal of Immunopathology and Pharmacology* 28, no. 4 (2015): 521–531.

Sato, K., M. Tsuchiya, J. Saldanha, et al. "Reshaping a human antibody to inhibit the interleukin 6-dependent tumor cell growth." *Cancer Research* 53 (1993): 851–856.

Sicras-Mainar, Antoni, Carlos Tornero-Tornero, Francisco Vargas-Negrín, Isabel Lizarraga, and Javier Rejas-Gutierrez. "Health outcomes and costs in patients with osteoarthritis and chronic pain treated with opioids in Spain: the OPIOIDS real-world study." *Therapeutic Advances in Musculoskeletal Disease* 12 (2020): 1759720X20942000.

Staurengo-Ferrari, Larissa, Kenji W. Ruiz-Miyazawa, Felipe A. Pinho-Ribeiro, Victor Fattori, Tiago H. Zaninelli, Stephanie Badaro-Garcia, Sergio M. Borghi et al. "Trans-chalcone attenuates pain and inflammation in experimental acute gout arthritis in mice." *Frontiers in Pharmacology* 9 (2018): 1123.

Thanh, Huyen Do Thi, Thi, Phuong Ngo, Thu, Hien Nguyen Thi, Thi, Nga Nguyen, Ngoc, Hung Le, Thi, Thao Do, and Minh, Ha Le "*Polygonatum kingianum* rhizome extract alleviates collagen antibody-induced arthritis by modulating proinflammatory cytokine production in mice." *Asian Pacific Journal of Tropical Biomedicine* 10 (2020): 490–495.

Tong, Li, Siddaraju M. Nanjundaiah, Shivaprasad H. Venkatesha, Brian Astry, Hua Yu, and Kamal D. Moudgil. "Pristimerin, a naturally occurring triterpenoid, protects against autoimmune arthritis by modulating the cellular and soluble immune mediators of inflammation and tissue damage." *Clinical Immunology* 155, no. 2 (2014): 220–230.

Wang, C., Y. Gao, Z. Zhang, C. Chen, Q. Chi, K. Xu, and L. Yang. "Ursolic acid protects chondrocytes, exhibits anti-inflammatory properties via regulation of the NF-κB/NLRP3 inflammasome pathway and ameliorates osteoarthritis." *Biomedicine & Pharmacotherapy* 130 (2020): 110568.

Wang, Q., X. Zhou, L. Yang, Y. Zhao, Z. Chew, J. Xiao, C. Liu, X. Zheng, Y. Zheng, Q. Shi, Q. Liang, Y. Wang, and H. Wang. "The natural compound notopterol binds and targets JAK2/3 to ameliorate inflammation and arthritis." *Cell Reports* 33, no. 8 (2020): 108442.

Wang, Zhigang, Zhe Chen, Sisi Yang, Yu Wang, Zhaoyi Huang, Jianfei Gao, Shenghao Tu, and Zhiguo Rao. "Berberine ameliorates collagen-induced arthritis in rats associated with anti-inflammatory and anti-angiogenic effects." *Inflammation* 37, no. 5 (2014): 1789–1798.

Weng, Weifeng, Fuyong Wang, Xiaojian He, Kaihua Zhou, Xiaotian Wu, and Xiao Wu. "Protective effect of corynoline on the CFA induced rheumatoid arthritis via attenuation of oxidative and inflammatory mediators." *Molecular and Cellular Biochemistry* 476, no. 2 (2021): 831–839.

WHO. Scientific Group on the Burden of Musculoskeletal Conditions at the Start of the New Millennium, "The burden of musculoskeletal conditions at the start of the new millennium." *World Health Organization Technical Report Series* 919 (2003): i.

WHO. 2021. "Chronic Rheumatic Conditions." Accessed May 5. http://www.who.int/chp/topics/rheumatic/en/.

Xiao, Y., M. Shi, Q. Qiu, M. Huang, S. Zeng, Y. Zou, Z. Zhan, L. Liang, X. Yang, and H. Xu. "Piperlongumine suppresses dendritic cell maturation by reducing production of reactive oxygen species and has therapeutic potential for rheumatoid arthritis." *Journal of Immunology* 196, no. 12 (2016): 4925–4934.

Xu, Renguo, Zhen Liu, Jiande Hou, Tao Huang, and Ming Yang. "Osthole improves collagen-induced arthritis in a rat model through inhibiting inflammation and cellular stress." *Cellular & Molecular Biology Letters* 23, no. 1 (2018): 1–11.

Xue, Xinghe, Yunlin Chen, Ye Wang, Jingdi Zhan, Bin Chen, Xiangyang Wang, and Xiaoyun Pan. "Troxerutin suppresses the inflammatory response in advanced glycation end-product-administered chondrocytes and attenuates mouse osteoarthritis development." *Food & Function* 10, no. 8 (2019): 5059–5069.

Yang, Y., X. Hu, L. Cheng, W. Tang, W. Zhao, Y. Yang, and J. Zuo. "Periplocoside A ameliorated type II collagen-induced arthritis in mice via regulation of the balance of Th17/Treg cells." *International Immunopharmacology* 44 (2017): 43–52.

Zhang, Lanyue, Xiaoxin Liang, Zhirong Ou, Mao Ye, Yaohui Shi, Yubin Chen, Jiawei Zhao, Danxia Zheng, and Hongping Xiang. "Screening of chemical composition, anti-arthritis, antitumor and antioxidant capacities of essential oils from four *Zingiberaceae* herbs." *Industrial Crops and Products* 149 (2020): 112342.

Zhai, Ke feng, Hong Duan, Lin Luo, Wen gen Cao, Fang kai Han, Ling ling Shan, and Xue mei Fang. "Protective effects of paeonol on inflammatory response in Il-1β-induced human fibroblast-like synoviocytes and rheumatoid arthritis progression via modulating NF-KB pathway." *Inflammopharmacology* 25, no. 5 (2017): 523–532.

Zhao, Hua, Aihua Liu, Longqiang Shen, Cuiping Xu, Ziwei Zhu, Jiaru Yang, Xinling Han, Fukai Bao, and Weimin Yang. "Isoforskolin downregulates proinflammatory responses induced by *Borrelia burgdorferi* basic membrane protein A." *Experimental and Therapeutic Medicine* 14, no. 6 (2017): 5974–5980.

4 Inside Alzheimer's Disease Diagnosis

Uncovering Nanoparticle Contributions and Perspective

Gomathi Rajkumar, Murugan Rajan, Mairim Serafini, Narendra Narain, Adriano A.S. Araujo, Lucindo José Quintans Júnior and Lijing Ke

CONTENTS

4.1 INTRODUCTION

Alzheimer's disease (AD), the most common form of dementia, is an irreversible neurodegenerative disorder of the elderly. The disease is characterized by loss of memory, especially episodic memory, cognitive decline, behavioural and physical disability and ultimately leads to death. Several biological pathways, molecules, heredity, environmental factors and others have been claimed towards its etiology (Hansen et al., 2008). The disruption of neurons observed even long before the onset of symptoms in AD patients urges the early diagnosis and the inability towards such an early detection of AD is in the limelight as a major threat across the globe. Reports suggest that the frequency of AD increases every year, and by 2050, it is expected that more than 100 million individuals will be affected worldwide (Prince et al., 2013). A considerable research effort has been made towards diagnosis and therapeutics for prevention and/or treatment of AD, however, the causes and etiology of AD remain elusive and AD therapy/prevention is considered to be the most high-risk approach in the pharmaceutical industry (Hajipour et al., 2017). Furthermore, several

DOI: 10.1201/9781003231745-4

drug candidates that were designed to ameliorate early-stage AD and prevent cognitive deficits, failed to improve conditions in clinical trials. Thus, new strategies in diagnosis and treatment are in high demand for combating AD.

In recent years, nanotechnology has offered engineering materials or devices with the smallest functional units on the nanometre scale (1–100 nm) using polymers, lipids, viruses and organometallic compounds (Gupta et al., 2019). This cutting-edge technology offers several exciting features for the diagnosis, treatment and drug delivery of AD (Paraskevaidi et al., 2018). Recently, several multifunctional nanoplatforms have emerged to provide diagnostic, therapeutic and theranostic solutions, offering newer insights in alleviating AD (Paraskevaidi et al., 2018; Gupta et al., 2019). The emphasis in this chapter focuses on diagnosis because an accurate diagnosis is the first step in enabling an early intervention, slowing down the progression of AD and improving the lives of patients (Paraskevaidi et al., 2018).

4.2 PATHOPHYSIOLOGY OF AD

During the past two decades, the cholinergic hypothesis, i.e. a dysfunction of acetylcholine-containing neurons in the brain, was associated with cognitive decline in Alzheimer's disease (Hansen et al., 2008). Acetylcholinesterase inhibitors, which inhibit the breakdown of acetylcholine into choline and acetate in the central nervous system (CNS), used to be the key focus in developing anti-AD drugs. However, the acetylcholinesterase inhibitors (i.e. donepezil) that once dominated the market are now declining, owing to their short effective period and limited efficiency (Hansen et al., 2008). In the past decade, several other hypotheses have emerged on the pathogenesis of AD. The major hallmark has been attributed to amyloid β (Aβ) deposition, neurofibrillary tangles induced by hyperphosphorylated tau proteins and neuronal death.

4.2.1 Amyloid β (Aβ)

The amyloid precursor protein (APP), is a central protein that plays a key role in the loss of neurons (cholinergic) in the hippocampus and cortex of the brain. Amyloid β (Aβ) is the product of proteolytic cleavage of APP. Under normal condition, α-secretases and γ-secretases cleave to APP-forming nonaggregating fragments. However, a dysfunction is observed when APP is cleaved by β-secretases and γ-secretases, resulting in aggregation of Aβ fragments in the intraneuronal network forming amyloid plaques. The oligomerization of Aβ plaques are found to cause neuronal cell death *via* intracellular neurofibrillary tangle (NFTs) formation and synaptic dysfunction, resulting in neurotoxicity in AD (Takahashi et al., 2017).

4.2.2 Hyperphosphorylation and Tau Protein Aggregation

Hyperphosphorylation occurs when there is an imbalance between the formation and clearance of Aβ in neurons. As a result, the tau protein loses its binding affinity for microtubule assembly, resulting in tau aggregates and NFTs, which lead to AD (Lippens et al., 2007).

4.2.3 Metal Ions

An abnormal breakdown of key metal ions required for brain function, such as copper, zinc and iron, revealed neurodegeneration and reactive oxygen species (ROS) production resulting in AD. Further studies also revealed that epigenetic changes caused by environmental exposure, inadequate diet and over expression of APP by ROS resulted in accumulation of Aβ and increased the severity of AD (Gupta et al., 2019).

4.2.4 Monoamine Oxidases

The enzyme monoamine oxidases (MAOs) found in the outer mitochondrial membrane of neuronal, glial, and other mammalian cells, act as catalysts in the oxidative deamination of amine neurotransmitters in the CNS. During an up-regulated activity of MAOs, an excess of neurotoxic byproducts such as hydrogen peroxides are produced, which subsequently promoted neuronal dysfunctions in both psychiatric disorders and neurodegenerative diseases (Gandhi and Abramov, 2012).

4.2.5 Astrocytes

Astrocytes perform several functions in the brain such as assisting endothelial cells in the formation of the blood brain barrier (BBB), transporting nutrients to the nervous system, maintaining ion balance and repairing the brain. Recently, investigations into the unusual behaviour of astrocytes along with the inflammation factor have been observed to be associated with the progression of AD (Gupta et al., 2019).

The last stage of AD is found to have drastic effects, despite its pathway, on both the physical and mental health of patients. Aside from loss of memory, cognitive impairment and behavioural instability in behaviour, AD patients were vulnerable to skin infections, sepsis, blood clots and severe internal inflammation, which led to organ failure (Ayaz et al., 2020). In AD, the main biomarkers are found to be Aβ peptide namely $A\beta_{1\text{-}40}$, $A\beta_{1\text{-}42}$, $A\beta_{42}$, tau protein in plasma, phosphorylated tau proteins, acetylcholine and others such as human apolipoprotein E (apoE), plasma fibrinogen and AAT protein in cerebrospinal fluid (CSF). $A\beta_{1\text{-}40}$, $A\beta_{1\text{-}42}$ and tau proteins are typical markers in blood-based detection (hemodiagnosis). Regardless of the prevalence of AD among older patients, it has been reported that the younger population may also be susceptible to early onset symptoms of AD pathology due to various genetic mutations in the primary genes involved in APP processing, such as APP, BACE1, PSEN1 and PSEN2. In addition, mutations in other genes, namely TREM2, clusterin, BIN1, ADAM10 and apolipoprotein E type 4 (APOEε4), have also been considered possible risk factors for AD (Ayaz et al., 2020).

4.3 NANOTECHNOLOGY-BASED AD DIAGNOSIS

Early detection is always desirable for treating diseases. Nanotechnology provides nanoparticles (NPs) with inherent valuable properties such as a high surface-to-volume ratio, ease of surface functionalization with desired ligands and ability to cross biological barriers, making them promising candidates for AD diagnosis and treatment. The nanodiagnostic approach provides potential application of imaging and fluid biomarkers owing to the ability of namely, (i) biosensing to detect disease-related biomarker and/or molecular processes and (ii) bioimaging to visualize specific sites in the human body (Nguyen and Zhao, 2015).

Generally, medical imaging techniques, cognitive tests and cerebrospinal fluid (CSF) analyses are used to diagnose AD. Commonly used medical imaging techniques are brain imaging with magnetic resonance imaging (MRI) and positron emission tomography (PET). For the early detection of AD, a sensitive, specific and preferably noninvasive biomarker would be extremely useful (Xu et al., 2021). Imaging probes made from nanomaterials such as fluorescent NPs, quantum dot/quantum rod NPs or multimodal NPs that could specifically bind to Aβ or tau proteins have been used to detect early pathological markers of AD *in vitro* and *in vivo* (Xu et al., 2021). To obtain effective imaging, NPs in the diagnosis of AD should meet the demands of sensitivity and specificity, as MRI records high resolution but low sensitivity while PET depicts low resolution but high sensitivity. CSF assays provide quantitative measurements of AD biomarkers such as Aβ and tau protein. Other strategies include monitoring circulating AD biomarkers employing mass spectrometry (MS), Western-blot and immunohistochemistry (IHC), and immuno-based assays such as enzyme-linked immuno sorbent assay (ELISA), flexible multi-analyte profiling xMAP® assays and the fully automated digital immunoassay analyser Simoa HD-1, accompanied with its own pros and cons (Serafín et al., 2020).

4.3.1 Metallic NPs

High stability and high reactivity offers numerous metallic functionalized NPs for AD diagnosis. Ultra-small Gadolinium-NPs were grafted with two peptides corresponding to the hydrophobic core of Aβ peptide favouring the formation of a β-sheet structure. The surface plasmon resonance spectroscopy experiments demonstrated that these NPs interacted successfully with $A\beta_{1\text{-}42}$ fibrils, supporting their role as an amyloid fibril marker in the diagnosis of AD (Gupta et al., 2019). In addition, immunological histochemical studies on Gadolinium-NPs revealed their hippocampus localization in mice. These NPs allowed for high positive MRI contrast, permitting MRI signal enhancement and, in turn, easier image interpretation in neuropathology.

Recently, Gao and co-workers developed an Aβ antibody with conjugated gold nanoparticles (AuNps-anti-Aβ complex) capable of immunoreaction efficiency with Aβ in the mouse brain. They used laser ablation inductively coupled plasma mass spectrometry (LA-ICP-MS) to establish quantitative imaging of AD pathology. The obtained images depicted a greater resolution than the conventional techniques, suggesting that the imaging tool also plays an important role in the diagnosis of AD pathologies (Gao et al., 2021).

Kang et al. (2009) suggested an ultrasensitive electrical detection approach for detecting $A\beta_{1-42}$ using scanning tunnelling microscopy (STM). Their research group used a gold surface to immobilize a monoclonal antibody fragment that had a high affinity for $A\beta_{1-42}$. The AuNPs antibody fragment complex was allowed to react with magnetic microparticles, which led to the formation of a "sandwich-like" structure. The resulting chip was analysed using STM. The frequency of Aβ-Au binding events correlated with the surface density of AuNPs, confirming an effective Aβ detection down to 10 fg/mL. The MRI signal images obtained using magnetic nanostructures could be used to effectively diagnose AD.

Garcia-Leis and Sanchez-Cortes (2021) used silver nanoparticles (AgNPs) to study the dynamic behaviour of the shortest fragment of Aβ i.e. $A\beta_{25-35}$, an active region of the amyloid peptide family. The TEM images obtained in their study demonstrated that NPs were subjected to high sensitivity and selective plasmon hybridization. They found that the self-aggregation of these peptides was restructured into β-sheet structures in the supra-molecular aggregates (spheroids and protofibrils). They were also adsorbed on NPs forming α-helix secondary structure. As a result, their findings suggest that the plasmon-based nanoplatforms could not only detect AD, but also trap and eliminate toxic peptides from biological systems, lowering neurotoxicity.

4.3.2 Fluorescent Dye-Loaded NPs

Fluorescent dye-loaded NPs are used in the diagnosis of AD. Thioflavin-T (ThT) is a widely used dye for *in vivo* amyloid detection. ThT-loaded nanocapsules containing polybutyl cyanoacrylate (PBCA) that were injected into the intracerebral region of mice *in vivo* were found to pass through the BBB and interact with fibrillary Aβ species after a biodegradation-induced release from the carriers, as evidenced in using confocal laser scanning microscopy studies. Despite the fact that ThT is hydrophilic, PBCA enhanced the ability of NPs to pass the BBB and favoured the visualization of Aβ peptide pathology in *in vivo* studies (Roney et al., 2005).

4.3.3 Copper Chelator-Containing NPs

Clioquinol, also known as 5-chloro-7-iodo-8-hydroxyquinoline (CQ), is a metal chelator and antimalarial drug (i.e., quinolone) derivative that has been employed in imaging studies. Roney et al. (2005) loaded NPs with radio-iodinated CQ and formulated as PBCA-CQ NPs that were capable of crossing the BBB. These NPs were capable of diverse biodistributions after intravenous administration, making them an ideal tool for bioimaging in AD studies.

4.3.4 Quantum Dot/Quantum Rod NPs

Quantum dots provide fluorescence for optical biosensing due to their semiconducting properties at the nanoscale level. Morales-Narváez et al. (2012) used a sandwich immunocomplex microarray to detect apolipoprotein E using CdSe/ZnS quantum dots with fluorescent dye Alex 647. From their results, it was evident that QDs could open new avenues in detecting various biomarkers of AD such as apolipoprotein E isoforms and others in microarrays. A sandwich fluoresce immunoassay based on tyrosinase (TYR)-induced tau aptamer-tau-tau antibody (anti-tau) was designed to detect tau protein. The resulting fluorescence intensity was formulated from fluorophore dopamine (DA)-functionalized $CuInS_2$/ZnS quantum dots and exhibited effective tau protein detection with a linear concentration ranging from 10 pM to 200 nM (Chen et al., 2019).

Despite the fact that hyperphosphorylated tau protein and Aβ-derived diffusible ligands (ADDLs) are well known biomarkers of AD, quantitation and detection of these species in bodily fluids are challenging owing to their low levels. However, the recent development of NPs with high sensitivity for these biomarkers as mentioned above could be promising in early diagnosis of AD.

4.3.5 Other NPs

Besides the above-mentioned NPs, several others are being studied for their ability to detect AD pathology. HRP-Au-gelsolin nanohybrid formulated uniting horseradish peroxidase (HRP), gelsolin and gold NPs when used as electrochemical nanoprobes was found to be highly sensitive in the assessment of Aβ levels both in normal and AD rat brains (Yu et al., 2018). When tested, Poly-L-lysine-mediated chemical signal-amplified nanobiosensors, which were activated by chemical agents, demonstrated precise Aβ detection sensitivity (Kang et al., 2019).

An electrochemical aptasensor design based on a superhydrophobic carbon filter with the AuPt alloy nanoparticle forming a nanocomposite and characterized with hydrophobic surface was developed for detecting Aβ oligomer by Liu et al. (2021). The hydrophobic surface promoted the sensor to resist the nonspecific adsorption, while the AuPt alloy boosted the electroactive area and electron transfer rate, thereby increasing the sensor's sensitivity and specificity. Additionally, the designed aptasensor also exhibited an antifouling ability and serum sensitivity. This research offered another candidature in AD diagnosis using electrochemical biosensors susceptible to interference in complex biological fluids.

As blood circulating biomarkers have the ability to detect AD, Hadjidemetriou et al. (2021) developed lipid-based NPs capable of monitoring prediseased stages of AD pathologies. The developed blood circulating NPs identified AD disease-specific protein fingerprints from the asymptomatic period to plaque deposition and cognitive decline using corona-coated liposome NPs (HSPC:Chol:DSPE-PEG2000) when delivered intravenously. Their study recommends the theranoomics enrichment strategy as a promising investigation in predicting the development and dynamics of a variety of neurodegenerative diseases.

Kim et al. (2016) developed a wafer-scale patterning approach for reduced graphene oxide (rGO) that may provide high resolution and mass productivity. Their technique of immobilizing anti-Aβ antibodies on the sensor's (rGO) reactions sites, detected Aβ peptides with ultra-high sensitivity and accuracy with detection limits as low as 100 fg/mL. A nanobiosensor prepared based on a composite of poly (diallyl dimethyl ammonium chloride) functionalized with single-walled carbon nanotubes (CNTs) and 2,2′-azinobis-(3-ethylbenzothiazoline-6-sulphonate) by Yu et al. (2018) exhibited detection limits of 0.04 μM for Cu^{2+} and 0.5 ng/mL for $A\beta_{1-42}$ in the plasma and hippocampus of AD rats, respectively. Similarly, an electrochemical hydrogel biosensor was developed by Sun et al. (2018) to monitor Aβ oligomer. An Aβ oligomer nanosensor was constructed by immobilizing the thiolated cellular prion protein (PrPC) peptide probe on gold NPs and graphene oxide. The biosensor presented a high accuracy (linear range for the detection-0.1 pM to 10 nM) in identification of oligomers in blood plasma and artificial CSF, and distinguished Aβ oligomers from fibrils or monomers of Aβ (Sun et al., 2018).

A competitive immunosensor based on gold nanostructured screen-printed carbon electrodes was developed for the quantification of unfolded p53 proteins (Amor-Gutiérrez et al., 2020). To ease selectivity, they constructed a monoclonal antibody that only reacts with the unfolded isoform of the protein as capture antibody. The sensor directly quantified the unfolded p53 in plasma without dilution. Their research group claimed that the developed biosensor could serve as a low-cost, simple, miniaturized and portable device that requires small amounts of sample and could contribute to future point-of-care applications favouring noninvasive diagnosis and monitoring progressive AD pathogenesis.

A paper-based lateral flow immunoassay for the determination of AD blood biomarkers, fetuin B and clusterin was developed by Brazaca et al. (2019). Their research group immobilized selective antibodies for the target biomarkers on AuNPs and deposited them on paper pads. When the blood sample was added on the paper-based device, the biofluid laterally flew towards the selective antibody, forming AuNP-antibody accumulation and causing a change from white to pink. Colour clustering image analysis was found to hold great potential for diagnosing AD in its early stages in their study. Similarly, an "aggregation-based colorimetric sensor array" was developed based on gelsolin-modified silver nanotriangles and silver nanorods for the simultaneous identification and detection of $A\beta_{40}$ and $A\beta_{42}$ with structural similarity (Liu et al., 2020). The fluctuations monitored from different aggregation behaviours of nanocomplex with biomarkers resulted in different colour and unique spectral changes offering a great opportunity for the early diagnosis of AD.

A variety of fluorescent cyclic peptide nanoparticles (c-PNPs) with negative, neutral and positive surface charges were designed, synthesized and detected for Aβ peptide aggregates through fluorescence wavelength and intensity changes in both the cerebrospinal fluid (CSF) and serum (Sun et al., 2021). Through the aromatic, hydrophobic and electrostatic interactions, their research group found binding of negative c-PNPs with Aβ aggregates in CSF and serum, as well as a constant increase in fluorescence intensity and wavelength blueshift. This finding demonstrates the importance of PNPs in the early detection of AD.

4.4 NPs AS THERANOSTIC AGENTS IN AD

Nanotheranostics is gaining in popularity owing to the potential for simultaneous diagnosis, monitoring and treatment of diseases by NPs. As a result of chemical modification on their large surface area, NPs serve a holistic role as a vehicle for delivering multimodal imaging agents and as therapeutic drugs to specific target sites (*in vivo*). In line with this, several NPs are under research as novel theranostic agents for AD pathology. An ultrasmall superparamagnetic iron oxide NP coupled to a phenothiazine-based near-infrared (NIR) fluorescent dye was developed as a novel theranostic agent for AD. These novel agents bonded strongly to Aβ species concomitant with a pronounced fluorescence enhancement in the NIR window, and blocked Aβ seeding mediated aggregation forming toxic Aβ species and reduced $A\beta_{1\text{-}42}$ *in vivo* (Cai et al., 2020). Several other forms are under various stages of research development.

4.5 FLUORESCENT MOLECULAR PROBES IN AD DIAGNOSIS

In addition to NPs, fluorescent small molecules targeting AD biomarkers are designed as new sensing strategies for potential biomarkers in the diagnosis of AD. Congo red was used as an Aβ plaque probe in the 19th century. They were later found to have low BBB permeability. Shortly thereafter, thioflavine, methoxy-04 and oxazine-derivate probes were used for the same. These cationic probes had a moderate binding affinity for Aβ plaque and slow clearance from the brain. The recently employed

Pittsburgh compound B (PiB) is the neutral version of the probe, which has improved binding affinity, BBB penetration ability and faster clearance kinetics than the above stated charged probes (Jun et al., 2019). It was found to offer strong fluorescence when bonded with Aβ plaques (Mathis et al., 2012). In another study, CRANAD compounds, which are curcumin derivatives, were prepared by converting the enolate moiety into the appropriate boron complex and replacing the aryl group of curcumin with different aromatic substrates (Heo et al., 2016). Their second-generation boron complexes, CRANAD-3 and CRANAD-28, displayed high binding affinity for Aβ species such as Aβ monomer, dimers and oligomers with excellent fluorescent responses depicting AD pathology.

The two main approaches for developing fluorescent probes for tau proteins were developed based on two major strategies: (i) intercalation of probes into p-tau aggregates and (ii) electrostatic binding of probes containing zinc ions with the phosphate groups in p-tau aggregates (Jun et al., 2019). In this regard, several probes such as aryl-quinoline derivatives, THK derivatives (Okamura et al., 2013), phenyldiazenyl-benzothiazole and styryl-benzimidazole scaffolds (Matsumura et al., 2011), thiohydantoin (Ono et al., 2011) and BODIPY-1 containing binuclear Zn(II)-2,2'-dipicolylamine moieties (Ojida et al., 2009), were all found to selectively detect p-tau proteins, demanding further clinical studies (Jun et al., 2019). Sulphorhodamine 101, a small molecular dye, has been reported to stain astrocyte, which is also a biomarker in AD pathology. Despite serving as a fast fluorescent in the red region, its application is still under further research owing to the limited BBB permeability (Nimmerjahn et al., 2004; Jun et al., 2019). Despite such progress, the fluorescent dyes, when formulated with NPs, could offer an enhanced detection strategy for AD biomarkers and pathology. This demands further research in this area to unleash the promising future in the diagnosis of AD.

4.6 CONCLUSION

The recent progress in nanotechnology has sped up the application of NPs in the diagnosis of AD, which demands an integrated strategy in detecting multiple pathological pathways. As such, nanomaterials play a key role in recognizing the biological attributes and can be used as promising diagnostic tools. This chapter provides a glimpse of the studies covering molecular mechanisms and functions of NPs developed for the future use in the diagnosis of AD. However, a wide gap lies between these laboratory discoveries and clinical translations and applications. To bridge the gap, current protocols and methods should be revisited and monitored for accuracy and consistency.

REFERENCES

Amor-Gutiérrez, Olaya, Estefanía Costa-Rama, Noemi Arce-Varas, Carmen Martínez-Rodríguez, Antonello Novelli, María Teresa Fernández-Sánchez, and Agustín Costa-García. "Competitive electrochemical immunosensor for the detection of unfolded p53 protein in blood as biomarker for Alzheimer's disease." *Analytica chimica acta* 1093 (2020): 28–34.

Ayaz, Muhammad, Muhammad Ovais, Irshad Ahmad, Abdul Sadiq, Ali Talha Khalil, and Farhat Ullah. "Biosynthesized metal nanoparticles as potential Alzheimer's disease therapeutics." In *Metal Nanoparticles for Drug Delivery and Diagnostic Applications*, pp. 31–42. Elsevier, 2020.

Brazaca, Laís C., José R. Moreto, Aída Martín, Farshad Tehrani, Joseph Wang, and Valtencir Zucolotto. "Colorimetric paper-based immunosensor for simultaneous determination of fetuin B and clusterin toward early Alzheimer's diagnosis." *ACS nano* 13, no. 11 (2019): 13325–13332.

Cai, Jing, Pascal Dao, Huixiong Chen, Longjia Yan, Yong Liang Li, Wanzheng Zhang, Li Li, Zhiyun Du, Chang-Zhi Dong, and Bernard Meunier. "Ultrasmall superparamagnetic iron oxide nanoparticles-bound NIR dyes: Novel theranostic agents for Alzheimer's disease." *Dyes and pigments* 173 (2020): 107968.

Chen, Li, Jianwei Lin, Jinquan Yi, Qinghua Weng, Ying Zhou, Zhizhong Han, Chunyan Li, Jinghua Chen, and Qian Zhang. "A tyrosinase-induced fluorescence immunoassay for detection of tau protein using dopamine-functionalized $CuInS_2$/ZnS quantum dots." *Analytical and bioanalytical chemistry* 411, no. 20 (2019): 5277–5285.

Gandhi, Sonia, and Andrey Y. Abramov. "Mechanism of oxidative stress in neurodegeneration." *Oxidative medicine and cellular longevity* 2012, (2012): 1–11.

Gao, Xue, Huijie Pan, Yachen Han, Liuxing Feng, Jinping Xiong, Shizhong Luo, and Hongmei Li. "Quantitative imaging of amyloid beta peptide (Aβ) in Alzheimer's brain tissue by laser ablation ICP-MS using gold nanoparticles as labels." *Analytica chimica acta* 1148, (2021): 238197.

Garcia-Leis, Adianez, and Santiago Sanchez-Cortes. "Label-free detection and self-aggregation of amyloid β-peptides based on plasmonic effects induced by Ag nanoparticles: Implications in Alzheimer's disease diagnosis." *ACS applied nano materials* 4, no. 4 (2021): 3565–3575.

Gupta, Jyoti, Munazza Tamkeen Fatima, Zeyaul Islam, Rizwan Hasan Khan, Vladimir N. Uversky, and Parveen Salahuddin. "Nanoparticle formulations in the diagnosis and therapy of Alzheimer's disease." *International journal of biological macromolecules* 130 (2019): 515–526.

Hadjidemetriou, Marilena, Jack Rivers-Auty, Lana Papafilippou, James Eales, Katherine A. B. Kellett, Nigel M. Hooper, Catherine B. Lawrence, and Kostas Kostarelos. "Nanoparticle-enabled enrichment of longitudinal blood proteomic fingerprints in Alzheimer's disease." *ACS nano* 15, no. 4 (2021): 7357–7369.

Hajipour, Mohammad Javad, Michelle R. Santoso, Farhad Rezaee, Haniyeh Aghaverdi, Morteza Mahmoudi, and George Perry. "Advances in Alzheimer's diagnosis and therapy: The implications of Nanotechnology." *Trends in biotechnology* 35 (2017): 937–953.

Hansen, Richard A., Gerald Gartlehner, Aaron P. Webb, Laura C. Morgan, Charity G. Moore, and Daniel E. Jonas. "Efficacy and safety of donepezil, galantamine, and rivastigmine for the treatment of Alzheimer's disease: A systematic review and meta-analysis." *Clinical interventions in aging* 3, no. 2 (2008): 211.

Heo, Cheol Ho, Avik Ranjan Sarkar, Sung Hoon Baik, Tae Sung Jung, Jeong Jin Kim, Hyuk Kang, Inhee Mook-Jung, and Hwan Myung Kim. "A quadrupolar two-photon fluorescent probe for in vivo imaging of amyloid-β plaques." *Chemical science* 7, no. 7 (2016): 4600–4606.

Jun, Yong Woong, Seo Won Cho, Junyang Jung, Youngbuhm Huh, Young Soo Kim, Dokyoung Kim, and Kyo Han Ahn. "Frontiers in probing Alzheimer's disease biomarkers with fluorescent small molecules." *ACS central science* 5, no. 2 (2019): 209–217.

Kang, Da-Yeon, Jin-Ho Lee, Byung-Keun Oh, and Jeong-Woo Choi. "Ultra-sensitive immunosensor for β-amyloid (1–42) using scanning tunneling microscopy-based electrical detection." *Biosensors and bioelectronics* 24, no. 5 (2009): 1431–1436.

Kang, Min-Sik, Kyu-Hyon Son, Tae-Hoon Lee, Sang-Mok Chang, Sung-Woong Han, and Hoon-Kyu Shin. "Development of chemically signal amplified nano-biosensor mediated by poly-l-lysine." *Journal of nanoscience and nanotechnology* 19, no. 3 (2019): 1786–1789.

Kim, Jinsik, Myung-Sic Chae, Sung Min Lee, Dahye Jeong, Byung Chul Lee, Jeong Hoon Lee, Young Soo Kim, Suk Tai Chang, and Kyo Seon Hwang. "Wafer-scale high-resolution patterning of reduced graphene oxide films for detection of low concentration biomarkers in plasma." *Scientific reports* 6, no. 1 (2016): 1–8.

Lippens, Guy, Alain Sillen, Isabelle Landrieu, Laziza Amniai, Nathalie Sibille, Pascale Barbier, Arnaud Leroy, Xavier Hanoulle, and Jean-Michel Wieruszeski. "Tau aggregation in Alzheimer's disease: What role for phosphorylation?" *Prion* 1, no. 1 (2007): 21–25.

Liu, Yibiao, Qing Xu, Yina Zhang, Bingyu Ren, Liumei Huang, Hong Cai, Tailin Xu, Qiong Liu, and Xueji Zhang. "An electrochemical aptasensor based on AuPt alloy nanoparticles for ultrasensitive detection of amyloid-β oligomers." *Talanta* 231 (2021): 122360.

Liu, Chang, Xinrui You, Dingkun Lu, Guoyue Shi, Jingjing Deng, and Tianshu Zhou. "Gelsolin encountering Ag nanorods/triangles: An aggregation-based colorimetric sensor array for in vivo monitoring the cerebrospinal $A\beta_{42}$% as an indicator of Cd^{2+} exposure-related Alzheimer's disease pathogenesis." *ACS applied bio materials* 3, no. 11 (2020): 7965–7973.

Mathis, Chester A., N. Scott Mason, Brian J. Lopresti, and William E. Klunk. "Development of positron emission tomography β-amyloid plaque imaging agents." *Seminars in nuclear medicine* 42, no. 6, (2012): 423–432. WB Saunders.

Matsumura, Kenji, Masahiro Ono, Shun Hayashi, Hiroyuki Kimura, Yoko Okamoto, Masafumi Ihara, Ryosuke Takahashi, Hiroshi Mori, and Hideo Saji. "Phenyldiazenyl benzothiazole derivatives as probes for in vivo imaging of neurofibrillary tangles in Alzheimer's disease brains." *MedChemComm* 2, no. 7 (2011): 596–600.

Morales-Narváez, Eden, Helena Montón, Anna Fomicheva, and Arben Merkoçi. "Signal enhancement in antibody microarrays using quantum dots nanocrystals: application to potential Alzheimer's disease biomarker screening." *Analytical chemistry* 84, no. 15 (2012): 6821–6827.

Nguyen, Kim Truc, and Yanli Zhao. "Engineered hybrid nanoparticles for on-demand diagnostics and therapeutics." *Accounts of chemical research* 48, no. 12 (2015): 3016–3025.

Nimmerjahn, Axel, Frank Kirchhoff, Jason N.D. Kerr, and Fritjof Helmchen. "Sulforhodamine 101 as a specific marker of astroglia in the neocortex in vivo." *Nature methods* 1, no. 1 (2004): 31–37.

Ojida, Akio, Takashi Sakamoto, Masa-aki Inoue, Sho-hei Fujishima, Guy Lippens, and Itaru Hamachi. "Fluorescent BODIPY-based Zn (II) complex as a molecular probe for selective detection of neurofibrillary tangles in the brains of Alzheimer's disease patients." *Journal of the American Chemical Society* 131, no. 18 (2009): 6543–6548.

Okamura, Nobuyuki, Shozo Furumoto, Ryuichi Harada, Tetsuro Tago, Takeo Yoshikawa, Michelle Fodero-Tavoletti, Rachel S. Mulligan et al. "Novel 18F-labeled arylquinoline derivatives for noninvasive imaging of tau pathology in Alzheimer disease." *Journal of nuclear medicine* 54, no. 8 (2013): 1420–1427.

Ono, Masahiro, Shun Hayashi, Kenji Matsumura, Hiroyuki Kimura, Yoko Okamoto, Masafumi Ihara, Ryosuke Takahashi, Hiroshi Mori, and Hideo Saji. "Rhodanine and thiohydantoin derivatives for detecting tau pathology in Alzheimer's brains." *ACS chemical neuroscience* 2, no. 5 (2011): 269–275.

Paraskevaidi, Maria, Pierre L. Martin-Hirsch, and Francis L. Martin. "Progress and challenges in the diagnosis of dementia: A critical review." *ACS chemical neuroscience* 9, no. 3 (2018): 446–461.

Prince, Martin, Renata Bryce, Emiliano Albanese, Anders Wimo, Wagner Ribeiro, and Cleusa P. Ferri. "The global prevalence of dementia: A systematic review and metaanalysis." *Alzheimer's & dementia* 9, no. 1 (2013): 63–75.

Roney, Celeste, Padmakar Kulkarni, Veera Arora, Peter Antich, Frederick Bonte, Aimei Wu, N. N. Mallikarjuana et al. "Targeted nanoparticles for drug delivery through the blood–brain barrier for Alzheimer's disease." *Journal of controlled release* 108, no. 2–3 (2005): 193–214.

Serafín, V., M. Gamella, M. Pedrero, A. Montero-Calle, C. A. Razzino, P. Yáñez-Sedeño, R. Barderas, S. Campuzano, and J. M. Pingarrón. "Enlightening the advancements in electrochemical bioanalysis for the diagnosis of Alzheimer's disease and other neurodegenerative disorders." *Journal of pharmaceutical and biomedical analysis* 189 (2020): 113437.

Sun, Leming, Dingchang Liu, Dongjie Fu, Tao Yue, Douglas Scharre, and Lianbing Zhang. "Fluorescent peptide nanoparticles to detect amyloid-beta aggregation in cerebrospinal fluid and serum for Alzheimer's disease diagnosis and progression monitoring." *Chemical engineering journal* 405 (2021): 126733.

Sun, Liping, Yong Zhong, Jie Gui, Xianwu Wang, Xiaorong Zhuang, and Jian Weng. "A hydrogel biosensor for high selective and sensitive detection of amyloid-beta oligomers." *International journal of nanomedicine* 13 (2018): 843.

Takahashi, Reisuke H., Toshitaka Nagao, and Gunnar K. Gouras. "Plaque formation and the intraneuronal accumulation of β-amyloid in Alzheimer's disease." *Pathology international* 67, no. 4 (2017): 185–193.

Xu, Yilan, Manna Zhao, Dongming Zhou, Tingting Zheng, and Heng Zhang. "The application of multifunctional nanomaterials in Alzheimer's disease: A potential theranostics strategy." *Biomedicine & pharmacotherapy* 137 (2021): 111360.

Yu, Yanyan, Peng Wang, Xiaodan Zhu, Qiwen Peng, Yi Zhou, Tianxiao Yin, Yixin Liang, and Xiaoxing Yin. "Combined determination of copper ions and β-amyloid peptide by a single ratiometric electrochemical biosensor." *Analyst* 143, no. 1 (2018): 323–331.

5 Nanophytomedicine-Based Therapies for Topical Applications

Sankar Veintramuthu and Nithyananth Muniswamy

CONTENTS

5.1 INTRODUCTION

Throughout world history, viz Babylonian and Egyptian civilizations, the topical use of medicines formulated from natural plant origins for treatment of skin diseases has been well documented. *Ebers Papyrus* (1525 BC), an ancient Egyptian text on herbal medicine, chronicles the use of topical ointments and salves for various ailments. Traditional Indian practices like Ayurveda and Siddha extensively utilized antimicrobial therapeutic metals like lead, tin, gold, copper, iron, silver and zinc in treating various skin infections. Phytomedicines are considered safe due to their low side effects and synergistic actions. Despite the advantages, their pharmacological response is significantly weakened by reduced plasma concentrations, although *in vitro* studies have shown promise. Such limited systemic bioavailability in phytoconstituents

DOI: 10.1201/9781003231745-5

like flavonoids, tannins, and terpenoids are mostly due to their reduced lipid permeability and minimal drug solubility.

The use of nanotechnology in phytomedicine administration has become widespread worldwide to address limitations. Researchers find nanotechnology an attractive platform due to its ability to deliver the required dose of the active drug to the site of action and provoke a significant action.

Gera et al. (2017) reported *Curcuma longa* nanoparticles had significant improvement in water solubility and enhanced systemic bioavailability of curcumin. Synthesized using biodegradable polymers and biocompatible lipids, nanoparticles are nontoxic, can deliver drugs at high concentrations in a sustained manner and have high penetration and retention-enhancing effects that significantly increase drug penetration across the skin (Sungthongjeen et al., 1999).

5.2 SKIN

Skin is a first-line defence against micro-organisms, a natural barrier against loss of fluids and salts and also regulates the body temperature (homeostasis). With a molecular size up to 500 Da, lipophilic (logP1-3) substances can easily permeate skin without any assistance. But larger molecules like peptides or proteins will find it difficult to permeate (Badenhorst et al. 2014). Histologically, human skin is made up of three layers, the epidermis, dermis and subcutaneous layer (see Figure 5.1). The dermis contains the first layer of living cells made up of four to five layers of dermal fibroblasts in an extracellular matrix of structural proteins, mainly collagen, elastin and keratinocytes. It also contains sweat glands and hair follicles, sensory nerve endings, sebaceous glands, lymphatic vessels and blood capillaries that extend up to dermal side of dermo-epidermal junction. Dermis regulates

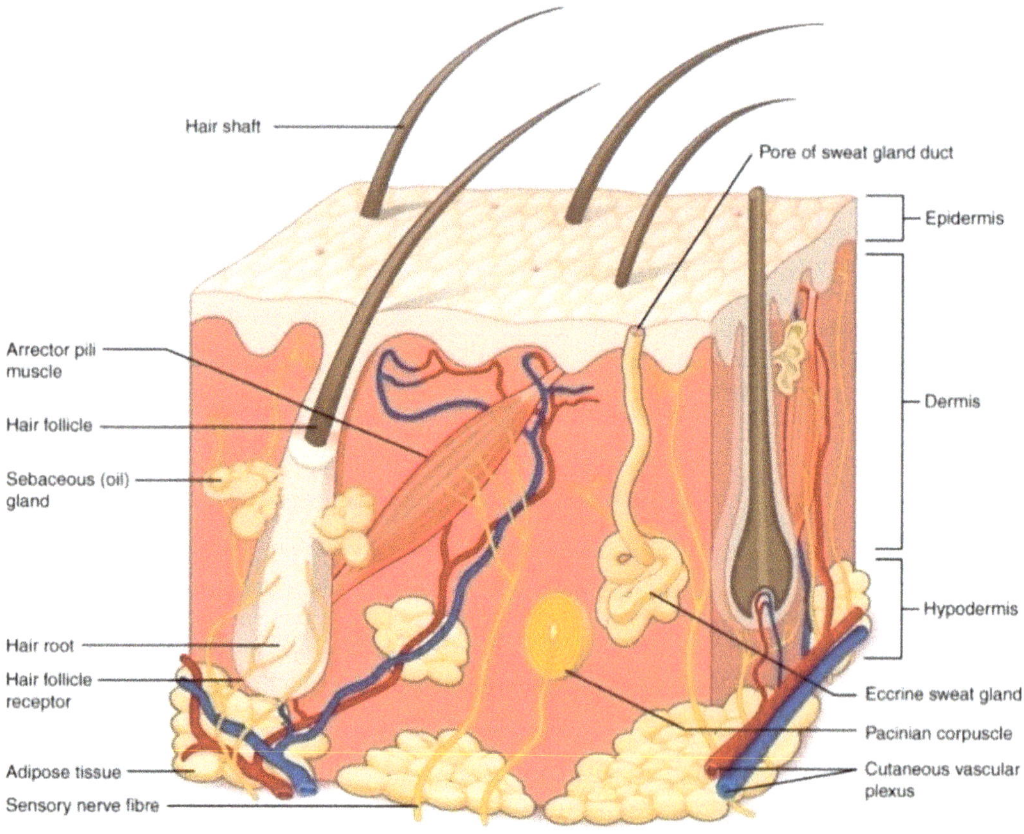

FIGURE 5.1 Anatomy of human skin.

nutrients and oxygen delivery inside and also aids in toxic removal from the vascular epidermis by diffusion throughout the dermo-epidermal junction (Montagna et al., 2020). Through topical delivery, a drug is mostly administered for local effect. Topical dermatological products include sunscreens, antiseptics, anti-fungal creams, keratolytic agents and anti-inflammatory agents used for the treatment of skin diseases like psoriasis and acne. In transdermal delivery, drugs transverse the skin reaching systemic circulation and incite a pharmacological response. Nicotine patches used for cessation of smoking are administrated through the transdermal route (Karunanidhi et al., 2014).

5.3 CHALLENGES IN DEVELOPING NANOPHYTOMEDICINE

The success of nanophytomedicine depends on how the phytoconstituents are processed. Insignificant levels of active phytoconstituents are present in plant sources, making screening, isolation, and identification extremely difficult in order to reduce variation in finished herbal products and demonstrate potency by biological assay. The anatomical part of the plant (leaf, flower or stem) from which phytoconstituents are extracted is critically important, as it can define the quality of active constituents. Uncertain bioavailability of active phytoconstituents can also be the result of plant-related factors like lack of standardization between batches or adulteration of botanical materials.

The lack of appropriate *in vivo* animal models to investigate the topical delivery of medicine and establish pharmacological and toxic profiles is still one of the major challenges in transforming promising delivery into successful clinical treatment. There are no established guidelines on methods of extraction from plant sources. Many herbal formulations have shown promising results in *in vitro* studies but failed to show the same in *in vivo* studies.

Studies have shown the potential of nanophytomedicine in treating various diseases but there are no specific guidelines or study protocols for nanoformulations intended for topical delivery. To successfully transform these technologies into relatable therapies is still a major challenge for many pharmaceutical companies.

Nanotechnologies have proven their ability to deliver drugs to the site of action, but they were also found to have significant interaction between nanoparticles and living organisms. Studies have shown nanoparticles adsorb protein portions of plasma and can significantly disturb the defence mechanisms of the body. Unexpected immune responses after delivery are relevant with nanoformulations. The fate of the nanoparticles after absorption must be investigated thoroughly before being used in any clinical practice. In comparison, developing nanoformulations containing phytoconstituents involves sophisticated techniques and expensive instruments in nanophytomedicine. These financial considerations drive many pharmaceutical companies to be pragmatic in their approach.

5.4 FACTORS TO BE CONSIDERED IN DEVELOPING NANODELIVERY

5.4.1 Particle Size, Particle-Size Distribution

Particle morphology of nanoparticles influences the behaviour characteristics of nanoparticles like drug release, stability, permeability and fate of nanoparticles inside the biological milieu. Manufacturing conditions, choice of excipients, temperature and viscosity of aqueous and organic solvents affect the particle size and their stability (Gollwitzer et al., 2016).

5.4.2 Zeta Potential

Zeta potential is a key parameter in evaluating the stability of nanoparticles. Zeta potential is a measure of electrokinetic potential at the shear plane. As a general rule, higher zeta potential (higher electric charge) on the nanoparticles will reduce or prevent aggregation in a buffer solution because of the high repellent forces among particles (Attama et al., 2007).

5.4.3 Surface Properties

Surface charges present in the nanoparticles play a critical role in their affinity toward cell membranes. Surface engineering of nanoparticles could greatly impact their uptake and restrict the cellular uptake of nanoparticles at intracellular targets viz mitochondria, lysosomes and cytoplasm. Positively charged nanoparticles greatly enhance cellular uptake. Similarly, having a large negatively charged domain could make cell membranes repel. Negatively charged sulphated proteoglycans and phosphatidylcholine on the cellular surface influences migration, cellular proliferation and motility. Nonetheless, non-specific adsorption (at sites having a positive charge) and the formation of nanoparticle clusters can increase the cellular uptake of negatively charged nanoparticles (Chang et al., 2006).

5.5 Techniques Used to Develop a Nanoparticulate System for Topical Delivery

5.5.1 High-Pressure Homogenization

High-pressure homogenization works when coarsely formed liquids are forced through a narrow nozzle at high pressure (100–200 bars) to breakdown a drug to nanosize. This high-energy stream of liquid passes into a disruption unit where it collides at an extremely high velocity. Here, particles are reduced to the submicron range due to the collision and cavitations in the disruption unit (Floury et al., 2000). High-pressure homogenizers provide a lot of advantages like reproducibility, homogenous particle size distribution and continuous production.

5.5.2 Solvent Emulsification-Evaporation Techniques

Organic solvent is used to dissolve lipophilic polymers and hydrophobic drugs that are completely immiscible in water. This mixture is emulsified in an aqueous phase using a high-speed homogenizer. Finally, the organic solvent is removed by evaporation at room temperature and reduced pressure, resulting in solid lipid nanoparticles (Iqbal, 2018).

5.5.3 Solvent Emulsification-Diffusion Technique

Lipids are dissolved partially in a water-miscible solvent previously saturated with water at a controlled temperature. By the addition of an aqueous solution of a stabilizing agent (saturated with solvent), the organic solvent is emulsified by mechanical stirring at a temperature used to dissolve lipids. The formed oil-water emulsion is then diluted with an excess of water in a ratio of 1:5 to 1:10 at a controlled temperature, resulting in diffusion from the internal phase into the external phase, which induces lipid aggregation to form solid lipid nanoparticles (Patravale et al., 2019).

5.5.4 Coprecipitation Technique

Coprecipitation is another popular method to synthesize nanoparticles. Nucleation, growth, coarsens, and agglomeration occur at the same time. In Coprecipitation process, precipitation of Fe^{2+} and Fe^{3+} salts (e.g., chlorides, sulphates and nitrates) in aqueous solutions occur by the addition of a base (e.g., NaOH). Particles can be reduced to nano- or micron-sized particles by adjusting the pH, precipitating agent, temperature and solvents (Mohammadi et al., 2021).

5.6 Nanoformulations for Topical Drug Delivery

5.6.1 Liposomes

Liposomes are amphilic bilayer lipid membranes made up of phospholipids as shown in Figure 5.2a and b. Many studies (Table 5.1) have shown their ability to deliver both hydrophilic and hydrophobic drugs at the site of action due to their biodegradable and biocompatible properties (Sachan et al., 2015).

TABLE 5.1
Nanoformulations Incorporating Phytoconstituents for Topical Application

S. No	Nanoformulations	Advantages	Indications	Method of Preparation	References
			A. Liposomes		
1.	Colchicine liposome	Improved skin accumulation, prolonged drug release	Antigout	Rotary evaporation sonication method	Saraf (2010)
2.	Quercetin liposomes	Improved skin permeation	Antioxidant and protective against UV-β radiation	Ethanol injection technique	Liu et al. (2013)
3.	Catechin liposomes	Increased bioavailability	Antioxidant, antiobesity, anti-inflammatory, antidiabetic	Reverse phase evaporation	Fang et al. (2006)
4.	Mixed liposomes of quercetin and curcumin	Improved skin penetration	Antioxidant	Antisolvent precipiation	Di Marzio et al. (2016)
5.	Aloe vera liposomes	Improved Bioavailability	Enhanced proliferation and production of collagen	Bangham method	Takahashi et al. (2009)
			B. Nanoparticles		
6.	Quercetin nanoparticles	Improved skin uptake	Antioxidant	Ethanol injection technique	Tan et al. (2011)
7.	Curcumin nanoparticles	IIncreased cellular uptake, better *in vitro* and *in vivo* bioavailability	Wound healing	Sonication method	Krausz et al. (2015)
8.	Silver nanoparticles of *Citrullus colocynthis* extract	Improved penetration	Wound healing	Chemical reduction method	Satyavani et al. (2011)
9.	Silver nanoparticles of *Indigofera aspalathoides* extract	Improved penetration	Wound healing	Green synthesis	Arunachalam et al. (2013)
			C. Solid Liquid Nanoparticles		
10.	Solid lipid nanoparticles of *Kaempferia parviflora* extract	Enhancement in skin permeability	Anti-inflammatory	Microemulsification method	Sutthanut et al. (2009)
11.	Solid lipid nanoparticles of triptolide	Better skin penetration	Anti-inflammatory, immunosuppressive	Emulsification-sonication	Mei et al. (2005)
12.	Quercetin nanoparticles	Improved skin penetration	Antioxidant	Nanoprecipitation technique	Wu et al. (2008)
13.	Curcumin nanoparticles	Better retention and Improved skin uptake	Antipsoriatic	Supercritical solution expansion	Mulik et al. (2009)
			D. Phytosomes		
14.	Polyherbal phytocomplex of *Trichosanthes curcumerina* extract and *Abrus precatorius* extract	Improved hair growth promoting effect	Antialopecia	Rotary evaporation technique	Sandhya et al. (2012)

(Continued)

TABLE 5.1 (Continued)
Nanoformulations Incorporating Phytoconstituents for Topical Application

S. No	Nanoformulations	Advantages	Indications	Method of Preparation	References
			E. Nanoemulsions		
15.	Genistein nanoemulsion	Enhancement of skin penetration	Anti-inflammatory	Spontaneous emulsification	Silva et al. (2009)
16.	Quercetin nanoemuslion	Improved penetration into stratum corneum and epidermis	Antioxidant activity	High speed homogenization method	Fasolo et al. (2009)
17.	Nanoemulsions of hydroalcoholic extracts of *Vellozias quamata*	Improved skin penetration	Antioxidant activity	Phase reversal technique	Quintão et al. (2013)
18.	*Artocarpus incises* extracts	Improved depigmenting action	Reduces hyperpigmentation	Rotary evaporation technique	Buranajaree et al. (2011)
19.	Glycyrrhetic acid nanoemulsion	Better stability and better transdermal penetration	Anti-inflammatory, antitumour, antiviral, antiulcer	Phase inversion temperature method	Puglia et al. (2010)
20.	*Cassia fistula* nanoemulsion	Increased stratum corneum hydration level	Antiageing, antioxidant	Macroemulsifi-cation method	Khan et al. (2015)
			F. Transferosomes		
21.	Curcumin transferosome gel	Improved skin penetration	Antipsoriatic	Desolvation method	Patel et al. (2009)
			G. Ethosomes		
22.	Matrine ethosome	Improves the percutaneous permeation	Anti-inflammatory, antibacterial, antirheumatic and antitumour properties	Ethanol injection –sonication method	Zhaowu et al. (2009)
23.	Ammonium glycyrrhizinate ethosomes	Increased *in vitro* percutaneous penetration	Anti-inflammatory	Solvent evaporation technique	Paolino et al. (2005)
24.	*Sesbania grandiflora* ethosomes	Increased transdermal penetration	Antiglycation	Solvent dispersion method	Nirved et al. (2012)
25.	*Sophora alopecuroides* ethosomes	Improved penetration through skin	Anticancer, antiendotoxic, anti-inflammatory	Passive loading method	Zhou et al. (2010)
			H. Beta-Cyclodextrin		
26.	Naringenin beta-cyclodextrin	Improved amorphous nature, dissolution profile and water solubility	Antimicrobial activity	Solvent evaporation technique	Semalty et al. (2014)
27.	Silymarin beta-cyclodextrin	Improved solubility, dissolution rate	Hepatoprotective	Kneading method	Ghosh et al. (2011)
			I. Niosomes		
28.	*Withania somnifera* niosomes	Significant enhancement of withaferin A and withanolide A delivery through human skin	Antimelanoma	Rotary evaporation method	Chinembiri et al. (2017)
			J. Polyvesicular nanoformulations		
29.	Curcumin ethosomes, liposomes, transferosomes	Better skin permeation	Wound healing, UV protectant	Shaking method	Kaur and Saraf (2011)
30.	*Capsicum frutescens* nanovesicle	Reduces the irritant effect of capsaicinoids	Antiarthritic	Dispersion method	Sarwa et al. (2014)
31.	*Panax ginseng* ethosomes and transferosomes	Greater skin permeation of ginsenoside Rh1	Anti-inflammatory, antitumour effects	Solvent evaporation method	Choi et al. (2015)

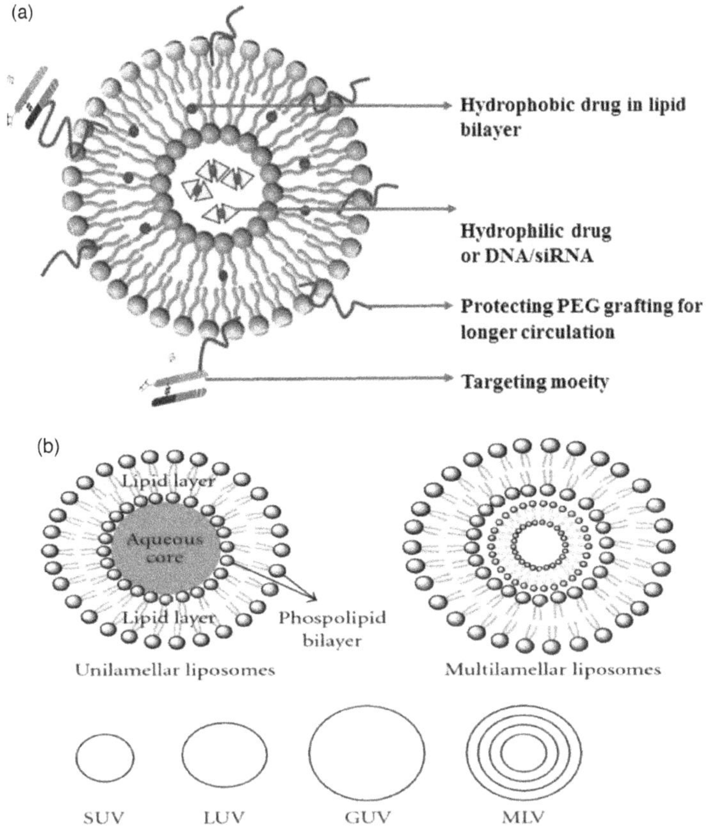

FIGURE 5.2 (a) Liposomes. (Adapted from Suthar et al., 2019.) (b) Types of liposomal drug delivery systems. (Adapted from Mishra et al., 2011.)

5.6.2 Ethosomes

Ethosomes are liposomes with a high concentration of ethanol (20–45%). Ethosomes generally contain phospholipids, cholesterol, stabilizers and a high concentration of ethanol. Due to the high concentration of ethanol, ethosomes are known to have high stability and permeability and can deliver the active constituents to the deeper layer of the skin (see Figure 5.3). Ethosomes provide

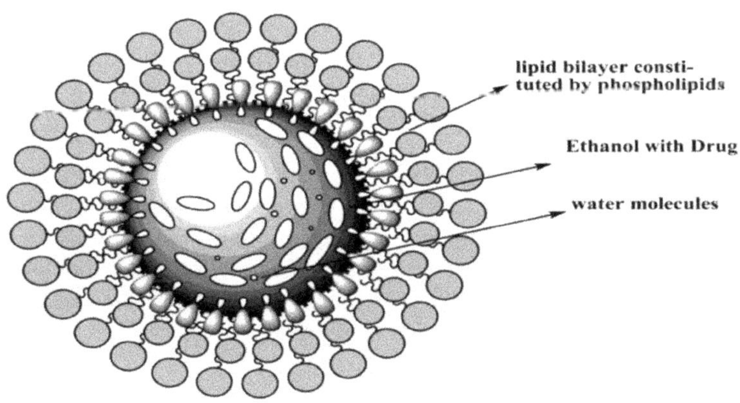

FIGURE 5.3 Ethosomes. (Adapted from Pandey, 2014.)

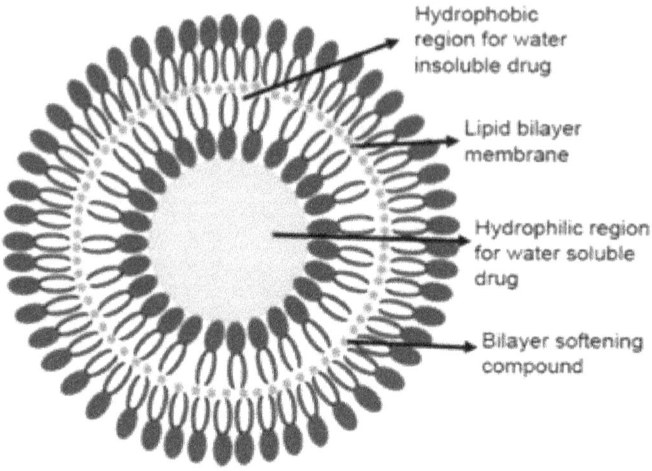

FIGURE 5.4 Transfersomes. (Adapted from Karunanidhi et al., 2014.)

excellent intracellular delivery of hydrophobic drugs and increased absorption of anti-inflammatory herbs (Sachan et al., 2015).

5.6.3 Transfersomes

Transfersomes are soft and deformable vesicular systems containing phospholipids, stabilizers, edge activators, ethanol and a high concentration of hydrophilic modulators like organic ions. Like ethosomes, transfersomes have better penetration and deliver drugs deeper into skin layers without losing drug integrity (see Figure 5.4).

5.6.4 Transethosomes

Transethosomes are modified versions of classical ethosomes. Along with ethosomes, edge activators such as Tween 80 are included. Transethosomes are deformable vesicles that can deliver active drugs deeper into skin. Both hydrophilic and lipophilic forms can be formulated (Ascenso et al., 2015).

5.6.5 Niosomes

Niosomes are unique microscopic bilayer lamellar vesicular systems ranging from 10 to 1000 nm made of a nonionic surfactant. Niosomes have better chemical stability in comparison to liposomes, which are prone to oxidation and have better permeation through the skin. Like liposomes, both polar and nonpolar medicines can be successfully incorporated as described in Figure 5.5.

5.6.6 Phytosomes

The phytosome (herbasome) is a patented delivery-system of the Italian pharmaceutical company Indena. Phytosomes are a standardized method to incorporate phospholipids into plant extracts to enhance absorption and bioavailability (see Figure 5.6). In contrast to liposomes, they form a molecular complex at a ratio of 1:1; 1:2 involving H-chemical bonds, which vastly improve chemical stability (Chivte et al., 2017). Phytosomes can be successfully encapsulated into different dosage forms like creams, lotions, gels and emulsions etc. (see Table 5.2).

5.6.7 Polymeric Nanomicelles

Polymeric nanomicelles are a self-assembly of amphiphilic block co-polymers in an aqueous medium. Polymeric nanomicelles contain both a hydrophobic outer shell and a hydrophilic inner shell in which

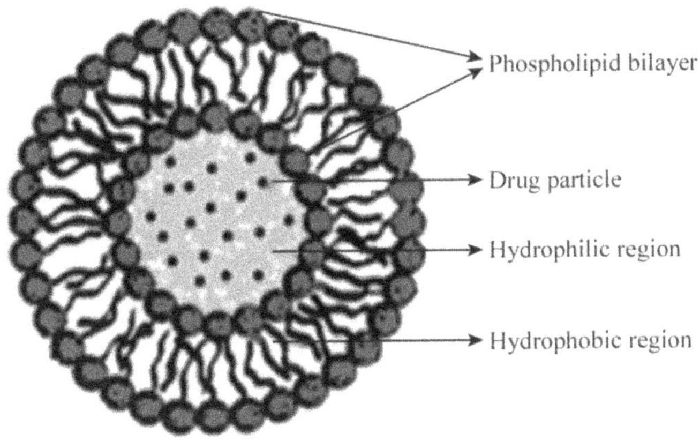

FIGURE 5.5 Niosome. (Adapted from Ravalika et al., 2017.)

the drug is incorporated into the core (see Figure 5.7). Drugs encapsulated in polymeric nanomicelles exhibit good solubility, increased circulation time and enhanced metabolic stability.

5.6.8 β-Cyclodextrin Complexes

Cyclodextrins (CDs) are biocompatible, nontoxic compounds that can form colloidal dispersions of nanoparticles between cyclodextrin derivatives. Their nature of forming inclusion complexes with hydrophobic drugs has provided a unique platform. Most of the CD is amphiphilic, noncytotoxic and nonhemolytic. Modified CDs are highly preferred since they display increased affinity with a biological membrane, which is an excellent platform for topical delivery. There are three major CDs: α-CD, β-CD, γ-CD, of which β-CD is widely popular because of it sexcellent cavity that incorporates drugs with a molecular weight between 200 and 800 g/mol. They are readily available and are reasonably priced. CD-conjugated nanoparticles have several advantages including improved drug solubility, targeted drug delivery, improved encapsulation efficiency and drug loading.

5.7 PHYSIOCHEMICAL CHARACTERIZATION OF NANOPARTICLES

There are no specific guidelines from the U.S. Food and Drug Administration to characterize nanoparticles based on the plant for topical applications. But the agency has issued specific draft guidance documents to address the application of nanotechnology in the food and cosmetics industries. So far, many scientific methods have been used to characterize the nanomaterials (see Table 5.3).

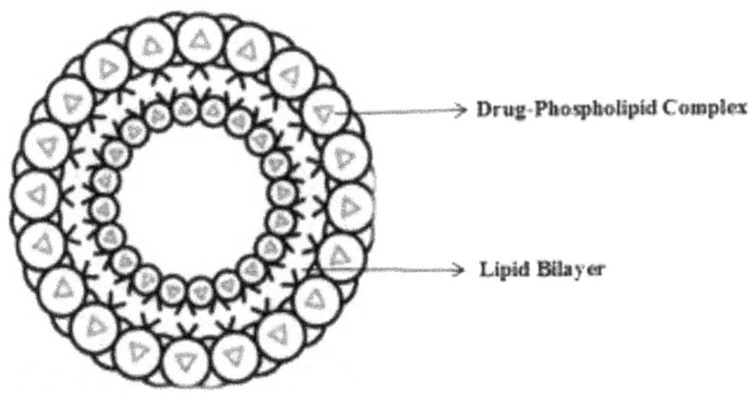

FIGURE 5.6 Phytosome. (Adapted from Kumar et al., 2020.)

TABLE 5.2
Marketed Phytosomal Products

S. No	Natural Sources	Phytoconstituents	Phytosomal Products	Dosage Form	Indications
1.	*Terminalia serica* (Silver cluster leaf)	Sericoside	Sericoside Phytosome™	3% gel, cream, emulsion, lotion	Antiageing, skin restructuring, wound healing, antioedema, anti-inflammatory
2.	*Glycyrrhiza glabra* (Mulethi)	Glycyrrhetinic acid	Glycyrrhetinic acid Phytosome	–	Anti-inflammatory, antierythemic, anti-irritant, skin infection
3.	*Centella asiatica* (Brahmi)	Asiatic acid, madecassic acid	Centella triterpenoid Phytosome	60–120 mg	Skin diseases, effective against ulcer, wound cure, effective against hair loss
4.	*Pinus maritima* (Pine)	Procyanidins	Pycnogenol Phytosome	–	Anti-inflammatory, antiwrinkle, antiallergic
5.	*Ruscus aculeatus* (Butchers' broom)	Ruscogenin, neoruscogenin	Ruscogenin Phytosome	Topical preparation	Anti-inflammatory, antiageing, sunscreen agent
6.	*Panicum miliaceum* (Millet)	Mineral salts, vitamins, unsaturated fatty acids, amino acids	Millet Phytosome	Topical preparation	Reduce stress, nutritional supplement for skin, nails and hair
7.	*Zanthoxylum bungeanum* (Tumburu)	Hydroxy-a-sanshool	Zanthalene Phytosome	Emulsion and lotion	Soothing and antireddening
8.	*Polygonum cuspidatum*	Resveratrol	Rexatrol	–	Antioxidant, antiageing

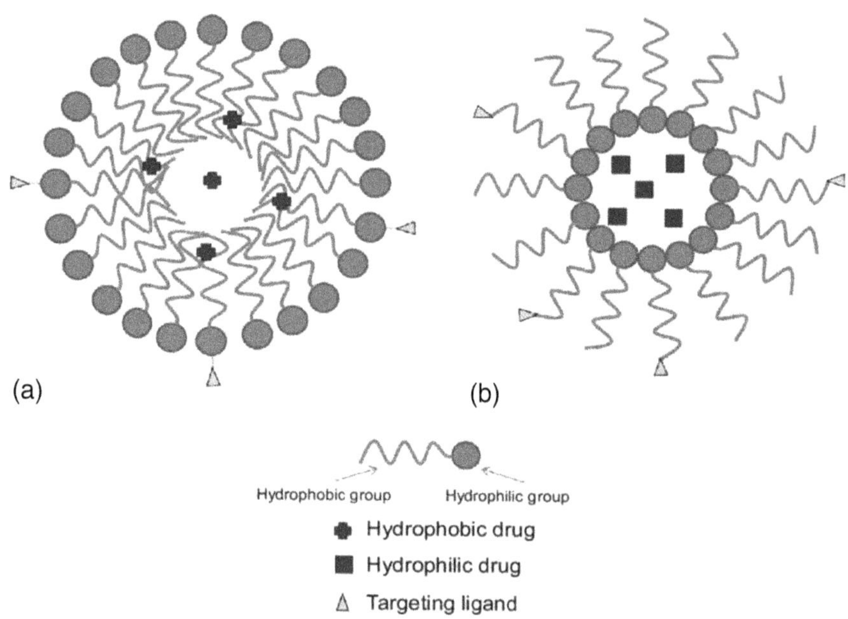

FIGURE 5.7 Polymeric nanomicelles. (Adapted from García et al., 2018.)

TABLE 5.3
Physiochemical Characterization of Nanoformulations Intended for Topical Delivery

S. No	Physiochemical Parameters	Purpose	Techniques
1.	Drug solubility	a. Instabilities arise due to result of super saturation and lead to metastable formulation during storage, which affects the drug release. b. To understand how the drug will behave in skin matrix after administration. c. Prevent the crystallization of active constituents, which could destabilize the nanoparticulate system.	UV spectroscopy, high-performance liquid chromatography
2.	Particle-size, shape, zeta potential	a. Particle size influences the drug release and penetration across the skin barriers. b. Zeta potential helps the formulators to understand their tendency to aggregate during storage.	Light scattering techniques
3.	Imaging	a. Understand the skin anatomy and skin perturbations. b. Information on the specific interaction between the nanoparticulate system and the skin. c. Drug permeation process in the epidermis.	Transmission microscopy, freeze fracturing microscope
4.	Stability determination	a. Drug leakage due to hydrolysis or oxidation of the phospholipid molecules could cause alterations in the vesicle size during storage. b. Ethosomes found to have good stability in their multilamellar and large unilamellar vesicles in terms of drugs entrapped and particle size but small unilamellar vesicles found to have aggregation due to increased surface area.	Electronic microscope, dynamic light scattering techniques Optical analyser Turbiscan Lab® Expert
5.	Chromatographic studies	Characterizing and quantify the impurities present in the drugs and formulation excipients.	High-pressure liquid chromatography Liquid chromatography-mass spectrometry
6.	Drug-skin interactions	a. Chemical stability of the drug in formulations. b. Structural organization of the SC layers. c. Permeation behaviour. d. Structural integrity of the lipids in human skin substances in stratum corneum.	Fourier-Transform infrared spectroscopy Attenuated total reflectance FTIR Small angle X-ray diffraction
7.	Thermal stability	Thermal integrity at various transition temperatures.	Differential scanning calorimetry
8.	Spreadability	Spread nature of the formulation influences the pharmacological response of the drug.	Contact angle of the formulation on standardized substrate and measure of its lubricity
9.	Dermal irritation assay	Skin compatibility with nanoformulations.	Animal *in vivo* models
10.	Rheology	a. Optimize the formulation parameters. b. Prolong the shelf life of the preparations.	Flow viscometer, oscillatory rheometer transient measurement.

5.8 SAFETY CONSIDERATIONS

While evaluating topical nanoformulations containing phytoconstituents, safety and toxic profiles should be established for nanoparticle systems and for the herbal drug itself. Many studies have reported that nanoparticles can gain access into systemic circulation and trigger the immune system, resulting in production of free radicals. Metallic nanoparticles like gold, silver and titanium along with carbon and silicon have the potential for toxicity towards the cells of skin membranes viz keratinocyctes and fibroblasts in addition to immunotoxicity. Because of the topical application, various nanoparticles containing formulations have caused considerable irritation, dermal inflammation and hyperpigmentation. Apart from the proven benefits of nanoparticulate systems, they have a profound impact on ecosystems. Nanoparticles have significant effects on plants, aquatic ecosystems and humans as they leach from the formulation into the soil and water. Nanoparticles containing titanium oxide and silver are reported to have ecotoxic effects. Since there are no uniform guidelines in assessing the safety and efficacy of nanophytoformulation, development is still a challenge. Developing *in vitro* models for evaluating safety and toxic profiles should be of primary concern for the successful transition from promising drug delivery to commercially viable product.

5.9 CONCLUSION

Phyto-based nanoformulations are a promising strategy to overcome poor solubility and permeability in formidable skin barriers, subsequently increasing the bioavailability of plant constituents. Nanotechnology continues to evolve in the field of medicine, providing an excellent platform to topically deliver drugs in a targeted and sustained manner to exert desired local or systemic effects.

REFERENCES

Arunachalam, Kantha D., Sathesh Kumar Annamalai, Aarrthy M. Arunachalam, and Subashini Kennedy. "Green Synthesis of Crystalline Silver Nanoparticles Using *Indigofera aspalathoides*—Medicinal Plant Extract for Wound Healing Applications." *Asian Journal of Chemistry* 25 (2013).

Ascenso, Andreia, Sara Raposo, Cátia Batista, Pedro Cardoso, Tiago Mendes, Fabíola Garcia Praça, Maria Vitória Lopes Badra Bentley, and Sandra Simões. "Development, Characterization, and Skin Delivery Studies of Related Ultra-Deformable Vesicles: Transfersomes, Ethosomes, and Transethosomes." *International Journal of Nanomedicine* 10 (2015): 5837.

Attama, A. A., B. C. Schicke, T. Paepenmüller, and C. C. Müller-Goymann. "Solid Lipid Nanodispersions Containing Mixed Lipid Core and a Polar Heterolipid: Characterization." *European Journal of Pharmaceutics and Biopharmaceutics* 67, no. 1 (2007): 48–57.

Badenhorst, T., Darren Svirskis, and Zimei Wu. "Pharmaceutical Strategies for the Topical Dermal Delivery of Peptides/Proteins for Cosmetic and Therapeutic Applications." *Austin Journal of Pharmacology and Therapeutics* 2 (2014): 1–10.

Buranajaree, Supasiri, Piyaporn Donsing, Rattima Jeenapongsa, and Jarupa Viyoch. "Depigmenting Action of a Nanoemulsion Containing Heartwood Extract of *Artocarpus incisus* on UVB-induced Hyperpigmentation in C57BL/6 Mice." *Journal of Cosmetic Science* 62, no. 1 (2011): 1–14.

Chang, Jeong Ho, Min Ae Cho, Hong Ha Son, Cheon Koo Lee, Myung Seok Yoon, Hyung Hwan Cho, Dong Sung Seo, and Kyung Ja Kim. "Characterization and Formation of Phospholipid Nanoemulsion Coatings on Mg-Modified Sericite Surface." *Journal of Industrial and Engineering Chemistry* 12, no. 4 (2006): 635–638.

Chinembiri, Tawona N., Minja Gerber, Lissinda H. Du Plessis, Jan L. Du Preez, Josias H. Hamman, and Jeanetta Du Plessis. "Topical Delivery of *Withania somnifera* Crude Extracts in Niosomes and Solid Lipid Nanoparticles." *Pharmacognosy Magazine* 13, no. Suppl 3 (2017): S663.

Chivte, Prajkta Satish, Vinal Shivram Pardhi, Vineeta Anil Joshi, and Ajitha Rani. "A Review on Therapeutic Applications of Phytosomes." *Journal of Drug Delivery and Therapeutics* 7, no. 5 (2017): 17–21.

Choi, Jae-Hwan, Sun-Hang Cho, Je-Jung Yun, Young-Beob Yu, and Cheong-Weon Cho. "Ethosomes and Transfersomes for Topical Delivery of Ginsenoside Rh1 from Red Ginseng: Characterization and in Vitro Evaluation." *Journal of Nanoscience and Nanotechnology* 15, no. 8 (2015): 5660–5662.

Di Marzio, Luisa, Cinzia Anna Ventura, Donato Cosco, Donatella Paolino, Antonio Di Stefano, Rosanna Stancanelli, Silvana Tommasini, Carmela Cannava, Christian Celia, and Massimo Fresta. "Nanotherapeutics for Anti-Inflammatory Delivery." *Journal of Drug Delivery Science and Technology* 32 (2016): 174–191.

Fang, Jia-You, Tsong-Long Hwang, Yen-Ling Huang, and Chia-Lang Fang. "Enhancement of the Transdermal Delivery of Catechins by Liposomes Incorporating Anionic Surfactants and Ethanol." *International Journal of Pharmaceutics* 310, no. 1–2 (2006): 131–138.

Fasolo, D., V. L. Bassani, and H. F. Teixeira. "Development of Topical Nanoemulsions Containing Quercetin and 3-O-methylquercetin." *Die Pharmazie—An International Journal of Pharmaceutical Sciences* 64, no. 11 (2009): 726–730.

Floury, Juliane, Anne Desrumaux, and Jérémie Lardières. "Effect of High-Pressure Homogenization on Droplet Size Distributions and Rheological Properties of Model Oil-in-Water Emulsions." *Innovative Food Science & Emerging Technologies* 1, no. 2 (2000): 127–134.

García, M. C., C. Aloisio, R. Onnainty, and G. Ullio-Gamboa. "Self-assembled nanomaterials." In *Nanobiomaterials*, pp. 41–94. Woodhead Publishing, 2018.

Gera, Meeta, Neelesh Sharma, Mrinmoy Ghosh, Do Luong Huynh, Sung Jin Lee, Taesun Min, Taeho Kwon, and Dong Kee Jeong. "Nanoformulations of Curcumin: An Emerging Paradigm for Improved Remedial Application." *Oncotarget* 8, no. 39 (2017): 66680.

Ghosh, A., S. Biswas, and T. Ghosh. "Preparation and Evaluation of Silymarin β-Cyclodextrin Molecular Inclusion Complexes." *Journal of Young Pharmacists* 3, no. 3 (2011): 205–210.

Gollwitzer, Christian, Dorota Bartczak, Heidi Goenaga-Infante, Vikram Kestens, Michael Krumrey, Caterina Minelli, Marcell Pálmai et al. "A Comparison of Techniques for Size Measurement of Nanoparticles in Cell Culture Medium." *Analytical Methods* 8, no. 26 (2016): 5272–5282.

Iqbal, Babar, Javed Ali, and Sanjula Baboota. "Recent Advances and Development in Epidermal and Dermal Drug Deposition Enhancement Technology." *International Journal of Dermatology* 57, no. 6 (2018): 646–660.

Karunanidhi, Priyanka, and Sanjay Singh. "A Review on Skin Targeted Delivery of Bioactives as Ultradeformable Vesicles: Overcoming the Penetration Problem." *Current Drug Targets* 15, no. 2 (2014): 184–198.

Kaur, Chanchal Deep, and Swarnlata Saraf. "Topical Vesicular Formulations of *Curcuma longa* Extract on Recuperating the Ultraviolet Radiation–Damaged Skin." *Journal of Cosmetic Dermatology* 10, no. 4 (2011): 260–265.

Khan, Barkat Ali, Naveed Akhtar, Abder Menaa, and Farid Menaa. "A Novel *Cassia fistula* (L.)-Based Emulsion Elicits Skin Anti-Aging Benefits in Humans." *Cosmetics* 2, no. 4 (2015): 368–383.

Krausz, Aimee E., Brandon L. Adler, Vitor Cabral, Mahantesh Navati, Jessica Doerner, Rabab A. Charafeddine, Dinesh Chandra et al. "Curcumin-Encapsulated Nanoparticles as Innovative Antimicrobial and Wound Healing Agent." *Nanomedicine: Nanotechnology, Biology and Medicine* 11, no. 1 (2015): 195–206.

Kumar, Dinesh, Nitin Vats, Kamal Saroha, and Avtar Chand Rana. "Phytosomes as Emerging Nanotechnology for Herbal Drug Delivery." In *Sustainable Agriculture Reviews* 43, pp. 217–237. Springer, Cham, 2020.

Liu, Dan, Haiyang Hu, Zhixiu Lin, Dawei Chen, Yongyuan Zhu, Shengtao Hou, and Xiaojun Shi. "Quercetin Deformable Liposome: Preparation and Efficacy against Ultraviolet B Induced Skin Damages In Vitro and In Vivo." *Journal of Photochemistry and Photobiology B: Biology* 127 (2013): 8–17.

Mei, Zhinan, Qunrong Wu, Sheng Hu, Xiaokuan Lib, and Xiangliang Yang. "Triptolide Loaded Solid Lipid Nanoparticle Hydrogel for Topical Application." *Drug Development and Industrial Pharmacy* 31, no. 2 (2005): 161–168.

Mishra, Gyan P., Mahuya Bagui, Viral Tamboli, and Ashim K. Mitra. "Recent Applications of Liposomes in Ophthalmic Drug Delivery." *Journal of Drug Delivery* 2011 (2011).

Mohammadi, Hamidreza, Elahe Nekobahr, Javad Akhtari, Majid Saeedi, Jafar Akbari, and Fereshteh Fathi. "Synthesis and Characterization of Magnetite Nanoparticles by Co-Precipitation Method Coated with Biocompatible Compounds and Evaluation of In Vitro Cytotoxicity." *Toxicology Reports* 8, (2021): 331–336.

Montagna, W., F. Ebling, and G. John. Human skin. *Encyclopedia Britannica* 2020. https://www.britannica.com/science/human-skin.

Mulik, Rohit, Kakasaheb Mahadik, and Anant Paradkar. "Development of Curcuminoids Loaded Poly (butyl) Cyanoacrylate Nanoparticles: Physicochemical Characterization and Stability Study." *European Journal of Pharmaceutical Sciences* 37, no. 3–4 (2009): 395–404.

Nirved U., V. Lokesh, G. Prasad Mishra, and Hemant M. Joshi. "Formulation and Evaluation of Ethosomes of *Sesbania grandiflora* Linn. Seeds." *Novel Science International Journal of Pharmaceutical Science* 1 (2012): 274–275.

Pandey, S., M. Goyani, V. Devmurari, and J. Fakir. "Transferosomes: A Novel Approach for Transdermal Drug Delivery." *Der Pharmacia Lettre* 1, no. 2 (2009): 143–150.

Pandey, V., Golhani, D., & Shukla, R. "Ethosomes: Versatile vesicular carriers for efficient Transdermal Delivery of Therapeutic agents." *Drug Delivery* 22, no. 8 (2014): 988–1002.

Paolino, Donatella, Giuseppe Lucania, Domenico Mardente, Franco Alhaique, and Massimo Fresta. "Ethosomes for Skin Delivery of Ammonium Glycyrrhizinate: In Vitro Percutaneous Permeation through Human Skin and In Vivo Anti-Inflammatory Activity on Human Volunteers." *Journal of Controlled Release* 106, no. 1–2 (2005): 99–110.

Patel, R., S. K. Singh, S. Singh, N. R. Sheth, and R. Gendle. "Development and Characterization of Curcumin Loaded Transfersome for Transdermal Delivery." *Journal of Pharmaceutical Sciences and Research* 1, no. 4 (2009): 71.

Patravale, Vandana B., and Amit G. Mirani. "Preparation and characterization of solid lipid nanoparticles-based gel for topical delivery." In *Pharmaceutical Nanotechnology*, pp. 293–302. Humana, New York, NY, 2019.

Puglia, Carmelo, Luisa Rizza, Markus Drechsler, and Francesco Bonina. "Nanoemulsions as Vehicles for Topical Administration of Glycyrrhetic Acid: Characterization and In Vitro and In Vivo Evaluation." *Drug Delivery* 17, no. 3 (2010): 123–129.

Quintão, Frederico J. O., Renata S. N. Tavares, Sidney A. Vieira-Filho, Gustavo H. B. Souza, and Orlando D. H. Santos. "Hydroalcoholic Extracts of *Vellozia squamata*: Study of its Nanoemulsions for Pharmaceutical or Cosmetic Applications." *Revista Brasileira de Farmacognosia* 23, no. 1 (2013): 101–107.

Ravalika, Veldurthi, and A. Sailaja. "Formulation and Evaluation of Etoricoxib Niosomes by Thin Film Hydration Technique and Ether Injection Method." *Nano Biomedicine and Engineering* 9, no. 3 (2017): 242–248.

Sachan, Anupam Kumar, and Ankita Gupta. "A Review on Nanotized Herbal Drugs." *International Journal of Pharmaceutical Sciences and Research* 6, no. 3 (2015): 961.

Saraf, S. "Applications of Novel Drug Delivery System for Herbal Formulations." *Fitoterapia* 81, no. 7 (2010): 680–689.

Sarwa, Khomendra K., Pranab J. Das, and Bhaskar Mazumder. "A Nanovesicle Topical Formulation of *Bhut jolokia* (hottest capsicum): A Potential Anti-Arthritic Medicine." *Expert Opinion on Drug Delivery* 11, no. 5 (2014): 661–676.

Satyavani, K., T. Ramanathan, and S. Gurudeeban. "Plant Mediated Synthesis of Biomedical Silver Nanoparticles by Using Leaf Extract of *Citrullus colocynthis*." *Research Journal of Nanoscience and Nanotechnology* 1, no. 2 (2011): 95–101.

Semalty, Ajay, Yuveraj Singh Tanwar, and Mona Semalty. "Preparation and Characterization of Cyclodextrin Inclusion Complex of Naringenin and Critical Comparison with Phospholipid Complexation for Improving Solubility and Dissolution." *Journal of Thermal Analysis and Calorimetry* 115, no. 3 (2014): 2471–2478.

Silva, A. P., B. R. Nunes, M. C. De Oliveira, L. Scherer Koester, P. Mayorga, V. Linck Bassani, and H. Ferreira Teixeira. "Development of Topical Nanoemulsions Containing the Isoflavonegenistein." *Die Pharmazie—An International Journal of Pharmaceutical Sciences* 64, no. 1 (2009): 32–35.

Sungthongjeen, S., T. Pitaksuteepong, A. Somsiri, and P. Sriamornsak. "Studies on Pectins as Potential Hydrogel Matrices for Controlled-Release Drug Delivery." *Drug Development and Industrial Pharmacy* 25, no. 12 (1999): 1271–1276.

Suthar, Shreya Mahesh, and Bhargavi Ankit Rathva. "Development of Liposomal Formulation: From Formulation to Sterilization." *World Journal of Pharmaceutical Research* 8, no. 3 (2019): 1561–1571.

Sutthanut, K., X. Lu, B. Sripanidkulchai, C. Yenjai, and M. Jay. "Solid Liquid Nanoparticles for Transdermal Delivery of *Kaempferia parviflora* Extracts." *Journal of Biomedical Nanotechnology* 5, no. 2 (2009): 224–232.

Takahashi, Makoto, Dai Kitamoto, Yonathan Asikin, Kensaku Takara, and Koji Wada. "Liposomes Encapsulating Aloe Vera Leaf Gel Extract Significantly Enhance Proliferation and Collagen Synthesis in Human Skin Cell Lines." *Journal of Oleo Science* 58, no. 12 (2009): 643–650.

Tan, Qi, Weidong Liu, Chenyu Guo, and Guangxi Zhai. "Preparation and Evaluation of Quercetin-Loaded Lecithin-Chitosan Nanoparticles for Topical Delivery." *International Journal of Nanomedicine* 6 (2011): 1621.

Wu, Tzu-Hui, Feng-Lin Yen, Liang-Tzung Lin, Tong-Rong Tsai, Chun-Ching Lin, and Thau-Ming Cham. "Preparation, Physicochemical Characterization, and Antioxidant Effects of Quercetin Nanoparticles." *International Journal of Pharmaceutics* 346, no. 1–2 (2008): 160–168.

Zhaowu, Zeng, Wang Xiaoli, Zhang Yangde, and Li Nianfeng. "Preparation of Matrine ethosome, its Percutaneous Permeation In Vitro and Anti-Inflammatory Activity In Vivo in Rats." *Journal of Liposome Research* 19, no. 2 (2009): 155–162.

Zhou, Yan, Yuhui Wei, Huanxiang Liu, Guoqiang Zhang, and Xin'an Wu. "Preparation and In Vitro Evaluation of Ethosomal Total Alkaloids of *Sophora alopecuroides* Loaded by a Transmembrane pH-gradient Method." *AAPS PharmSciTech* 11, no. 3 (2010): 1350–1358.

6 Patents and Nanobiotechnological Trends in the Treatment of Diabetes

J.A.C. Nascimento Júnior, A.M. Santos, T. de A. Andrade, S. Shanmugam, L.A. Frank and M.R. Serafini

CONTENTS

6.1 INTRODUCTION

Until the early 1920s, there was no effective treatment for diabetes mellitus (DM), which resulted in the death of many young people. Diabetes is characterized by the nonproduction of insulin (INS). The body does not absorb glucose from the blood, which causes a lack of energy for the cells.

Type 1 diabetes can occur at any age, although its highest prevalence is among children and young people. Daily, patients with this disease should receive exogenous insulin delivery.

Type 2 diabetes usually begins in adulthood as progressive insulin resistance, with pancreatic involvement (Inzucchi 2002). Therapy consists of an adequate diet, physical activity, controlled blood glucose levels and often the administration of antihyperglycaemic medication (Kalakotla et al. 2015).

One way to make medication more effective is the use of nanostructured carrier systems (Frank et al. 2015). The use of nanoparticles (NPs) containing drugs and biomolecules for the treatment of diabetes has been a modern approach. The ideas and researchers behind the creation of something new, the development of a manufacturing process or an improvement of an existing technique or product are protected through patents (Kimura, da Silva, and Costa 2019).

These documents are an incentive to technological development, encouragement to scientific research and to meeting the needs of society (Fialkoski, Ricardo, and Malfatti 2019).

We carried out a patent survey considering the subject "diabetes and nanotechnology" published in different databases over a period of five years. The European Patent Office (EPO) and the World Intellectual Property Organization (WIPO) databases were searched using the title and summary fields "nano*", code A61P 3/10 (hypoglycaemic drugs) in International Patent Classification (IPC) field and publication date between January 2015 and May 2020.

DOI: 10.1201/9781003231745-6

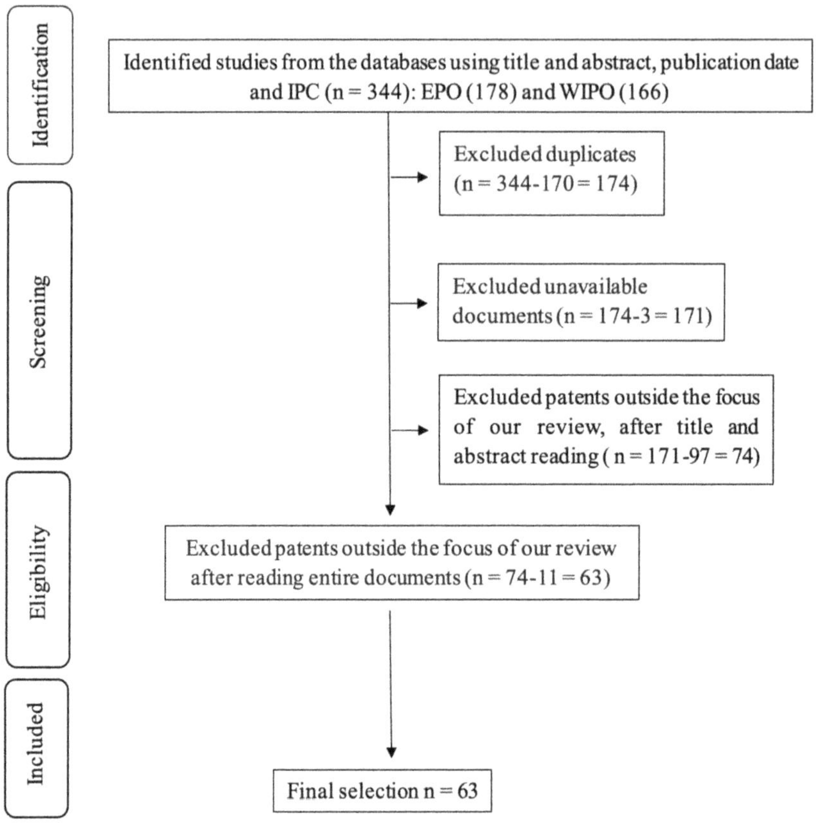

FIGURE 6.1 Flowchart of patent search and screening.

A total of 344 patents were identified for preliminary assessment of the database, of which 170 were excluded due to duplication, three because the text was not available, 97 were out of focus from our review after reading the title and summary and another 11 patents were excluded after reading the full patent for being out of focus in the review. This resulted in the selection of 63 patents for our critical analysis. Figure 6.1 illustrates the systematic guidelines based on the PRISMA methodology.

6.2 ANALYSES, CORRELATIONS AND INNOVATIONS ON THE TOPIC

6.2.1 Patent Search and Screening

During the period of 2015 to 2020 (Table 6.1), the largest number of patent documents occurred in 2018, with 20 cases, represented in Figure 6.2. From 63 patents selected for evaluation, China led the ranking with 51 patents filed, showed in Figure 6.3. Also, with a significant difference in quantity compared to the second place in the ranking, six patents were deposited via the Patent Cooperation Treaty (PCT) by WIPO.

Universities filed the most significant number with 44 patents, pharmaceutical industry with 12 and hospitals with two and independent researchers with three orders. Partnerships can be made between entities as noted between universities and industries, as well as with hospitals, presenting each patent as shown in Figure 6.4.

TABLE 6.1
Patents Published in Relation to the Treatment of Diabetes with Nanotechnology

Inventor(s) and Year	Active	Particle Type	Pharmaceutical Form	Route
Kang et al. 2015	Dogwood	Gold nanoparticle	Powder	Oral
Liu, Gu, and Chen 2019	Insulin	Cyclodextrin	Suspension	Oral
Guan, Li, and Huang 2019	Artemisinin	Polymeric nanoparticle	Suspension	Oral
van Endert, Motte, and Dubreil 2020	Insulin	Nanoparticle	Suspension	Parenteral
Shao, Zou, and Lin 2018	Insulin	Polymeric nanoparticle	Suspension	Parenteral
Astudillo De La Vega 2019	Natural compounds	Nanoparticle	Suspension	Oral
Geho 2019	Insulin	Nanoparticle	Suspension	Parental, oral, buccal or inhalable
Han and Sun 2016	Natural compounds	No specified	Powder	NS
Gu 2018	Insulin	Nanoparticle	NS	Oral
Liu, Sun, et al. 2018	Insulin	Nanoparticle and microparticle	Suspension	Oral
Liu, Zhou, and Xu 2015	Insulin	Functionalized nanoparticle	Suspension	Oral
Shao, Guo, and Li 2018	Ginsenoside	Nanoparticle	Gel	Transdermal
Fu et al. 2016	Glibenclamide	Nanoparticle	Suspension	Parental, oral or inhalable
Han et al. 2016	Glucagon	Nanoparticle with peptide	Suspension	Parenteral
Li 2018	Glucose	Nanoparticle	Hydrogel	Parental
Zhu, Wu, Li, et al. 2016	Glucose	Nanoparticle	Suspension	NS
Gu and Yu 2019	Insulin	Nanoparticle	Suspension	NS
Lei et al. 2018	Insulin	Gold nanoparticle	NS	Transdermal
Zhen and Xiuli 2019	Insulin	Nanoparticle	Suspension	Transdermal
Anderson et al. 2016	Insulin	Nanoparticle with gel	Suspension	Parental
Zeng et al. 2015	Insulin	Nanoparticle	Suspension	Parental
Yao and Zhang 2017	Insulin	Nanoparticle	Suspension	Oral
Zhang, Wu, and Deng 2016	Insulin	Nanoparticle	Suspension	Oral
Liu, Sun, et al. 2019	Insulin	Eudragit nanogranules	Powder	Oral
Guan, Zhang, and Qiu 2018	Insulin	Nanoparticle	Suspension	Oral
Zeng, Yang, et al. 2017	Insulin	Nanomicelles	Suspension	Oral
Li and Zhou 2018	Insulin	Nanoparticle	Suspension	NS
Whitehead and Lamson 2020	Insulin	Nanoparticle	Gel	Transdermal
Ke and Fan 2019	Insulin	Nanoparticle	Suspension	Parental
Zhu, Gao, and Yu 2018	Insulin	Nanoparticle	Tablet	Oral
Wang 2018	Insulin	Nanoparticle	Tablet	Oral
Chen and Zeng 2018	Empagliflozin	Nanoparticle	Tablet	Oral
Yu 2020	Hybrid Cas9 protein	Liposome	Suspension	Transdermal
Shao, Lin, and Guo 2018	Insulin	Nanoparticle	Adhesive transdermal	Transdermal
Liu, Tian, et al. 2019	Insulin	Nanoparticle	Suspension	Oral
He et al. 2019	Insulin	Nanoparticle	Suspension	Oral
Rademacher and Williams 2015	Insulin	Nanoparticle	Suspension	Transbuccal
Zhao, Yang, et al. 2017	Natural product	Nanoparticle	Solution	NS
Chen, Xu, et al. 2017	STR1720	Nanoparticle	Suspension	NS
Hua et al. 2019	Bifidobacterium	Nanoparticle with selenium	NS	Parental
Wu et al. 2018	Insulin	Nanoparticle	NS	Oral
Long 2018	Insulin	Nanoparticle	Suspension	Oral

(Continued)

TABLE 6.1 *(Continued)*
Patents Published in Relation to the Treatment of Diabetes with Nanotechnology

Inventor(s) and Year	Active	Particle Type	Pharmaceutical Form	Route
Liu, Tian, et al. 2018	Insulin	Nanogranules	Powder	Oral
Zhang and Meng 2018	Insulin	Nanoparticle	Suspension	Oral
Sun 2015	Insulin	Nanoparticle	Tablet	Sublingual
Zeng, Wu, et al. 2017	Insulin	Nanoparticle	Suspension	Oral
Wu et al. 2019	Insulin	Nanoparticle	Gel	Transdermal
Hou et al. 2016	Insulin	Nanoparticle	Solution	NS
Xiong et al. 2018	Insulin	Polymeric nanoparticle	Suspension	Oral
Sun et al. 2017	Insulin	Nanocomplex	Solution	Oral
Lu 2019	Insulin	Nanoparticle	Solution	Oral
Chen et al. 2015	Insulin	Nanoparticle	Suspension	Oral
Chen, Wang, et al. 2017	Insulin	Nanoparticle	Suspension	Oral
Shao, Zou, and Guo 2018	Insulin	Mesoporous silica nanoparticles	Adhesive transdermal	Transdermal
Hong et al. 2017	Insulin	Nanocomposite	Suspension	Parental
Wang and Zhang 2015	Insulin	Nanoparticle	Capsule	Oral
Sun et al. 2015	Insulin	Nanoparticle	Capsule	Oral
Lei, Liu, and Zhang 2020	Insulin	Gold nanoparticle	NS	Transdermal
Fu et al. 2019	Roxburgh rose	Functionalized nanoparticle	Suspension	NS
Zhu, Wu, Wu, et al. 2016	Insulin	Nanoparticle	Suspension	Oral
Bornal and Quintana 2016	Insulin	Nanoparticle	Suspension	Oral
Shao, Li, and Guo 2018	Insulin	Nanoparticle	Transdermal	Transdermal
Shao et al. 2019	Insulin	Nanoparticle	Transdermal	Transdermal

Finally, the type of pharmaceutical nanotechnology compositions were quantified. The suspensions showed the most significant number since they are the form in which insulin is available for treatment. Oral insulin administration was also present in the inventions, highlighting the use of capsules, tablets, powders and solutions. The application of dermal insulin was also addressed, with the use of transdermal gels and adhesives, as represented in Figure 6.5.

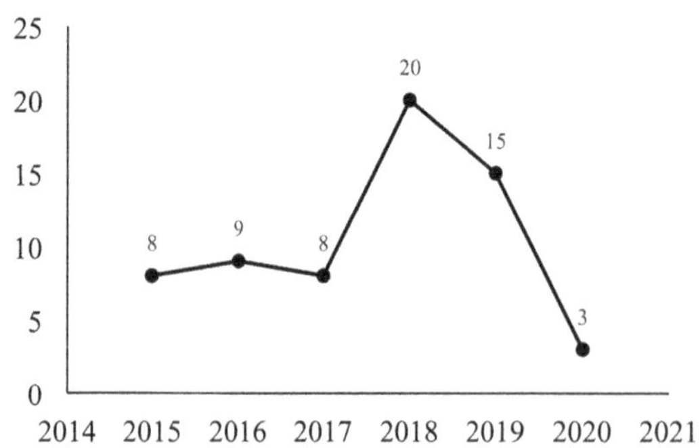

FIGURE 6.2 Patent publication numbers per period after final selection.

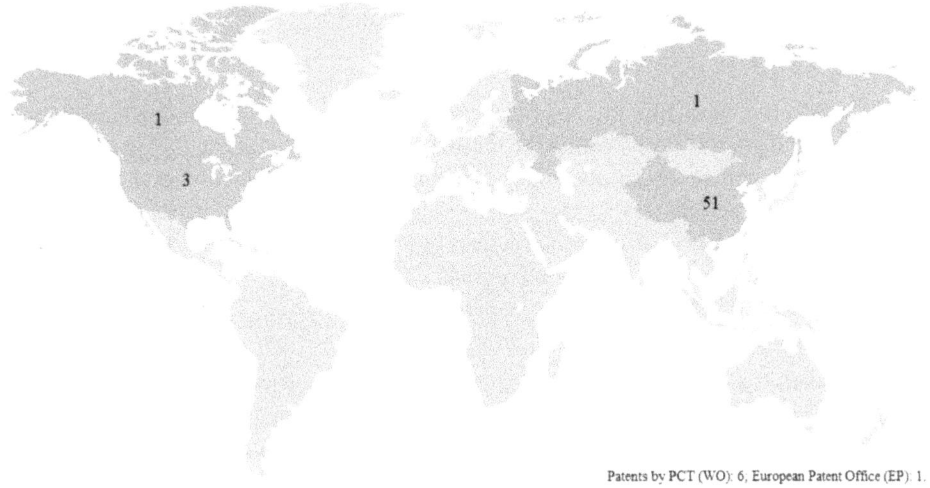

FIGURE 6.3 Number of patent filings by country.

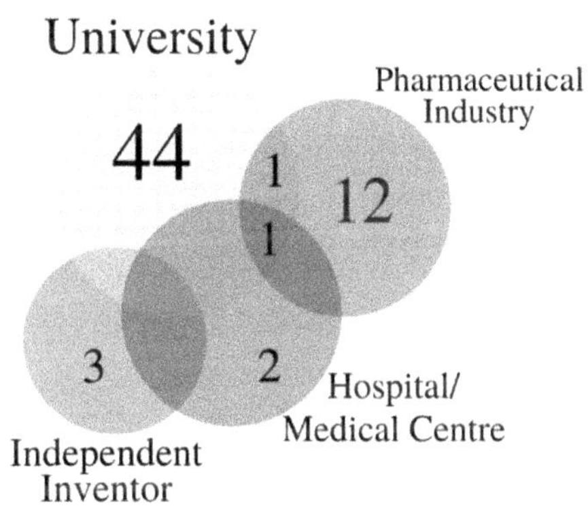

FIGURE 6.4 Final selection of patents: type of applicant.

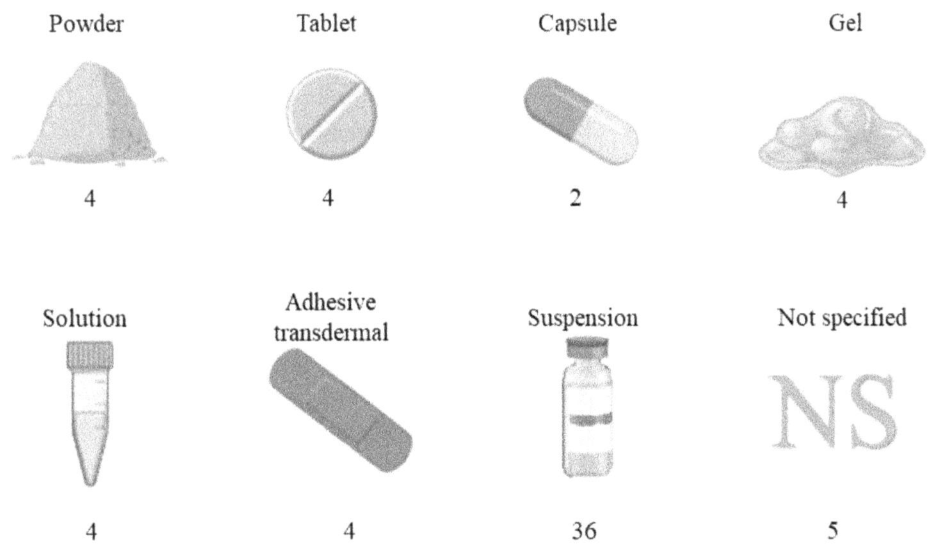

FIGURE 6.5 Final selection of patents: pharmaceutical form.

6.3 TECHNOLOGICAL SCENARIO OF PATENTS

6.3.1 Insulin Therapy

Subcutaneous injection of insulin has been the standard treatment of insulin-dependent patients to maintain their blood glucose levels within the normal range. Nonetheless, daily injections of insulin have disadvantages, such as suboptimal control, fluctuation in blood glucose levels, hypoglycaemic events, unwanted pain, peripheral hyperinsulinemia leading to hypertension and atherosclerosis and other side effects (Momoh et al. 2020). Alternatives have been developed to achieve effectiveness in other types of administration, which will be presented next.

6.3.1.1 Oral Insulin

Insulin administered orally alleviates the issue of injection pain, provides better glucose homeostasis, mimics the physiological fate of insulin (INS) and can improve patient compliance (Yu et al. 2015). However, the bioavailability of oral delivery is affected by extensive degradation of the protein and peptide drugs through enzymatic and chemical activity (Khaleel Basha et al. 2020).

To address these problems, materials such as NPs have been used to deliver insulin and improve its activity and administration, which can achieve biocompatibility, biodegradability and the ability to protect insulin from degradation (Han et al. 2016; Mumuni et al. 2020).

In 2015, Wang and Zhang invented a soluble starch nanoparticle insulin with a sustained release and a small particle size. In this invention, the nanoparticle has a large specific area and advantages such as a high embedding rate on transported medicines and becoming a safe, nontoxic and side-effect free option with small hygroscopicity.

In another invention by Chen, Wang et al., insulin was protected from the influence of the protease and acidic environment of the gastrointestinal tract. This causes the intestinal epithelium to absorb the insulin through the chitosan and the carboxymethyl chitosan, consequently reducing blood glucose (Chen, Wang et al. 2017).

To improve the oral bioavailability of insulin, nanomedicine technologies might employ an insulin-polymer conjugation, such as micelles and microparticles (MPs) (Woldu and Lenjisa 2014). That type of technology is seen in Liu, Sun et al. (2018), which resulted in a stable formulation and a fast-dissolving process of the drug. The invention has advantages such as higher entrapment efficiency, small intestine mucus adhesion and penetration functions (Liu, Sun, et al. 2018).

Polymeric micelles as the most typical and potential delivery vectors have showed significant advances in biomedical area due to their versatility of functions and clinical success (Zhang et al. 2020). Thus, Zeng, Yang et al. used de micelles to develop an insulin-loaded nanomicelle with pH sensibility. The invention provides the proper release and absorption of insulin in the intestinal tract, having excellent stability and dispersibility in an aqueous solution (Zeng, Yang et al. 2017).

Moreover, Liu, Zhou and Xu (2015) developed an amino polyethylene glycol-functionalized nanographene oxide insulin to treat type 2 DM. The formulation can inhibit islet amyloid protein aggregation in the pancreas. The use of nanographene oxide brings higher stability and solubility to the invention.

6.3.1.2 Parenteral Insulin

Parenteral administration is the major route for insulin delivery due to its large molecular weight, poor membrane permeability and susceptibility to enzymatic proteolysis (Zhang et al. 2019). The majority of diabetes patients have resistance to initiate treatment, due to their fear and anxiety about the injection and its side effects. For this reason, nanoparticles and hydrogels raise the acceptance of INS therapy (Ji et al. 2020).

Zeng et al. (2015) used nanoparticles as a carrier and the insulin as an active constituent; the formulation can release insulin for a long period of time, consequently reducing the frequency of drug administration.

Moreover, hydrogels associated with nanoparticles can be administrated by using minimally invasive approaches (Zhao, Wu et al. 2017). Thus, Li (2018) used this skill of the hydrogel to control the slow release of insulin. The co-delivery ensured that all the compounds of the formulation were delivered at the same time and presented in the same environment (van Endert, Motte, and Dubreil 2020). For this reason, Hong et al. (2017) used a short amylase insulin, where the nanoparticle can tightly adhere to membrane in the small intestine, prolonging the time of the compounds in the stomach and intestine.

6.3.1.3 Transdermal Insulin

Transdermal delivery can avoid chemical and enzymatic degradation in the digestive tract as well as hepatic first-pass metabolism. The advantage of this route occurs since the skin is the most extensive organ and receives one-third of the total blood supplied throughout the body (Fathima, Begum, and Fatima 2017).

The liposome is also an alternative to improve transdermal administration, as seen in Yu (2020), in which a nanoliposome carrier encapsulates a hybrid protein and a guide RNA to treat diseases such as type 2 DM by gene editing. The formulation can suppress the expression of a DNA target, resulting in high drug delivery efficiency.

Shao, Guo and Li (2018) disclosed hypoglycaemic drug insulin combined with a ginsenoside-insulin nanogel. The formulation can slow the release of insulin, being long in the drug effect period, and avoid the side effects caused by the injection. The area of the diffusion is 2.80 cm^2 to achieve a relatively stable transmission state.

Whitehead and Lamson (2020) provided a delivery drug based on transepithelial nanoparticles, which are made porous to allow the permeation of the compounds, eluting from the nanoparticles to adjust the release profile of the ingredient.

Shao, Li and Guo (2018) invented a transdermal-delivered ursolic acid/insulin nanoscale slow-release preparation. The ursolic acid/insulin nano- and microneedles have high drug loading capacity, which can penetrate the endothelial blood vessel to increase power and not be damaged.

Another invention deposited by Shao et al. (2019) brought a formulation based on a vitamin D-insulin nanometre slow-release transdermal preparation. The vitamin D is taken as a carrier and subjected to co-assembling with insulin to obtain a nanometre solution. The insulin and vitamin D can improve pancreatic β-cell function.

6.3.1.4 Sublingual Delivery

Pharmaceutical formulations for the sublingual route provide convenient and easy methods for drug administration. Sun (2015) developed a safe, effective, controlled in slow release and stable sublingual tablet. The invention exhibits a small particle size, uniform distribution, a high encapsulation rate, promotes mucosal absorption, inhibits enzymatic hydrolysis and shows a controllable release. The formulation also can reduce the order of the lateral chains of lipid molecules and increase the mobility of membrane lipids.

6.3.2 Other Antidiabetic Therapy

The treatment of diabetes can also be performed using other therapies as antidiabetic drugs working through different mechanisms. Kang et al. (2015) disclosed an invention using a nano-level homogeneous ultrafine powder of dogwood as a medicine against type 2 DM. Compared to the regular oral treatment, the dogwood nanopowder is safer, more effective, mild, long-lasting, nontoxic and without side effects. The dogwood nanopowder can regulate genes in the skeletal muscle to enhance the use of glucose and increase the genes' expression level.

Guan, Li and Huang (2019) developed a polymeric nanoparticle to improve insulin resistance and pancreatic β-cell injury in type 2 DM at the same time causing the transformation of the pancreatic α-cells into β-cell. This is accomplished through the combination of citrus pectin and artemisinin.

Horacio Astudillo De La Vega (2019) developed an invention focused on the treatment and protection of DM with natural compounds. The formulation of the treatment method is based on the development of responses in metabolism control and cell growth.

Meanwhile, glibenclamide has been incorporated into nanocrystals by Fu et al. (2016). The formulation has no physiological toxicity, but has good biocompatibility, exhibits high drug-loading capacity and is stable.

Chen and Zeng (2018) related an invention using empagliflozin in nanoassembled tablets, which can increase of the dissolution rate of the active ingredient. Moreover, the invention has good water absorption, expansion properties and nonphysiological toxicity.

6.4 CONCLUSIONS AND PERSPECTIVES

Each year, the number of people with DM increases around the world. However, the low adherence and possible side effects decrease the effectiveness of treatment. Among the medicines found in this review, the most widely used was insulin with nanotechnology for administration by several routes. Additionally, oral antiglycaemic drugs and natural products were often used. The use of nanoparticles sought to improve bioavailability and reduce adverse effects.

However, a large number of these inventions were created by universities, which reduces the availability of these drugs to the population for a few years. The emergence of automatic insulin infusion is now available as well as the stem cells are used, which aid the development of vaccines and new antiglycaemic drugs, bringing novel treatment of the disease. Thus, the control and possible cure of the disease gets closer and closer.

REFERENCES

Anderson, Daniel G., Zhen Gu, Alex Arthur Aimetti, and Robert S. Langer. 2016. Injectable nano-network gels for diabetes treatment. U.S. Patent Application 14/787,652, filed March 10, 2016.

Astudillo De La Vega, Horacio. 2019. Combination of bioenergy and nutra-epigenetic metabolic regulators, nutraceutical compounds in conventional and nanotechnology-based combinations, for reversing and preventing cellular senescence accelerated by chronic damage caused by diabetes and other. EP3470060, issued 2019.

Bornal, Ada Yeste, and Francisco J. Quintana. 2018. Tolerogenic nanoparticles for treating diabetes mellitus. U.S. Patent Application 15/560,276, filed March 15, 2018.

Chen, Qing, and Juntang Zeng. 2018. Nano-assembled empagliflozin tablet for diabetes, and preparation method thereof. CN109010293, issued 2018.

Chen, Xiao Dong, Fei Yu, Xue E. Wu, and Liming Che. 2015. Preparation method of insulin-loaded nanoparticle. CN104840948, issued 2015.

Chen, Xiguang, Juan Wang, Chao Feng, Xiaojie Cheng, Ming Kong, and Ya Liu. 2017. Preparation method of insulin-supporting carboxymethyl chitosan/chitosan nanometer preparation. CN106421807, issued 2017.

Chen, Yulong, Cangbao Xu, Ke Li, Yaping Zhang, Min Jia, Ying Yang, and Huanhuan Liu. 2017. Nano-preparation for treating type 2 diabetes mellitus and preparation method of nano-preparation. CN107468670, issued 2017.

Danne, Thomas, Lutz Heinemann, and Jan Bolinder. "New insulins, biosimilars, and insulin therapy." *Diabetes technology & therapeutics* 21, no. S1 (2019): S–57.

van Endert, P., L. Motte, and C. Dubreil. 2020. Biocompatible tolerogenic nanoparticle. WO2020025408, issued 2020.

Fathima, Syeda Ayesha, Shireen Begum, and Syeda Saniya Fatima. "Transdermal drug delivery system." *International journal of pharmaceutical and clinical research* 9, no. 01 (2017): 35–43.

Fialkoski, Diego, Carlos Ricardo, and Maneck Malfatti. "Nanotecnologia: uma prospecção tecnológica no âmbito nacional e internacional." *Cadernos de Prospecção* 12, no. 3 (2019): 590.

Frank, Luiza A., Renata V. Contri, Ruy C. R. Beck, Adriana R. Pohlmann, and Silvia S. Guterres. "Improving drug biological effects by encapsulation into polymeric nanocapsules." *Wiley interdisciplinary reviews: Nanomedicine and nanobiotechnology* 7, no. 5 (2015): 623–639.

Fu, Qiang, Zhonggui He, Wenqian Yang, Mengran Guo, Zhibin Guo, and Shijia Guan. 2016. Glibenclamide nanocrystal preparation and preparation method thereof. CN105878194, issued 2016.

Fu, Xiong, Lei Wang, Chao Li, and Qiang Huang. 2019. Roxburgh rose polysaccharide functionalized nanometer selenium compound, preparation method thereof and application of roxburgh rose polysaccharide functionalized nanometer selenium compound in blood glucose reducing drug. CN109303922, issued 2019.

Geho, W. B. 2019. Compositions comprising lipid-based nanoparticles for treating diabetes mellitus. WO2019136386, issued 2019.

Gu, Yan. 2018. Drug release rate rapid regulating nano insulin-chitosan produced by pressurizing coating method. CN107596352, issued 2018.

Gu, Zhen, and Jichen Yu. 2019. Glucose-responsive insulin delivery system using hypoxia-sensitive nanocomposites. RU2017139931, issued 2019.

Guan, Yanqing, Jian Li, and Jingmin Huang. 2019. Artemisinin-loaded citrus pectin oral nanoparticle. CN110200980, issued 2019.

Guan, Yanqing, Li Zhang, and Jiani Qiu. 2018. Insulin-loaded nanoparticles and application thereof. CN108721605, issued 2018.

Han, Haixia, and Xiaowei Sun. 2016. Diabetic foot treating nano-level pharmaceutical composition and preparation method thereof. CN105477492, issued 2016.

Han, Jing, Yan Tang, Yuanxin He, Yuhe Qian, Leilei Gong, and Yunkai Wan. 2016. Glucagon-like peptide-1 slow-released nano preparation, preparation method and application. CN106074376, issued 2016.

He, Zhiyu, Yongming Chen, Lixin Liu, Haiquan Mao, and Kam W. Leong. 2019. Nanoparticle loaded with therapeutic protein and microcapsule thereof. WO2019076125, issued 2019.

Hong, Yan, Na Ji, Zhengbiao Gu, Li Cheng, Zhaofeng Li, and Caiming Li. 2017. Preparation method of short amylase-insulin or short amylase-insulin-procyanidine nanocomposite. CN107213457, issued 2017.

Hou, Lin, Zhenzhong Zhang, Yazhen Zheng, Xiaoning Shan, Lisha Hao, Yinling Zhang, and Qianhua Feng. 2016. Preparation and applications of mesoporous silica/insulin nanoparticles modified by phenylboronic acid. CN106236734, issued 2016.

Hua, Zichun, Yan Lin, Yan Zhang, and Genxing Xu. 2019. Nano-selenium *Bifidobacterium longum* as well as preparation method and application thereof. CN109576176, issued 2019.

Inzucchi, Silvio E. "Oral antihyperglycemic therapy for type 2 diabetes: scientific review." JAMA 287, no. 3 (2002): 360–372.

Ji, Linong, Leili Gao, Liming Chen, Yangang Wang, Zhongshu Ma, Xingwu Ran, Zilin Sun et al. "Insulin delivery with a needle-free insulin injector versus a conventional insulin pen in Chinese patients with type 2 diabetes mellitus: A 16-week, multicenter, randomized clinical trial (the FREE study)." *EClinicalMedicine* 23 (2020): 100368.

Kalakotla, Shanker, Krishna Mohan Gottumukkala, M. Rani, and P. L. Pravallika. "Herbal drugs and herbal mediated silver nano particles as anti-diabetics: a new horizon." *International journal of pharmaceutical sciences review and research* 31 (2015): 142–148.

Kang, Jiefang, Yuan Bai, Shao Jun Wang, Hongju Wang, Ning Yang, and Zhezhi Wang. 2015. Application of dogwood nano powder to resist type-2 diabetes. CN104688799, issued 2015.

Ke, Tianyi, and Kelong Fan. 2019. Nano drug carrier used for treating diabetes and composite medicine thereof. CN110179997, issued 2019.

Khaleel Basha, S., M. Syed Muzammil, R. Dhandayuthabani, and V. Sugantha Kumari. "Development of nanoemulsion of Alginate/Aloe vera for oral delivery of insulin." *Materials today: Proceedings* (2020) May. Elsevier BV.

Kimura, Patrícia de Carvalho, Sandra Beserra da Silva, and Silvio Claudio da Costa. "Prospecção Tecnológica para Verificação do Potencial de Patenteabilidade de Alimento Formulado com Produtos de Estévia." *Cadernos de Prospecção* 12, no. 4 (2019): 890.

Lei, Yifeng, Sheng Liu, and Yujie Zhang. 2020. Responding type gold nano-cluster particles for blood glucose regulation and preparation method of responding type gold nano-cluster particles. CN108635570, issued 2020.

Lei, Yifeng, Sheng Liu, Yujie Zhang, Wubin Dai, and Ke Huang. 2018. Gold nano-cluster particle for intelligently releasing insulin for regulating blood glucose and preparation method thereof. CN108079282, issued 2018.

Li, Jianshu, and Anna Zhou. 2018. Method for preparing polypeptide-insulin nanoparticle. CN108159403, issued 2018.

Li, Yi. 2018. Glucose responsiveness injectable nano gel as well as preparation method and application thereof. CN109091453, issued 2018.

Liu, Jie, Xianbo Zhou, and Taoyuan Xu. 2015. Functionalized nano graphene oxide supported insulin derivative material, preparation method thereof and application in preparation of medicine for treating diabetes. CN104922692, issued 2015.
Liu, Yu, Bohan Gu, and Yong Chen. 2019. Application of sulfanilic acid-beta-cyclodextrin mediated supermolecular nanoparticles in controlled release of insulin. CN109908367, issued 2019.
Liu, Zhijia, Lilong Sun, Yongming Chen, Lixin Liu, Mao Haiquan, Jinrong Liang, and Chengxin Sun. 2018. Enteric nano-microparticles loaded with insulin as well as preparation method and applications of enteric nano-microparticles. CN108379560, issued 2018.
Liu, Zhijia, Lilong Sun, Yongming Chen, Lixin Liu, Haiquan Mao, Kam W. Leong, and Chengxin Sun. 2019. Insulin-loaded enteric-coated nanoparticles, preparation method therefor and uses thereof. WO2019148811, issued 2019.
Liu, Zhijia, Houkuan Tian, Yongming Chen, Haiquan Mao, Kam W. Leong, Lixin Liu, Zhiyu He, and Chengxin Sun. 2019. Nanoparticle formulation for oral insulin delivery and preparation method therefor. WO2019148810, issued 2019.
Liu, Zhijia, Houkuan Tian, Yongming Chen, Haiquan Mao, Jinrong Liang, Lixin Liu, Zhiyu He, and Chengxin Sun. 2018. Oral insulin nanometer granular preparation and method for preparing same. CN108371708, issued 2018.
Long, Qiaoming. 2018. Oral insulin nano composite and preparation method thereof. CN108236720, issued 2018.
Lu, Baocun. 2019. Preparation method of blood glucose reducing insulin nanoparticles. CN109276707, issued 2019.
Momoh, Mumuni A., Kenechukwu C. Franklin, Chinazom P. Agbo, Calister E. Ugwu, Musiliu O. Adedokun, Ofomatah C. Anthony, Omeje E. Chidozie, and Augustine N. Okorie. "Microemulsion-based approach for oral delivery of insulin: formulation design and characterization." *Heliyon* 6, no. 3 (2020): e03650.
Mumuni, Momoh A., Franklin C. Kenechukwu, Kenneth C. Ofokansi, Anthony A. Attama, and David Díaz. "Insulin-loaded mucoadhesive nanoparticles based on mucin-chitosan complexes for oral delivery and diabetes treatment." *Carbohydrate polymers* 229 (2020): 115506.
Mumuni, Momoh A., Frankilin C. Kenechukwu, Omeje C. Ernest, Adedokun M. Oluseun, Barikisu Abdulmumin, Darlington C. Youngson, Ofokansi C. Kenneth, and Attama A. Anthony. "Surface-modified mucoadhesive microparticles as a controlled release system for oral delivery of insulin." *Heliyon* 5, no. 9 (2019): e02366.
Rademacher, Thomas, and Phillip Williams. 2015. Nanoparticle-insulin and insulin analogue compositions. WO2015114341, issued 2015.
Shao, Jingwei, Yan Guo, and Chao Li. 2018. Ginsenoside-insulin nano gel as well as preparation method and application thereof. CN108273042, issued 2018.
Shao, Jingwei, Chao Li, and Yan Guo. 2018. Transdermally delivered ursolic acid/insulin nanoscale slow-release preparation and preparation method thereof. CN108272778, issued 2018.
Shao, Jingwei, Juanfang Lin, and Yan Guo. 2018. Nanometer transdermal slow-release preparation based on insulin loaded on PAMAM (polyamide-amine) dendrimers and preparation method thereof. CN108465105, issued 2018.
Shao, Jingwei, Junjie Zou, and Yan Guo. 2018. Preparation method of mesoporous silica-insulin nanometer slowly-released transdermal patch. CN108434125, issued 2018.
Shao, Jingwei, Junjie Zou, and Juanfang Lin. 2018. Chitosan/insulin nano sustained release percutaneous preparation and preparation method thereof. CN108553639, issued 2018.
Shao, Jingwei, Junjie Zou, Bingchen Zhang, Zixuan Chen, and Xiang Chen. 2019. Vitamin D-insulin nanometer slow-release transdermal preparation, and preparation method thereof. CN109289041, issued 2019.
Sun, Chengxin, Lixin Liu, Yongming Chen, Zhiyu He, and Zhijia Liu. 2017. Preparation method of apocynum acidic polysaccharide cell-penetrating peptide oral insulin nanocomposite. CN106924720, issued 2017.
Sun, Minfu. 2015. Oral sublingual tablet formula and method for preparing oral sublingual tablet by using insulin lipidosome, insulin microsphere (capsule), and insulin nanoparticle (capsule) as raw materials. CN105193751, issued 2015.
Sun, Qingjie, Liu Xiong, Xiaojing Li, Yang Qin, and Chao Qiu. 2015. Preparation method of waxy corn starch nano particle-insulin sustained-release capsules. CN104337795, issued 2015.
Wang, Chao, and Binglong Zhang. 2015. Preparation method of soluble starch nanoparticle-insulin sustained-release capsule. CN104667265, issued 2015.
Wang, Quanbao. 2018. Nano insulin-chitosan manufactured through pressurization and vibration packing method. CN107670025, issued 2018.

Whitehead, Kathryn Ann, and Nicholas George Lamson. 2020. Method of increasing epithelial permeability using nanoparticles. US2020129444, issued 2020.

Woldu, Minyahil A., and Jimma L. Lenjisa. "Nanoparticles and the new era in diabetes management." *International journal of basic & clinical pharmacology* 3, no. 2 (2014): 277–284.

Wu, Zhimin, Tengfei Lu, Manzhen Duan, Yuejin Liu, Di Xiong, and Rong Li. 2019. PH sensitive/membrane adhesive bifunctional insulin-loaded mixed micelle nanoparticle and preparation method thereof. CN110522739, issued 2019.

Wu, Zhimin, Tengfei Lu, Di Xiong, Yuejin Liu, Langsheng Pan, Chao Yu, Manzhen Duan, and Rong Li. 2018. Oral insulin micelle nanoparticles and preparation method thereof. CN108478785, issued 2018.

Xiong, Xiangyuan, Daoge Chen, Yanchun Gong, Ziling Li, and Yuping Li. 2018. Preparation method and applications of targeting nano vesicle. CN108514548, issued 2018.

Yao, Ping, and Zhe Zhang. 2017. Insulin oral nano-preparation and preparation method thereof. CN107362354, issued 2017.

Yu, Fei, Yang Li, Chang Sheng Liu, Qin Chen, Gui Huan Wang, Wei Guo, Xue E. Wu, Dong Hui Li, Winston Duo Wu, and Xiao Dong Chen. "Enteric-coated capsules filled with mono-disperse micro-particles containing PLGA-lipid-PEG nanoparticles for oral delivery of insulin." *International journal of pharmaceutics* 484, no. 1–2 (2015): 181–191.

Yu, Kyeong-Nam. 2020. Nano-liposome carrier composition containing hybrid of Cas9 protein and guide RNA. US2020146988, issued 2020.

Zeng, Fanchu, Qianqian Yang, Zhimin Wu, Yuejin Liu, and Dan Li. 2017. Method for preparing oral insulin-loaded nano-micelles. CN107519484, issued 2017.

Zeng, Qingbing, Yiting Wang, Weiwei Li, Mei Shen, and Qingyuan Chen. 2015. Insulin controlled-release nanometer particle and preparing method thereof. CN104922694, issued 2015.

Zeng, Qingbing, Lei Wu, Wenting Zhu, Jun Wang, and Jie Liu. 2017. PH response insulin slow release nanoparticle, and preparation method and application thereof. CN106581647, issued 2017.

Zhang, Ting, Jingwen Luo, Qiang Peng, Jianxia Dong, Yunuo Wang, Tao Gong, and Zhirong Zhang. "Injectable and biodegradable phospholipid-based phase separation gel for sustained delivery of insulin." *Colloids and surfaces B: Biointerfaces* 176 (2019): 194–201.

Zhang, Xingwang, Baojian Wu, and Wenji Deng. 2016. Insulin oral selenium nano preparation and preparation method thereof. CN106075403, issued 2016.

Zhang, Xuefei, and Xiangyang Meng. 2018. Oral insulin nano-particles and preparation method thereof. CN108434120, issued 2018.

Zhang, Yi, Jian Li, Zhen Wang, Ming-Ze Xu, Zhi Zeng, Jin-Peng Huang, and Yan-Qing Guan. "Natural plant-derived polygalacturonic acid-oleanolic acid assemblies as oral-delivered nanomedicine for insulin resistance treatment." *Chemical engineering journal* 390 (2020): 124630.

Zhao, Chao, Chengfeng Yang, Bin Liu, Xucong Lyu, Yuqing Chen, and Lu Zhang. 2017. Nanoparticles capable of assisting in lowering blood sugar and preparation method thereof. CN106420879, issued 2017.

Zhao, Fuli, Di Wu, Dan Yao, Ruiwei Guo, Weiwei Wang, Anjie Dong, Deling Kong, and Jianhua Zhang. "An injectable particle-hydrogel hybrid system for glucose-regulatory insulin delivery." *Acta biomaterialia* 64 (2017): 334–345.

Zhen, Gu, and Hu Xiuli. 2019. H_2O_2-responsive nanoparticles and uses thereof. US20190298659, issued 2019.

Zhu, Limin, Dunzi Wu, Heyu Li, Huanling Wu, and Kaichun Zhou. 2016. Glucose temperature-sensibility self-assembly nano particles and preparing method and application thereof. CN106038513, issued 2016.

Zhu, Limin, Dunzi Wu, Huanling Wu, Heyu Li, and Kaichun Zhou. 2016. Self-assembled nanoparticles having glucose sensitivity and temperature sensitivity and preparation method of nanoparticles. CN106075404, issued 2016.

Zhu, Shengwei, Peibei Gao, and Lin Yu. 2018. Nano insulin-chitosan capable of rapidly adjusting drug releasing speed. CN107550889, issued 2018.

7 The Antidiabetic Potential of *Chenopodium album* Linn. (Chenopodiaceae) in *In Vitro* and *In Vivo* Studies

Govindasamy Hariharan, Ren-You Gan, Sahayaraj Jasmin Vinitha, Kesavan Keerthana, Gurunagarajan Sridharan, Hua-Bin Li, Nayagham Agnel Arul John, Gopalsamy Rajiv Gandhi, Ricardo Queiroz Gurgel and Gnanasekaran Jothi

CONTENTS

7.1 INTRODUCTION

Diabetes mellitus is a disorder categorized by increased levels of blood sugar caused by an insufficiency in insulin action or insulin production. Type 1 diabetes mellitus (T1DM) is a chronic autoimmune disease characterized by increased blood glucose and macrovascular and microvascular complications (Yatoo et al., 2017). According to the International Diabetes Federation (IDF), T1DM worldwide accounts for 10–15% of all diabetic cases. Natural products are novel sources of

DOI: 10.1201/9781003231745-7

alternative or complementary medicine for the treatment of various ailments with lower side effects, easier availability and more reasonable costs than the synthetic oral hypoglycaemic agents that exhibit a variety of adverse effects (Sukalingam et al., 2015; Li et al., 2019; Shang et al., 2020a, b; Tao et al., 2020)

The World Health Organization (WHO) has recommended the use of traditional medicine-based plant treatments for diabetes (World Health Organization, 2017a, b). There are more than 200 species of plants that have been reported to exhibit hypoglycaemic properties (Sireesha and Raghunandan, 2016). *C. album* Linn. (Chenopodiaceae) also known as Paruppukirai or Chakaravarthikirai (Tamil) and Goosefoot or Fat Hen (English) has been widely cultivated in India and consumed as a food material (Kumar et al., 2015). Leaves have been reported to possess antidiabetic, antihelmintic, antiphlogistic, antirheumatic, laxative and odontalgic properties (Arora and Itankar, 2018).

A preliminary study on the hypoglycaemic effects of flavonoids, phenolics, saponin and glycosides rich in *C. album* (*Chenopodium album*) by Khatun et al. (2015) served as evidence, indicating the therapeutic potential of *C. album*. Therefore, the current investigation evaluates its antidiabetic effect both *in vitro* and *in vivo* methods.

7.2 MATERIALS AND METHODS

7.2.1 Chemicals and Reagents

Alloxan and other fine chemicals were purchased from Sisco Research Laboratories (SRL) Pvt. Ltd., India. Organic solvents were procured from Merck (Chennai, India). All other chemicals of the reagent grade were obtained from local suppliers (SRL, Himedia & Biosystems, Chennai, India).

7.2.2 Plant Materials

C. album was collected between January and March, 2014, at Srirangam, Tiruchirapalli, India. The plant was identified at Botanical Survey of India (Gamble, 1997) and authenticated by Dr. S. John Britto. A specimen (number: GH001) was deposited in the Rapinat Herbarium, Centre for Molecular Systematics, St. Joseph's College (Autonomous), Tiruchirapalli, India.

7.2.3 Preparation of Plant Extract

Fresh aerial parts of *C. album* were washed, shade dried and coarsely powdered using an electric blender. The plant powder (200 g) was mixed and boiled with 1200 mL of distilled water until reduced to one third and evaporated to dryness. The residue obtained was stored at 4°C and used for further experiments (Sharma et al., 2011).

7.2.4 *In Vitro* Antidiabetic Evaluation Assays

The assays of α-amylase (Narkhede et al., 2011), α-glucosidase enzyme inhibition (McCue et al., 2005), glucose adsorption capacity (Ou et al., 1999), glucose diffusion (Sathiavelu et al., 2013) and *in vitro* antioxidant activity were evaluated based on Diphenyl-2-picrylhydrazyl radical scavenging (Re et al., 1999), Ferric reducing antioxidant power assay (Singhal et al., 2011), Nitric oxide scavenging (Sreejayan and Rao, 1996) and 2,2′-Azinobis(3-ethylbenzothiazoline-6-sulphonic acid ammonium salt) radical scavenging (Wagner and Bladt, 1996) assays.

7.2.5 Experimental Animals

Healthy adult Wistar rats (two to three months old and 150–200 g) were obtained from Biogen, Bangalore and acclimatized for five days prior to the experiment. Animals were fed with a standard rat chow diet (Sai Durga Foods and Feeds, Bangalore, India), and water *ad libitum*. All the studies were approved by the Committee for the Purpose of Control and Supervision of Experiments on

Animals (CPCSEA), India on 23 January 2014, with the project entitled "antidiabetic activity of plant-based foods" (SAC/IAEC/BC/2014/M. Sc-001).

7.2.6 The Design of Animal Experiments

The rats were starved for 16 hours before the induction. Alloxan monohydrate was injected into rats intraperitoneally at a dosage of 150 mg/kg body weight in physiological saline (Szkudelski, 2001). Blood sugar levels were measured after four days of alloxan induction, and animals exhibiting hyperglycaemia (blood sugar level >200 mg/dL) were chosen and then grouped for further experiments.

Thirty rats were randomly divided into five groups with six rats in each group treated as follows for 45 days: Group I, normal control rats treated with tap water; Group II, diabetic control rats treated with alloxan (single dose); Group III, diabetic + Aqueous Extract of *Chenopodium album* (150 mg/kg); Group IV, diabetic + AECA (300 mg/kg); Group V, diabetic + AECA (600 mg/kg); and Group VI, diabetic + glibenclamide (1 mg/kg) (Rao et al., 2016) as the positive control. AECA was administered by oral gavage between 10 a.m. and 11 a.m., once a day for 45 days. After the experimental period, animals were sacrificed by cervical dislocation under diethyl ether anaesthesia and blood samples were collected (Parasuraman et al., 2010). Liver and pancreas tissues were collected for biochemical assays.

7.2.7 Biochemical Analysis

Biochemical parameters were carried by following standard protocols. The blood glucose was estimated by Folin and Wu (1919), HbA_1C (glycosylated haemoglobin) by Dodge et al. (1963) and Nayak and Pattabiraman (1981) and plasma insulin by Bakirel et al. (2008). The measurements of glucokinase activities by Brandstrup et al. (1957), glucose-6-phosphatase and fructose-1,6-bisphosphatase by King et al. (1965) and Gancedo and Gancedo (1971), respectively, were estimated in liver homogenate. Methods for measuring glycogen follows Carroll et al. (1956), protein in liver tissue by Lowry et al. (1951) and the extraction of lipids by Folch et al. (1957). Blood urea and creatinine were estimated by Barker (1944) and Bonsnes and Taussky (1945), cholesterol by Zlatkis et al. (1953), triglycerides by Foster and Dunn, (1973), free fatty acids and phospholipids by Fiske and Subbarow (1925), Bartlett (1959) and Falholt et al. (1973). Lipoprotein cholesterol estimates were done with the Biosystem kit (Biosystems, The Precision Scientific Co, Coimbatore).

7.2.8 Histopathological Preparation

Histopathological examination was performed according to Suvarna Layton (2013). Pancreas specimens were taken serial transverse sections with 4–5 μm thickness, stained and examined under a microscope (200X).

7.2.9 High-Performance Liquid Chromatography (HPLC) Analysis

The identification and quantification of the phenolic compounds in *C. album* were conducted as described by Andrade et al. (2017) using a high-performance liquid chromatography (HPLC) system (model 1290 Infinity, Agilent Technologies, Bengaluru, India). The chromatographic separation was performed using an Ascentis express C18 column (2.7 μm, 150 × 2.1 mm) (THERMO, Sigma Aldrich, Bengaluru, India). For analysing gallic acid, the mobile phase consisted of 0.1% (v/v) ortho-pthalaldehyde in water (A) and acetonitrile (B), with a flow rate of 1.0 mL/min. It was eluted with the following gradient profile at 40°C: 0–5 min, 25% B; 5–10 min, 50% B; 10–25 min, 25% B. For analysing kaempferol, the mobile phase consisted of 0.5% (v/v) ortho-pthalaldehyde in water (A) and tetrahydrofuran/acetonitrile/methanol (300/10/10, v/v/v) (B), with a flow rate of 1.0 mL/min. It was eluted with the following gradient profile at 40°C: 0–10 min, 50% B; 10–15 min, 25% B; 15–25 min, 25% B. The injection volume was 20 mL. Peak qualification and quantification were determined based on the retention time, spectra and peak areas of compounds and compared with standard compound as reported by Chen et al. (2011).

7.2.10 Statistical Analysis

All the *in vitro* experiments were performed in triplicate, and the results were expressed as mean ± standard deviation (SD). The *in vivo* results are expressed as the mean ± standard error of the mean (SEM). The statistical significance (p value < 0.05) was evaluated by a one-way ANOVA plus the *post hoc* Dunnet method for multiple comparisons using SPSS 11.5 (Scribd, Trichy, India).

7.3 RESULTS

7.3.1 α-Amylase and α-Glucosidase Inhibitory Effects of AECA

The *in vitro* results showed that AECA could dose-dependently increase the inhibitory activities against α-amylase (Figure 7.1a) and α-glucosidase enzymes (Figure 7.1b) compared to acarbose. The glucose diffusion retardation index (GDRI) of the AECA (1%) showed significantly higher inhibition at the end of a 60-minute treatment (Figure 7.1c) than the control and diminished over time.

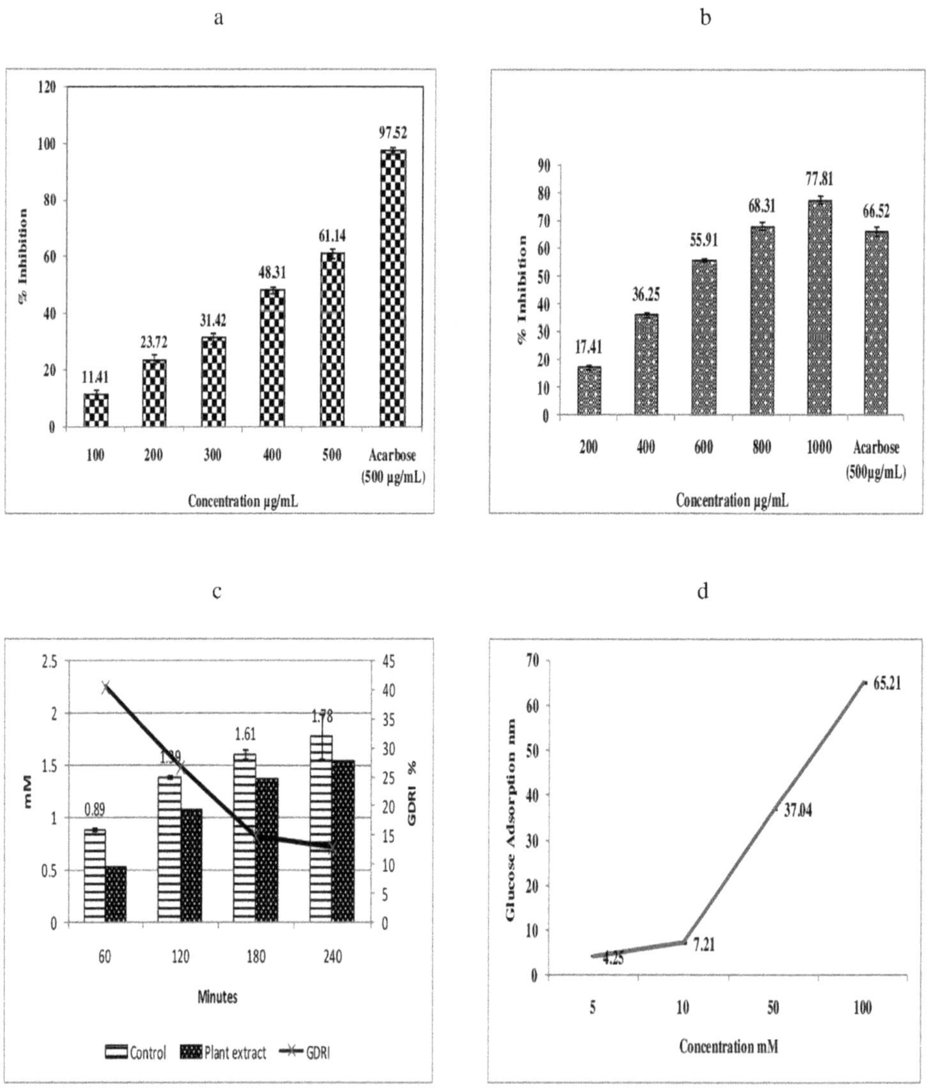

FIGURE 7.1 Effects of AECA on *in vitro* antidiabetic activities. (a) Alpha amylase inhibitory activity, (b) Alpha glucosidase inhibitory activity, (c) Glucose diffusion assay, (d) Glucose adsorption assay.

The glucose adsorption capacity of AECA is shown in Figure 7.1d. There was a significant increase ($p < 0.05$) in the glucose adsorption capacity observed at 50 mM (37.04%) and 100 mM (65.21%) under the utilization of glucose solution.

7.3.2 Antioxidant Effects of AECA

AECA were evaluated based on the DPPH, ABTS, FRAP and NO assays *in vitro* and its antioxidant effects were found to be concentration-dependent, with the IC50 410 µg/mL, 285 µg/mL, 1400 µg/mL and 100 µg/mL, respectively (Figure 7.2a–c), compared with ascorbic acid (standard).

7.3.3 Effects of AECA on *In Vivo* Biochemical Parameters and Glucose Metabolizing Enzymes

The antidiabetic effects of AECA *(in vivo)* are represented in Table 7.1. The diabetic control group showed a significant increase in the levels of blood glucose, HbA1C, blood urea and serum creatinine as well as reduction in the levels of insulin, hepatic and muscle glycogen and serum and

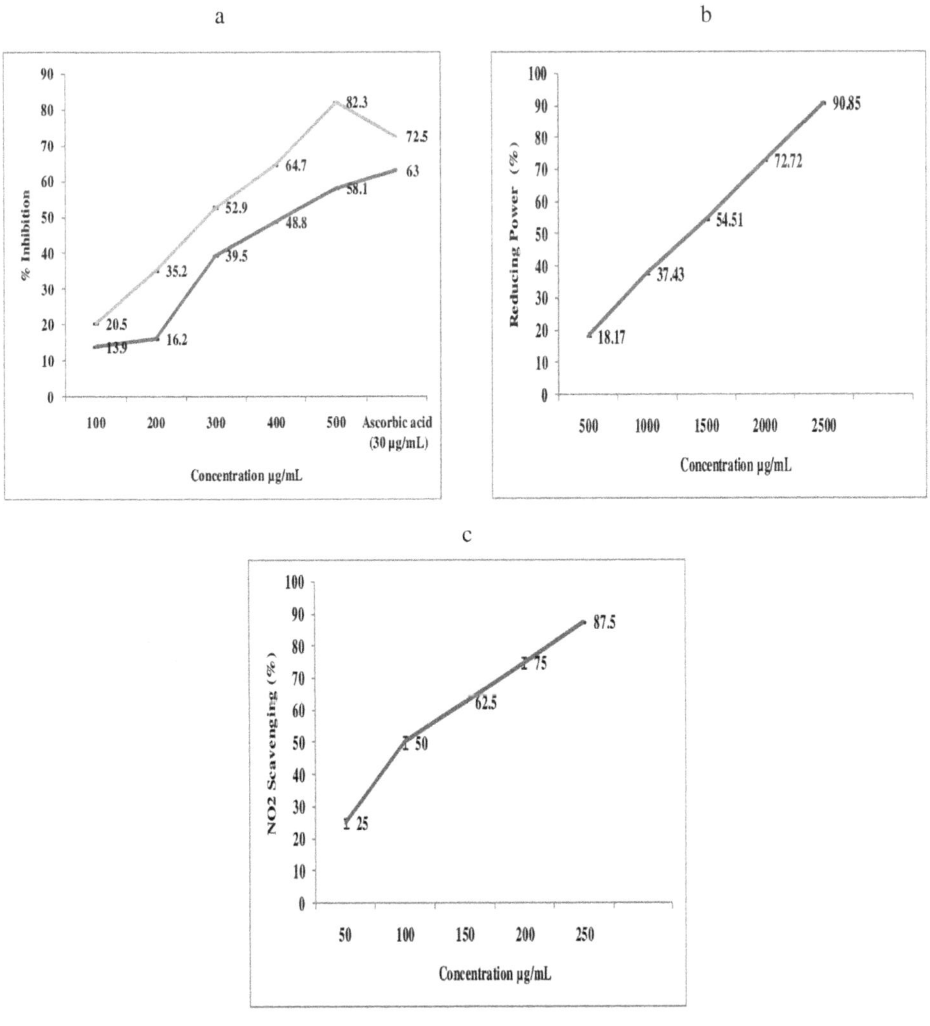

FIGURE 7.2 Effects of AECA on *in vitro* antioxidant activities. (a) DPPH and ABTS assay, (b) Reducing power assay, (c) Nitric oxide assay.

TABLE 7.1
Effects of AECA on Biochemical Parameters

Groups	Blood Glucose (mg/dL)	Insulin (µU/mL)	Glycosylated Hemoglobin (%)	Glycogen Liver (mg/g)	Glycogen Muscle (mg/g)	Protein Serum (mg/dL)	Protein Tissue (mg/100g)	Urea (mg/dL)	Creatinine (mg/dL)
Normal control	86.5 ± 3.56	5.2 ± 0.46	2.6 ± 0.41	32.5 ± 6.34	5.4 ± 0.56	36.9 ± 2.16	39.6 ± 4.21	35.4 ± 1.51	0.13 ± 0.06
Diabetic control	264.4 ± 19.41*	3.1 ± 0.32*	7.1 ± 0.23*	13.4 ± 3.14*	8.5 ± 0.36*	20.1 ± 1.74*	24.9 ± 3.16*	86.2 ± 3.91*	3.05 ± 0.11*
Diabetic + AECA (150 mg/kg)	201.5 ± 15.31**	3.9 ± 0.25**	5.8 ± 0.24**	21.3 ± 3.56**	7.3 ± 0.14**	24.2 ± 0.91**	28.4 ± 3.52**	60.3 ± 4.31**	2.14 ± 0.91**
Diabetic + AECA (300 mg/kg)	148.3 ± 11.52**	4.6 ± 0.11**	4.1 ± 0.16**	29.4 ± 2.11**	6.4 ± 0.52**	30.5 ± 2.52**	34.6 ± 4.52**	49.5 ± 4.61**	1.01 ± 0.41**
Diabetic + AECA (600 mg/kg)	88.9 ± 5.14**	5.4 ± 0.21**	2.8 ± 0.14**	36.5 ± 1.51**	5.3 ± 0.13**	35.6 ± 2.36**	39.5 ± 4.01**	33.4 ± 2.51**	0.23 ± 0.09**
Diabetic + Glibenclamide (1 mg/kg)	81.3 ± 6.32	4.9 ± 0.17	2.9 ± 0.17	34.4 ± 2.56	5.5 ± 0.24	32.1 ± 4.17	33.6 ± 3.51	39.6 ± 3.01	0.91 ± 0.06

Note: These values indicate the mean ± standard error of the mean (SEM) of six rats per group.
* $p < 0.05$, statistically significant when diabetic control is compared with normal control.
** $p < 0.05$, statistically significant when groups III, IV and V are compared with diabetic control.

TABLE 7.2
Effects of AECA on Activities of Glucose-Metabolizing Enzymes

Groups	Glucokinase (U/g protein)[2]	Glucose-6-Phosphatase (U/g protein)[2]	Fructose-1,6-Bisphosphatase (U/g protein)[2]
Normal control	124.7 ± 12.36	8.9 ± 1.31	32.5 ± 4.31
Diabetic control	88.5 ± 9.51*	17.5 ± 2.04*	96.2 ± 5.36*
Diabetic + AECA (150 mg/kg)	96.5 ± 10.14**	14.3 ± 1.91**	72.3 ± 7.46**
Diabetic + AECA (300 mg/kg)	108.8 ± 8.32**	11.5 ± 1.06**	52.4 ± 3.92**
Diabetic + AECA (600 mg/kg)	121.6 ± 7.56**	8.4 ± 0.94**	34.5 ± 2.56**
Diabetic + Glibenclamide (1 mg/kg)	116.6 ± 8.16	9.2 ± 1.52	38.9 ± 3.51

Note: The values indicate the mean ± standard error of the mean (SEM) of six rats per group.
* $p < 0.05$, statistically significant when diabetic control is compared with normal control.
** $p < 0.05$, statistically significant when groups III, IV and V are compared with diabetic control.

tissue proteins when compared to the normal control. There was a significant resumption ($p < 0.05$) of biochemical parameters in animals treated with AECA (300 and 600 mg/kg) compared to the diabetic control rats. Oral administration of AECA significantly reduced ($p < 0.05$) the activities of glucose-6-phosphatase and fructose-1,6-bisphosphatase and enhanced the glucokinase concentration compared to the diabetic control (Table 7.2).

7.3.4 Effects of AECA on Lipid Profiles

Table 7.3 shows the results of serum cholesterol, triglycerides, phospholipids and free fatty acid. Diabetic untreated rats have distorted lipid profiles. Treatment with AECA (150, 300 and 600 mg/kg)

TABLE 7.3
Effects of AECA on Serum Cholesterol, Triglycerides, Phospholipids and Free Fatty Acids

Groups	Cholesterol (mg/dL)	Triglycerides (mg/dL)	Phospholipids (mg/dL)	Free Fatty Acids (mg/dL)
Normal control	50.5 ± 3.01	54.2 ± 3.56	48.2 ± 4.01	36.7 ± 2.14
Diabetic control	132.6 ± 7.16*	156.3 ± 11.9*	116.3 ± 7.96*	101.8 ± 6.52*
Diabetic + AECA (150 mg/kg)	114.6 ± 5.52**	109.5 ± 8.52**	92.4 ± 7.32**	82.7 ± 4.31**
Diabetic + AECA (300 mg/kg)	86.5 ± 3.14**	98.4 ± 4.31**	76.5 ± 6.41**	66.5 ± 3.51**
Diabetic + AECA (600 mg/kg)	56.3 ± 2.79**	53.1 ± 3.92**	52.3 ± 6.51**	40.4 ± 3.92**
Diabetic + Glibenclamide (1 mg/kg)	61.5 ± 4.26	68.4 ± 4.15	68.4 ± 7.14	46.2 ± 4.61

Note: The values indicate the mean ± standard error of the mean (SEM) of six rats per group.
* $p < 0.05$, statistically significant when diabetic control is compared with normal control.
** $p < 0.05$, statistically significant when groups III, IV and V are compared with diabetic control.

TABLE 7.4
Effects of AECA on Serum HDL, LDL and VLDL Cholesterol Levels

Groups	HDL (mg/dL)	LDL (mg/dL)	VLDL (mg/dL)
Normal control	56.5 ± 4.16	16.8 ± 2.31	10.84 ± 1.56
Diabetic control	29.6 ± 3.21*	95.4 ± 6.44*	31.26 ± 3.12*
Diabetic + AECA (150 mg/kg)	34.2 ± 3.76**	58.5 ± 4.14**	21.92 ± 1.52**
Diabetic + AECA (300 mg/kg)	40.9 ± 4.52**	25.9 ±1.56**	19.68 ± 1.56**
Diabetic + AECA (600 mg/kg)	54.6 ± 3.14**	8.9 ± 0.91**	10.62 ± 10.62**
Diabetic + Glibenclamide (1 mg/kg)	48.3 ± 3.96	0.48 ± 0.6	13.68 ± 1.24

Note: The values indicate the mean ± standard error of the mean (SEM) of six rats per group.
* $p < 0.05$ statistically significant when diabetic control is compared with normal control.
** $p < 0.05$ statistically significant when groups III, IV and V are compared with diabetic control.

significantly ($p < 0.05$) reduced the serum lipid profile compared to the diabetic control. AECA treatment at 600 mg/kg generally reversed the lipoprotein profile to normal status (Table 7.4).

7.3.5 Effects of AECA on the Histology of the Pancreas

The alterations in the pancreatic tissues of diabetic control, AECA (150, 300 and 600 mg/kg) and standard drug (glibenclamide 1 mg/kg)-treated groups are depicted in Figure 7.3. Islets of Langerhans of normal control showed granulated cytoplasm and acini tissues (Figure 7.3a) and the diabetic rats exhibited irregular necrotic pancreatic acini and cytoplasmic degeneration (Figure 7.3b). AECA treatment could gradually increase the restoration of the β-cell histoarchitecture in the pancreatic tissue as shown in Figure 7.3c–e as compared with standard drug treated pancreatic tissue shown in Figure 7.3f.

7.3.6 Qualification and Quantification of Phenolic Compounds by HPLC

Phenolic compounds were analysed by HPLC. The major phenolic compounds in AECA were identified as gallic acid and kaempferol (Figure 7.4a,b), with the content of 200 and 10 μg/g, respectively. The HPLC peak of gallic acid and kaempferol are shown in Figure 7.5a–c.

7.4 DISCUSSION

The aqueous extract of *C. album* has significant antioxidant potential due to its rich polyphenolic content; hence, two phenolic compounds, gallic acid and kaempferol, were identified and quantified using HPLC. The inhibition of α-glucosidase and α-amylase enzymes by AECA was expounded by synergistic actions of the presence of flavonoids. The GDRI, which is determined based on the retardation of glucose diffusion, indicates the effect of fibre on the delay in glucose absorption in the gastrointestinal tract. The GDRI of the AECA was 40.45% at 60 minutes and gradually reduced to 26.68% at 120 minutes. *C. album* showed a strong inhibition towards the flow of glucose across

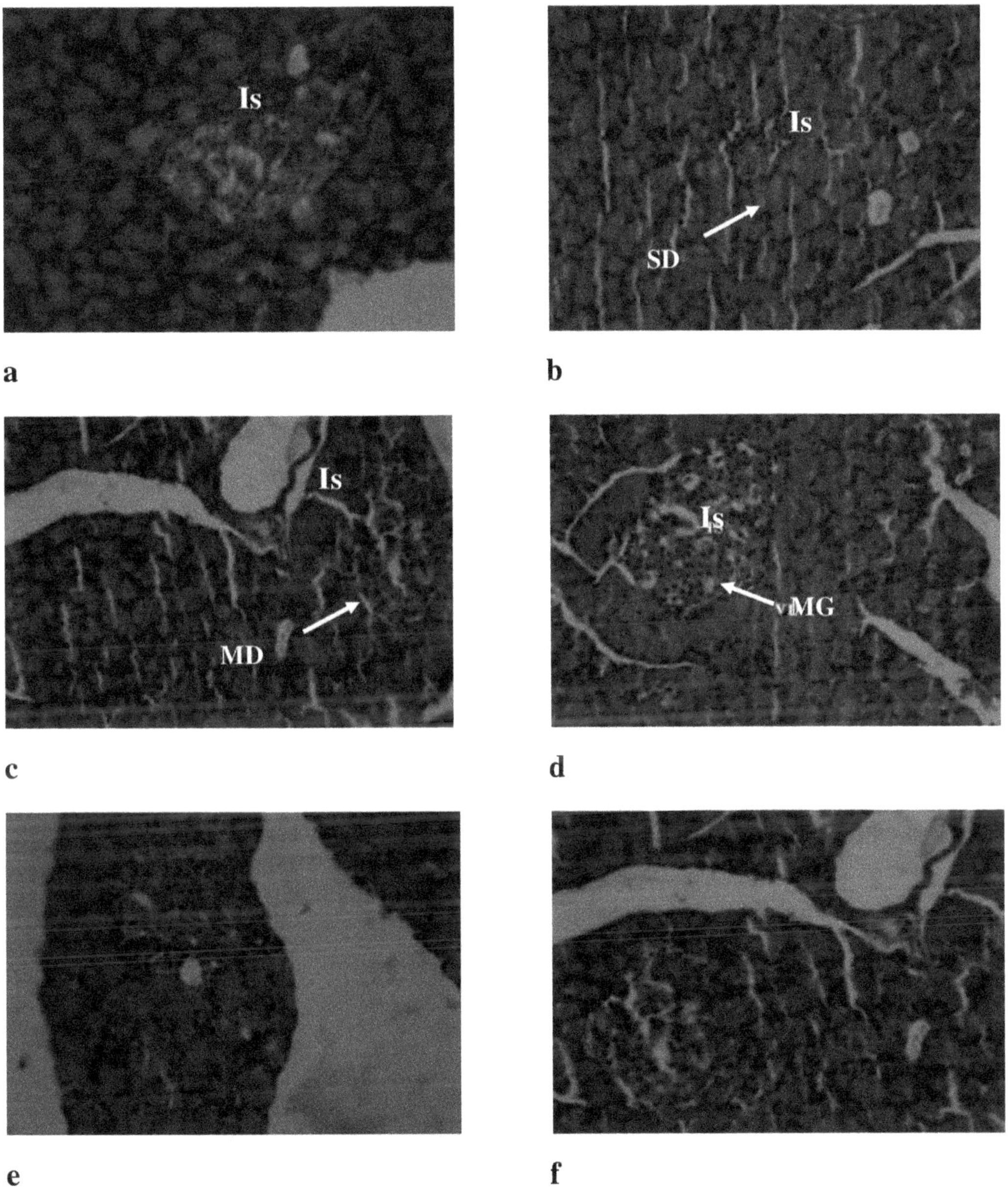

FIGURE 7.3 Effects of AECA on histology of pancreas (H & E 200 ×). (a) Normal control: round islets of Langerhans with granulated β-cells. (b) Diabetic control: small, shrunken islets and destruction of β-cells. (c) Diabetic + (AECA 150 mg/kg): mild granulated β-cells. (d) Diabetic + (AECA 300 mg/kg): mild granulation and rejuvenated β-cells. (e) Diabetic + (AECA 600 mg/kg): mild granulation to well granulated and rejuvenated β-cells. (f) Diabetic + Glibenclamide (1 mg/kg): mild granulation and rejuvenation of β-cells. *Abbreviations:* Is, Islets of Langerhans; SD, Severe degeneration; MD, Mild degeneration; MG, Mild granulation; WG, Well granulated islets.

the dialysis membrane. It was, therefore, speculated that AECA might help to retain glucose in the intestinal lumen, even at low glucose concentration levels, and exhibit good retardation potential of glucose adsorption and glucose diffusion, which might be due to the fibre content of the plant.

Alloxan damages the β cells (Boas et al., 2020), thereby reducing insulin secretion and consequently increasing the level of free circulating glucose. The observed reduction in plasma glucose

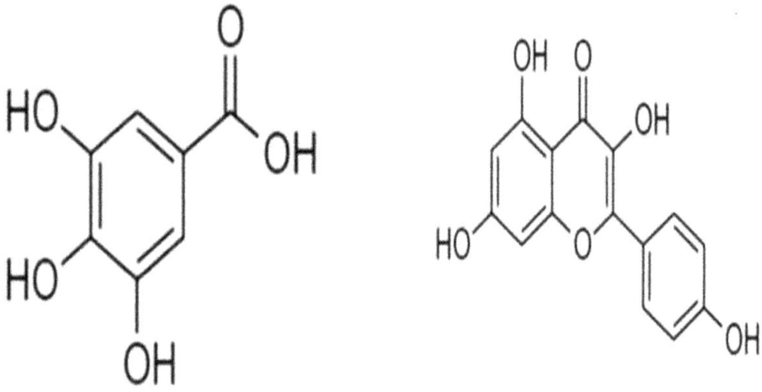

a
Gallic acid

b
Kaempferol

FIGURE 7.4 Chemical structures of phenolic compounds in AECA. (a) Gallic acid, (b) Kaempferol.

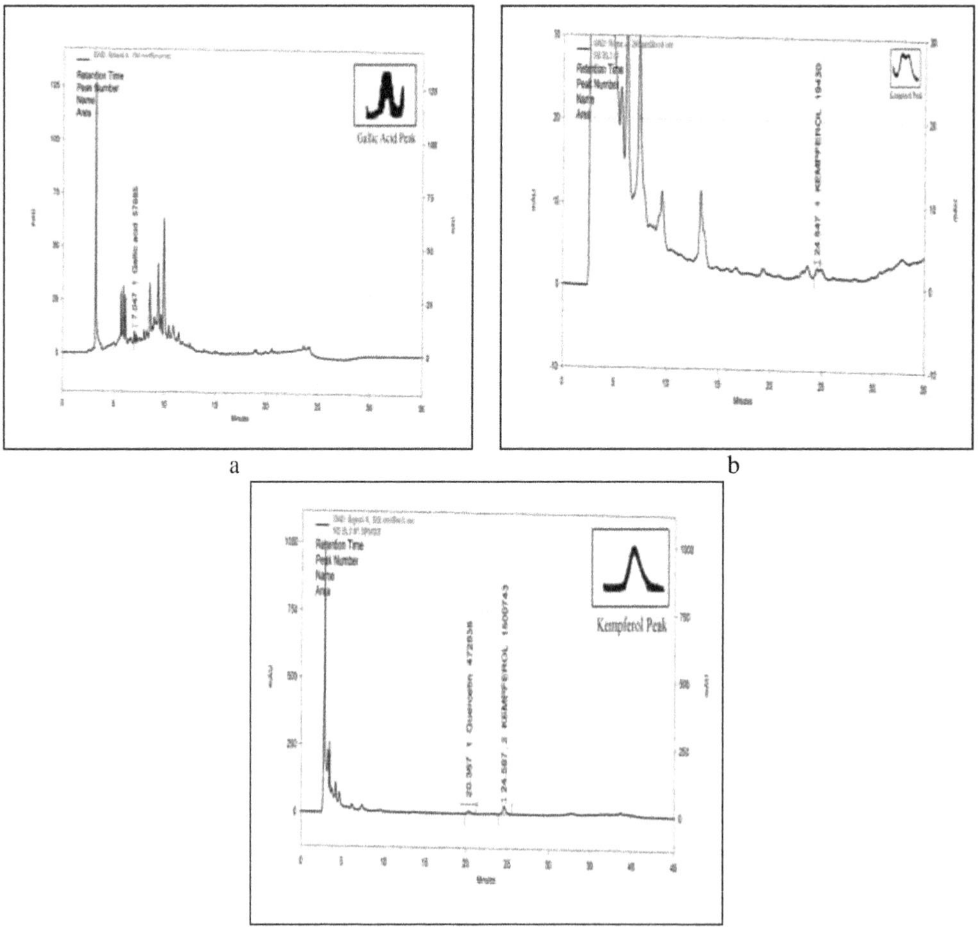

c

FIGURE 7.5 HPLC analysis. (a) AECA with gallic acid, (b) AECA with kaempferol, (c) AECA with kaempferol.

levels and increased insulin concentration post-treatment with AECA could be attributed to the glucose-lowering potential of the phenolic compounds as well as it has restored the structural and functional integrity of β-cells. Measurement of HbA_1C level provides information of long-term glycaemic status and correlates with various complications related to diabetes (Gupta et al., 1997). The AECA-treated group of animals was found to have restored HbA_1C levels.

Glycogen is the primary intracellular storage form of glucose, and its level in various tissues is a direct reflection of insulin activity. A reduction in glycogen in hepatic tissues was observed in the diabetic control. The glycogen was restored to a near-normal level with AECA treatment. Insulin is the prime hormone that regulates protein synthesis (James et al., 2017). The increased insulin level in the AECA-treated diabetic rats caused profound restoration of serum and tissue protein levels. The disturbance in protein metabolism, together with oxidative stress, affects renal function, elevating the levels of serum urea and creatinine in alloxanized animals. The levels of urea and creatinine decreased considerably in the AECA-treated rats, indicating the restoration of protein metabolism.

Glucose-6-phosphatase and fructose-1,6-bisphospatase are key enzymes in gluconeogenesis that increase due to insulin insufficiency in diabetic rats (Arathi and Sachdanandam, 2003). Treatment with AECA regulated the activity of glucokinase, glucose-6-phosphatase and fructose-1,6-bisphosphatase in a dose-dependent manner. The increase in total cholesterol (TC), triglycerides (TG) and LDL and the decline in high-density lipoprotein-C (HDL-C) are the most common abnormalities related to diabetes. In this study, the decreased levels of TC and TG, and the increased level of HDL-C in AECA (150, 300 and 600 mg/kg) in the treated groups are due to the increase in the insulin action on cholesterol and fatty acid biosynthesis and intestinal cholesterol absorption (Wang et al., 2014).

The histology of the alloxan-induced pancreas showed partial destruction of the pancreatic β-cell. Treatment with AECA (300 and 600 mg/kg) regenerated the β-cells and normalized their activity and restorations in the exocrine acinar cells. The increased cellularity of β-cells confirmed the regenerative capability. Overall, the present investigation supports the antidiabetic, antihyperlipidemic and antioxidant potential of AECA due to its high levels of flavonoids and phenols (Krishnasamy et al., 2016).

7.5 CONCLUSION

Chenopodium album Linn. is the most widely used green leafy plant in India. In the present study, the antioxidant and antidiabetic potentials of *C. album* were studied using *in vitro* and *in vivo* experimental approaches. The obtained data revealed the significant potential of AECA in reducing intestinal glucose-metabolizing enzymes as well as scavenging free radicals. Gallic acid and kaempferol were quantified by HPLC, which is well known for its antidiabetic activity. The study also confirmed that AECA (300 mg and 600 mg/kg) significantly restored the altered levels of blood glucose, serum insulin, glucose-metabolizing enzymes and protein, glycogen and lipid profiles. Histopathological analysis also substantiated the cell-protective efficacy of AECA against alloxan-induced pancreatic β-cell damage. Hence, further in-depth studies with a focus on isolating active compounds and evaluating their mechanisms of action are needed for the management of diabetes mellitus.

REFERENCES

Andrade, Julianna Karla Santana, Marina Denadai, Christean Santos de Oliveira, Maria Lucia Nunes, and Narendra Narain. "Evaluation of bioactive compounds potential and antioxidant activity of brown, green and red propolis from Brazilian northeast region." *Food Research International* 101, (2017): 129–138.

Arathi, G., and P. Sachdanandam. "Therapeutic effect of *Semecarpus anacardium* Linn. nut milk extract on carbohydrate metabolizing and mitochondrial TCA cycle and respiratory chain enzymes in mammary carcinoma rats." *Journal of Pharmacy and Pharmacology* 55, no. 9 (2003): 1283–1290.

Arora, Sumit, and Prakash Itankar. "Extraction, isolation and identification of flavonoid from *Chenopodium album* aerial parts." *Journal of Traditional and Complementary Medicine* 8, no. 4 (2018): 476–482.

Bakirel, Tülay, Utku Bakırel, Oya Üstüner Keleş, Sinem Güneş Ülgen, and Hasret Yardibi. "*In vivo* assessment of antidiabetic and antioxidant activities of rosemary (*Rosmarinus officinalis*) in alloxan-diabetic rabbits." *Journal of Ethnopharmacology* 116, no. 1 (2008): 64–73.

Barker, S. B. "The direct colorimetric determination of area in blood and urine." *Journal of Biological Chemistry* 152 (1944): 453–463.

Bartlett, Grant R. "Phosphorus assay in column chromatography." *Journal of Biological Chemistry* 234, no. 3 (1959): 466–468.

Boas, Villas Gustavo Roberto, Joao Marcos Rodrigues Lemos, Matheus William de Oliveira, Rafael Claudino Dos Santos, Ana Paula Stefanello da Silveira, Flávia Barbieri Bacha, Caren Naomi Aguero Ito et al. "Aqueous extract from *Mangifera indica* Linn. (Anacardiaceae) leaves exerts long-term hypoglycemic effect, increases insulin sensitivity and plasma insulin levels on diabetic Wistar rats." *Plos One* 15, no. 1 (2020): e0227105.

Bonsnes, Roy W., and Hertha H. Taussky. "On the colorimetric determination of creatinine by the Jaffe reaction." *Journal of Biological Chemistry* 158, no. 3 (1945): 581–591.

Brandstrup, N., J. E. Kirk, and C. Bruni. "The hexokinase and phosphoglucoisomerase activities of aortic and pulmonary artery tissue in individuals of various ages." *Journal of Gerontology* 12, no. 2 (1957): 166–171.

Carroll, Nicholas V., Robert W. Longley, and Joseph H. Roe. "The determination of glycogen in liver and muscle by use of anthrone reagent." *Journal of Biological Chemistry* 220, no. 2 (1956): 583–593.

Chen, Ying, Qing-Wen Zhang, Song-Lin Li, Yan Yi, Jing Zhao, Ying Wang, and Wen-Cai Ye. "*Psidium guajava*, a potential resource rich in corosolic acid revealed by high performance liquid chromatography." *Journal of Medicinal Plants Research* 5, no. 17 (2011): 4261–4266.

Dodge, James T., Carolyn Mitchell, and Donald J. Hanahan. "The preparation and chemical characteristics of hemoglobin-free ghosts of human erythrocytes." *Archives of Biochemistry and Biophysics* 100, no. 1 (1963): 119–130.

Falholt, K., B. Lund, and W. Falholt. "An easy colorimetric micromethod for routine determination of free fatty acids in plasma." *Clinica Chimica Acta* 46, no. 2 (1973): 105–111.

Fiske, Cyrus H., and Yellapragada Subbarow. "The colorimetric determination of phosphorus." *The Biochemical Journal* 66, no. 2 (1925): 375–400.

Folch, Jordi, Mark Lees, and G. H. Sloane Stanley. "A simple method for the isolation and purification of total lipids from animal tissues." *Journal of Biological Chemistry* 226, no. 1 (1957): 497–509.

Folin, O., and H. Wu. "A system of blood analysis." *Journal of Biological Chemistry* 38, no.1 (1919): 81–110.

Foster, Lowell B., and Ralph T. Dunn. "Stable reagents for determination of serum triglycerides by a colorimetric Hantzsch condensation method." *Clinical Chemistry* 19, no. 3 (1973): 338–340.

Gamble, J. S. Flora of presidency of Madras. Botanical Survey of India. Dehradun, 1997. 2, 1181.

Gancedo, Juana M., and Carlos Gancedo. "Fructose-1,6-diphosphatase, phosphofructokinase and glucose-6-phosphate dehydrogenase from fermenting and non-fermenting yeasts." *Archivfür Mikrobiologie* 76, no. 2 (1971): 132–138.

Gupta, B. L., M. Nehal, and N. Z. Baquer. "Effect of experimental diabetes on the activities of hexokinase, glucose-6-phosphate dehydrogenase and catecholamines in rat erythrocytes of different ages." *Indian Journal of Experimental Biology* 35, no. 7 (1997): 792–795.

James, Haleigh A., Brian T. O'Neill, and K. Sreekumaran Nair. "Insulin regulation of proteostasis and clinical implications." *Cell Metabolism* 26, no. 2 (2017): 310–323.

Khatun, M., S. Haque, M. A. Hassan, and S. N. Islam. "Medicinal value of some edible leafy vegetables from Bangladesh." *World Journal of Pharmaceutical Research* 4, no. 2 (2015): 139–151.

King, J., M. J. Princeton, and D. Van. *Practical clinical enzymology*. Nostrand Company, London, 1965. 363.

Krishnasamy, Gopinath, Karthikeyan Muthusamy, David Raj Chellappan, and Nagarajan Subbiah. "Antidiabetic, antihyperlipidaemic, and antioxidant activity of *Syzygium densiflorum* fruits in streptozotocin and nicotinamide-induced diabetic rats." *Pharmaceutical Biology* 54, no. 9 (2016): 1716–1726.

Kumar, P., S. Kumar, S. Kumar, and R. Kumar. "*In vitro* study of plant extract from *Chenopodium album* that inhibits a key enzyme in diabetes and its role in diabetic oxidative stress." *Pelagia Research Library* 6, no. 12 (2015): 48–61.

Li, Bang-Yan, Xiao-Yu Xu, Ren-You Gan, Quan-Cai Sun, Jin-Ming Meng, Ao Shang, Qian-Qian Mao, and Hua-Bin Li. "Targeting gut microbiota for the prevention and management of diabetes mellitus by dietary natural products." *Foods* 8, no. 10 (2019): 440.

Lowry, Oliver H., Nira J. Rosebrough, A. Lewis Farr, and Rose J. Randall. "Protein measurement with the Folin phenol reagent." *Journal of Biological Chemistry* 193 (1951): 265–275.

McCue, Patrick, Young-In Kwon, and Kalidas Shetty. "Anti-amylase, anti-glucosidase and anti-angiotensin I-converting enzyme potential of selected foods." *Journal of Food Biochemistry* 29, no. 3 (2005): 278–294.

Narkhede, M. B., P. V. Ajimire, A. E. Wagh, Manoj Mohan, and A. T. Shivashanmugam. "*In vitro* antidiabetic activity of *Caesalpina digyna* (R.) methanol root extract." *Asian Journal of Plant Science and Research* 1, no. 2 (2011): 101–106.

Nayak, S. S., and T. N. Pattabiraman. "A new colorimetric method for the estimation of glycosylated hemoglobin." *Clinica Chimica Acta* 109, no. 3 (1981): 267–274.

Ou, Shiyi, Kongrong Gao, and Yan Li. "An *in vitro* study of wheat bran binding capacity for Hg, Cd, and Pb." *Journal of Agricultural and Food Chemistry* 47, no. 11 (1999): 4714–4717.

Parasuraman, S., R. Raveendran, and R. Kesavan. "Blood sample collection in small laboratory animals." *Journal of Pharmacology & Pharmacotherapeutics* 1, no. 2 (2010): 87.

Rao, Nalamolu Koteswara, Krupavaram Bethala, Sreenivas Patro Sisinthy, and Sugumaran Manickam. "Antihyperglycemic and in vivo antioxidant activities of *Phyllanthus watsonii* A. shaw roots in streptozotocin induced type 2 diabetic rats." *International Journal of Pharmacognosy and Phytochemical Research* 8 (2016): 335–340.

Re, Roberta, Nicoletta Pellegrini, Anna Proteggente, Ananth Pannala, Min Yang, and Catherine Rice-Evans. "Antioxidant activity applying an improved ABTS radical cation decolorization assay." *Free Radical Biology and Medicine* 26, no. 9–10 (1999): 1231–1237.

Sathiavelu, Arunachalam, Sundaramoorthy Sangeetha, Rastogi Archit, and Sathiavelu Mythili. "In vitro anti-diabetic activity of aqueous extract of the medicinal plants *Nigella sativa, Eugenia jambolana, Andrographis paniculata* and *Gymnema sylvestre*." *International Journal of Drug Development and Research* 5, no. 2 (2013): 323–328.

Shang, A., R. Y. Gan, X. Y. Xu, Q.-Q. Mao, P.-Z. Zhang, and H.-B. Li. "Effects and mechanisms of edible and medicinal plants on obesity: An updated review." *Critical Reviews in Food Science and Nutrition* (2020a): Advance Online Publication. doi:10/1080/10408398.2020.1769548.

Shang, Ao, Hong-Yan Liu, Min Luo, Yu Xia, Xiao Yang, Hang-Yu Li, Ding-Tao Wu et al. "Sweet tea (*Lithocarpus polystachyus* rehd.) as a new natural source of bioactive dihydrochalcones with multiple health benefits." *Critical Reviews in Food Science and Nutrition* (2020b): 1–18.

Sharma, Ganesh N., Susheel K. Dubey, Piush Sharma, and Nitin Sati. "Medicinal values of bael (*Aegle marmelos*) (L.) Corr.: A review." *International Journal of Current Pharmaceutical Research* 2, no. 1 (2011): 12–22.

Singhal, Manmohan, Arindam Paul, Hemendra Pratap Singh, Sushil Kumar Dubey, and Kalpesh Gaur. "Evaluation of reducing power assay of chalcone semi-carbazones." *Journal of Chemical and Pharmaceutical Research* 3 (2011): 639–645.

Sireesha, K., and N. Raghunandan. "Evaluation of *in vivo* antidiabetic and antioxidant activity of *Artocarpus hirsutes* seeds in streptozotocin-induced diabetic rats. *Asian Journal of Pharmaceutical and Clinical Research* 9, no. 1 (2016): 170–173.

Sreejayan, N., and M. N. Rao. "Free radical scavenging activity of curcuminoids." *Arzneimittel-forschung* 46, no. 2 (1996): 169–171.

Sukalingam, Kumeshini, Kumar Ganesan, and Kumar Ponnusamy. "Evaluation of antidiabetic activity of polyherbal formulations on type 2 diabetic patients: A single blinded randomized study." *International Journal of Integrative Medical Sciences* 2, no. 3 (2015): 90–98.

Suvarna, S. K., and C. Layton. *Bancroft's theory and practice of histological techniques.* 7th edn. Churchill Livingstone Elsevier, United Kingdom. 2013.

Szkudelski, T. "The mechanism of alloxan and streptozotocin action in B cells of the rat pancreas." *Physiological Research* 50, no. 6 (2001): 537–546.

Tao, Jun, Sha Li, Ren-You Gan, Cai-Ning Zhao, Xiao Meng, and Hua-Bin Li. "Targeting gut microbiota with dietary components on cancer: Effects and potential mechanisms of action." *Critical Reviews in Food Science and Nutrition* 60, no. 6 (2020): 1025–1037.

Wang, Wangen, Yanran He, Pei Lin, Yunfei Li, Ruifen Sun, Wen Gu, Jie Yu, and Ronghua Zhao. "*In vitro* effects of active components of *Polygonum multiflorum* radix on enzymes involved in the lipid metabolism." *Journal of Ethnopharmacology* 153, no. 3 (2014): 763–770.

Wagner, Hildebert, and Sabine Bladt. *Plant drug analysis: A thin layer chromatography atlas.* Springer Science & Business Media, Germany. 1996.

World Health Organization. 2017a. Diabetes. Available at http://www.who.int/mediacentre/ factsheets/fs312/en/ (accessed on 26 May 2017).

World Health Organization. 2017b. Global reports on diabetes. Available at http://apps.who.int/iris/bitstream/10665/204871/1/9789241565257-eng.pdf (accessed on 22 December 2017).

Yatoo, Mohd Iqbal, Archana Saxena, Arumugam Gopalakris, Mahmoud Alagawany, and Kuldeep Dhama. "Promising antidiabetic drugs, medicinal plants and herbs: An update." *International Journal of Pharmacology* 13, no. 7 (2017): 732–745.

Zlatkis, Albert, Bennie Zak, and Albert J. Boyle. "A new method for the direct determination of serum cholesterol." *Journal of Laboratory and Clinical Medicine* 41, no. 3 (1953): 486–492.

8 Green Synthesis of Nanoparticles from Herbals for Biomedical Applications

A Systematic Review

Saikumar Sathyanarayanan, Benedict Mathews Paul and Parimelazhagan Thangaraj

CONTENTS

8.1 INTRODUCTION

Herbal medicine is considered the most effective pharmacological system of the 21st century and it is the future medicine for health care (Ahmad Khan and Ahmad 2019). Plants produce several types of bioactive molecules, making them rich sources of different medications. Most of the drugs today are obtained from natural sources or semisynthetic derivatives of natural products used in the traditional systems of medicine (Yuan et al. 2016). Thus, it is a logical approach in drug discovery to screen conventional natural products. Approximately 20% of the plants found in the world have been submitted to pharmaceutical or biological tests, and a sustainable number of new medicines introduced in the market are obtained from natural or semisynthetic resources (Mothana and Lindequist 2005). Medicinal plants are finding their way into pharmaceuticals, cosmetics and nutraceuticals. Medicinal plants are primarily used in the pharmaceutical field for the wide range of substances present in plants used to treat chronic and infectious diseases (Okigbo, Anuagasi, and Amadi 2009). According to the World Health Organization (WHO), medicinal plants are the best source to obtain various drugs (Nascimento et al. 2000). Many ethnic groups use several plant species to treat various ailments ranging from minor infections to dysentery, skin diseases, asthma, malaria and a horde of other indications (Dahanukar, Kulkarni, and Rege 2000). About 80% of the people in developing countries use traditional medicines for their health care (Ghosh 2003). However, in less developed/developing countries, 80% of people still rely only on conventional medicine obtained from local plants and 85% of traditional therapy involves plant extracts (Yuan et al. 2016).

The action of plant medicines depends on the function of the various active components. Each plays a vital role in the action and they are all related to each other. All the components provide a combined effort, thus enhancing the therapeutic value. However, most herbal medicines are

DOI: 10.1201/9781003231745-8

insoluble, which leads to low viability and requires systematic administration and repeated administration of high doses, making drug use a flawed candidate. Recently, in phytomedicine research, nanoscale-level drug design and delivery has attracted many researchers. Adaptation of nanotechnology by herbal medicines has numerous benefits, including solubility and vitality, protection from toxicity, improving pharmacological function, improving stability and prolonged delivery and safety from physical and chemical degradation.

Nanotechnology is a division of implemented science and technology that creates devices sized in the 1–100 nm range. Nanotechnology applications for managing and diagnosing diseases and scrutinizing and controlling biological systems are referred to as nanomedicine (Ansari, Islam, and Sameem 2012). Nanotechnology is used to improve inadequately water-soluble plant drugs and targeted distribution in a single cell or tissue and release large plant molecules to the affected area of tissue (Gopi and Amalraj 2016). Therefore, nanoscale drug delivery systems of herbal medicines have the potential to improve function and overcome problems associated with herbal medicines. Consequently, it is necessary to integrate nanocarriers into the traditional medical system to combat asthma, diabetes, cancer and other chronic diseases (Sandhiya and Ubaidulla 2020). In this review, we try to bring to light the latest documents focusing on various nanoparticles synthesized by herbal plants and their different applications and biological activities. The review opens the door for the importance of green synthesis of nanomaterials with phytocompounds and plans for future treatment options.

8.2 LITERATURE REVIEW METHOD

PubMed, Science Direct, Medline, Scopus, Google Scholar, Iran Medex and Web of Science databases were searched from February 2021 to June 2021 to collect relevant data for this review study. The search terms or keywords, herbal drugs, nanotechnology, novel drug delivery systems, plant extract, green synthesis, silver, gold, zinc, copper and platinum, were used to search the articles. The publications with available abstracts and full text were reviewed for this study. The green synthesis of nanomaterials from the different plant parts and their extracts was characterized by evaluating various parameters, including size, shape and biological activity. Medicinal plants are defined in this review as crude extracts. Although this study is not limited to a specific geographical area, most of the plants included in the study are native to Asian countries.

8.3 RESULTS AND DISCUSSION

Many studies showed nanomaterials such as silver, gold, zinc oxide, copper oxide, titanium and platinum are synthesized using herbal plants. The botanicals present in medicinal plants induce the synthesis of nanoparticles. A wide variety of nonvolatile and volatile phytochemicals present in various parts of plants such as leaves, stem bark, roots, fruits and seeds may reduce the size of the metal particles during green synthesis. Many scientific reports proved that nanomaterials synthesized from the plants could suppress many free radicals, which are the primary causes for many noncommunal diseases, and possess anticancer and antimicrobial properties. As per our search, most of the studies were either *in vitro* or *in vivo* studies; very few were conducted at the preclinical level. This review discusses the various green synthesized nanoparticles from medicinal plants and their biological activity favourable for humanity.

8.4 NANOPARTICLE SYNTHESIS USING PLANTS

A broad spectrum of metal nanoparticles has been produced using different plants listed in Table 8.1. These nanoparticles have unique physical as well as chemical properties compared to their corresponding mass material, with plentiful applications (Zhang et al. 2020). Various methods are used for nanoparticle synthesis. Green synthesis nanomaterials from herbal materials show less harmful and beneficial biological activity. The leaf extracts of *Curcuma longa* have exhibited the capability

TABLE 8.1
Green Synthesis of Nanoparticles from Herbal Plants

Plant Name	Parts Used	Type of Nanoparticles	Biological Activity	Nanoparticle Size	Reference
Abutilon indicum	Leaves	Manganese oxide	Antibacterial and cytotoxicity	80 ± 0.5 nm	(Khan et al. 2020)
Adinandra poilanei	Leaves, twigs	Silver	Antioxidant capability and anticancer activity	14.8 nm	(Ahn and Park 2020)
Alpinia officinarum	Rhizome	Silver	Anti-inflammatory	100 nm	(Zhang et al. 2019b)
Artocarpus hirsutus	Seed	Silver	Antibacterial activity	25–40 nm	(Shobana et al. 2020)
Atropa acuminata	Leaves	Silver	Antioxidant	5–20 nm	(Rajput, Kumar, and Agrawal 2020)
Berberis vulgaris	Root	Silver	Antibacterial activity	30–70 nm	(Behravan et al. 2019)
Chenopodium quinoa	Leaves	Titanium dioxide	Antifungal activity	6–8 nm	(Irshad et al. 2020)
Cissus vitiginea	Leaves	Copper	Antibacterial activity	5–20 nm	(Wu et al. 2020)
Crataegus pentagyna	Fruit	Silver	Antibacterial activity	25–45 nm	(Ebrahimzadeh et al. 2020)
Curcuma longa	Leaves	Silver	Wound-healing activity	15–40 nm	(Maghimaa and Alharbi 2020)
Deverra tortuosa	Aerial part	Zinc oxide	Anticancer activity	9.26–31.18 nm	(Selim et al. 2020)
Dysphania ambrosioides	Leaves	Zinc oxide	Antibacterial	5–30 nm	(Álvarez-Chimal et al. 2021)
Ficus palmate	Leaves	Zinc oxide	Antidiabetic activity	34 nm	(Sati et al. 2020)
Galega officinalis	Leaves	Silver	Antimicrobial activity	8–34 nm	(Manosalva et al. 2019)
Hypotrachyna laevigata	Barks	Silver	Anticancer activity	21.3 ± 4.9 nm	(Ahn and Park 2020)
Kalanchoe blossfeldiana	Aerial part	Zinc oxide	Cytotoxic activity	94 nm	(Aldalbahi et al. 2020)
Lysiloma acapulcensis	Stem and roots	Silver	Antimicrobial activity	1.2–62 nm	(Garibo et al. 2020)
Matricaria chamomilla	Flower	Zinc oxide	Antibacterial effect	51.2±3.2 nm	(Ogunyemi et al. 2019)
Millettia pinnata	Leaves	Gold	Cytotoxicity	37 nm	(Kumar, Ghosh, and Pandey 2019)
Ocimum basilicum	Leaves	Sulphur nanoparticles	Potential to limit the deleterious effects of Mn stress	23 nm	(Ragab and Saad-Allah 2020)
Peganum harmala	Leaf and seed	Gold	Antibacterial activity	52.04 nm	(Moustafa and Alomari 2019)
Phyllanthus niruri	Stem	Silver	Antiviral activity	70– 120 nm	(Sharma et al. 2018)
Quercus semecarpifolia	Leaves	Silver	Cytotoxic activity	20–50 nm	(Khattak et al. 2019)
Rheum ribes	Aerial part	Silver	Anticancer l activity	18.2 ± 3.6 nm	(Aygün et al. 2020)
Rubia tinctorum	Aerial part	Silver	Anticancer activity	12 nm	(Ghandehari et al. 2019)
Salvia miltiorrhiza	Leaves	Silver	Anticancer	100 nm	(Zhang et al.2019a)
Silybum marianum	Seeds	Silver	Cytotoxic	13.20 nm	(Shah et al. 2020)
Terminalia mantaly	Leaf, root	Gold	Antitumor activity	22.5–43 nm	(Majoumouo et al. 2020)
Trigonella foenum-graecum	Leaf	Titanium dioxide	Antimicrobial activity	20–90 nm	(Subhapriya and Gomathipriya 2018)
Veronica multifida	Leaves	Zinc oxide	Antimicrobial and antibiofilm	10–100 nm	(Doğan and Kocabaş 2020)

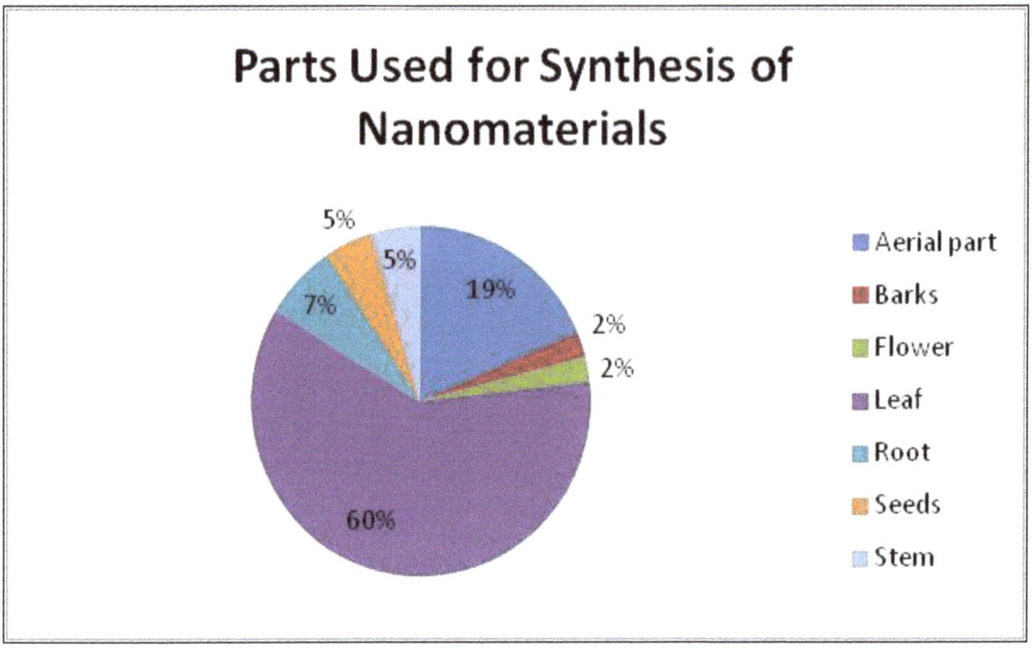

FIGURE 8.1 Plant parts used for synthesis of nanomaterials.

to synthesize silver nanoparticles (AgNPs). The size of the AgNPs obtained became extensively uniform in shape and ranged from 15 to 20 nm (Maghimaa and Alharbi 2020). According to Khan et al. (2020), the leaf extract of *Abutilon indicum* can also reduce manganese oxide, resulting in nanoparticles of 80 nm in size.

Crataegus pentagyna fruit has been used to produce AgNPs (Ebrahimzadeh et al. 2020). *Trigonella foenum-graecum* extract from leaves produced titanium dioxide nanoparticles ranging in size from 20 to 90 nm (Subhapriya and Gomathipriya 2018); gold nanoparticles of around 43 nm were produced by the *Terminalia mantaly* leaf and root extract (Majoumouo et al. 2020) and leaf extract from *Ocimum basilicum* produced sulphur nanoparticles of 23 nm (Ragab and Saad-Allah 2020). The *Alpinia officinarum* rhizome may be used to produce both AgNPs (Zhang et al. 2019b). From our search results, the leaf was used most commonly for nanoparticles, followed by aerial parts. The percentage of plant parts utilized for synthesizing nanoparticles is graphically represented in Figure 8.1.

8.5 BIOLOGICAL ACTIVITY

Biological activities of nanoparticles from search results were analysed and a graphical diagram is presented in Figure 8.2. The antifungal activity of titanium dioxide from the extract of the *Chenopodium quinoa* leaf exhibits reliable activity that may be helpful in biomedical applications (Irshad et al. 2020). Studies have shown that the plant extracts used in zinc oxide from *Matricaria chamomilla* promote greater antibacterial efficacy against bacterial strains. The increase in the antibacterial efficacy of zinc oxide has been linked with the biomolecule found in the extract during the synthesis process (Ogunyemi et al. 2019). The antiviral analysis of AgNPs obtained from the *Phyllanthus niruri* showed the efficacy of destroying the tested chikungunya virus' growth. The bactericidal effectiveness of nanoparticles has been traced to the development of highly reactive oxygen species on the surface of the nanoparticles, which causes viral death (Sharma et al. 2018).

The anticancer investigation of zinc oxide nanoparticles biosynthesized using the extract of *Deverra tortuosa* against Caco-2 cancer cell lines and human lung adenocarcinoma A549 showed significant

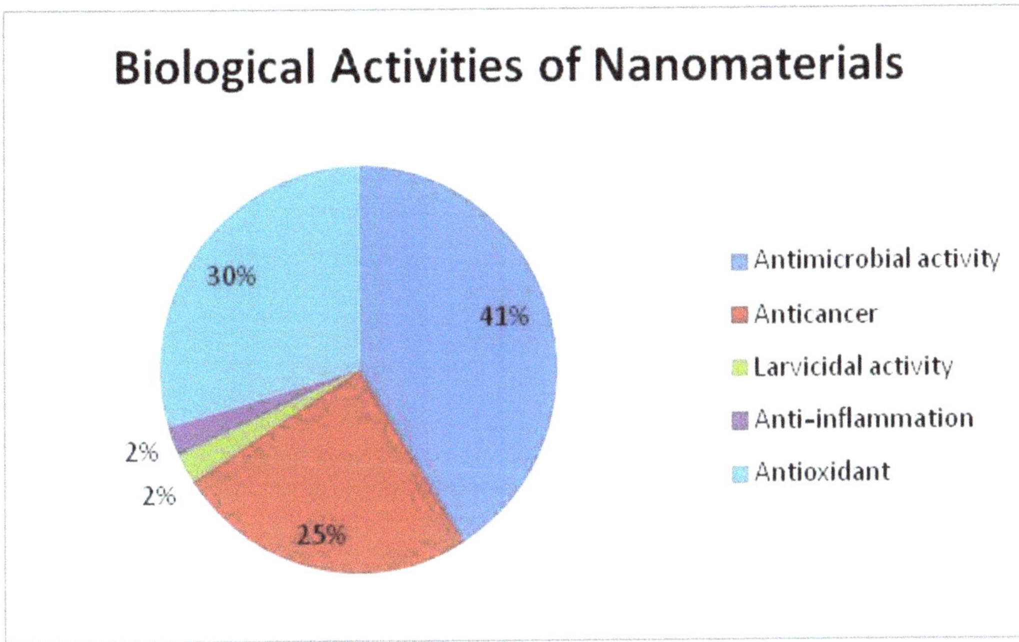

FIGURE 8.2 Biological activities of nanomaterials.

cytotoxicity against the two examined cancer cell lines (Selim et al. 2020). Gold nanoparticles synthesized from *Terminalia mantaly* were investigated in Caco-2, MCF-7 and HepG2 cancer cell lines. The nanoparticles show more toxicity to the cancer cells (Majoumouo et al. 2020). Reports have shown that AgNPs from *Salvia miltiorrhiza* had potential anticancer efficiency, and study results show them as ecofriendly with antimicrobial and anticarcinogenic effects (Zhang et al. 2019a).

8.6 CONCLUSION AND FUTURE PERSPECTIVE

Around the globe, research is being done on herbal remedies and natural products. The development of herbal medicines in the pharmaceutical distribution system in many companies is carried out at basic and clinical trial stages. It is only necessary to develop the best techniques for the proper delivery of such drugs to the point that they do not affect existing sites or treatment of the whole body. Not only does it provide relief from side effects such as toxicity and hypersensitivity reactions, but it is also highly desirable to increase the patient's strength internally.

In the future, the concept of herbal nanoparticles for drug distribution may attract some potential research groups and produce attention-grabbing results. Plants and other biological entities are rich in various natural compounds that can be deposited in nanoparticles during synthesis. Nanoparticles coated with bioactive compounds can be used as promising antibacterial, anticancer and antifungal agents.

Despite efforts to create biological nanoparticles for potential use in nanomedicine, there are still many challenges. Supplementary *in vitro* and animal studies are needed to understand the toxicity and efficacy of biologically synthesized nanoparticles. Therefore, using "herbal remedies" in nanocarriers will increase the effectiveness of treatment for various chronic diseases and health benefits.

There are many successful examples among us with empirical evidence in the direction of nano research. Herbal medicines are rich sources of beneficial compounds that contain antioxidants and components that can be used in targeted foods. We discuss the biological method of integrating nanoparticles from plants, which offers the best opportunity for medical institutions and other industries due to its biological function and synthesis method.

The effectiveness of biological nanoparticles as anticancer, antioxidant, antibacterial and effluent treatment has been well discussed. This type of collaborative research between traditional "herbal remedies" and new approaches to modern drug delivery, i.e., nanotechnology, has established attractive therapies for medicine that will improve people's health in the future. The practical and valuable applications of natural products and herbal medicines are used to develop nanoparticles to enhance their anticancer, antioxidant and antibacterial activities. To improve the biological applications of nanoparticles, more research should focus on possible ways to reduce the toxicity of nanoparticles.

CONFLICT OF INTEREST

The authors have no conflict of interest.

ACKNOWLEDGEMENT

The author Saikumar Sathyanarayanan is thankful to the management authorities of PSG College of Arts and Science, Coimbatore, that provided him with support throughout the study.

REFERENCES

Ahmad Khan, Mohd Sajjad, and Iqbal Ahmad. 2019. "Herbal Medicine." In *New Look to Phytomedicine*, 3–13. Elsevier. doi:10.1016/B978-0-12-814619-4.00001-X.

Ahn, Eun-Young, and Youmie Park. 2020. "Anticancer Prospects of Silver Nanoparticles Green-Synthesized by Plant Extracts." *Materials Science and Engineering: C* 116 (November): 111253. doi:10.1016/j.msec.2020.111253.

Aldalbahi, Ali, Seham Alterary, Ruba Ali Abdullrahman Almoghim, Manal A. Awad, Noura S. Aldosari, Shouq Fahad Alghannam, Alhanouf Nasser Alabdan, et al. 2020. "Greener Synthesis of Zinc Oxide Nanoparticles: Characterization and Multifaceted Applications." *Molecules* 25 (18): 4198. doi:10.3390/molecules25184198.

Álvarez-Chimal, Rafael, Víctor Irahuen García-Pérez, Marco Antonio Álvarez-Pérez, and Jesús Ángel Arenas-Alatorre. 2021. "Green Synthesis of ZnO Nanoparticles Using a *Dysphania ambrosioides* Extract. Structural Characterization and Antibacterial Properties." *Materials Science & Engineering. C, Materials for Biological Applications* 118 (January): 111540. doi:10.1016/j.msec.2020.111540.

Ansari, S. H., Farha Islam, and Mohd Sameem. 2012. "Influence of Nanotechnology on Herbal Drugs: A Review." *Journal of Advanced Pharmaceutical Technology and Research* 3: 142–46. Wolters Kluwer – Medknow Publications. doi:10.4103/2231-4040.101006.

Aygün, Ayşenur, Fulya Gülbağça, Mehmet Salih Nas, Mehmet Hakkı Alma, Mehmet Harbi Çalımlı, Buket Ustaoglu, Yasemin Celik Altunoglu, Mehmet Cengiz Baloğlu, Kemal Cellat, and Fatih Şen. 2020. "Biological Synthesis of Silver Nanoparticles Using *Rheum ribes* and Evaluation of Their Anticarcinogenic and Antimicrobial Potential: A Novel Approach in Phytonanotechnology." *Journal of Pharmaceutical and Biomedical Analysis* 179 (February): 113012. doi:10.1016/j.jpba.2019.113012.

Behravan, Mahmoodreza, Ayat Hossein Panahi, Ali Naghizadeh, Masood Ziaee, Roya Mahdavi, and Aliyar Mirzapour. 2019. "Facile Green Synthesis of Silver Nanoparticles Using *Berberis vulgaris* Leaf and Root Aqueous Extract and its Antibacterial Activity." *International Journal of Biological Macromolecules* 124 (March): 148–54. doi:10.1016/j.ijbiomac.2018.11.101.

Dahanukar, S. A., R. A. Kulkarni, and N. N. Rege. 2000. "Pharmacology of Medicinal Plants and Natural Products." *Indian Journal of Pharmacology* 32: 81–188. https://pdfs.semanticscholar.org/5b28/cccaa13ef8660cb7d19c87252448da272262.pdf.

Doğan, S. Şahin, and A. Kocabaş. 2020. "Green Synthesis of ZnO Nanoparticles with *Veronica multifida* and Their Antibiofilm Activity." *Human & Experimental Toxicology* 39 (3): 319–27. doi:10.1177/0960327119888270.

Ebrahimzadeh, Mohammad Ali, Ali Naghizadeh, Omid Amiri, Mina Shirzadi-Ahodashti, and Sobhan Mortazavi-Derazkola. 2020. "Green and Facile Synthesis of Ag Nanoparticles Using *Crataegus pentagyna* Fruit Extract (*CP*-AgNPs) for Organic Pollution Dyes Degradation and Antibacterial Application." *Bioorganic Chemistry* 94 (January): 103425. doi:10.1016/j.bioorg.2019.103425.

Garibo, Diana, Hugo A. Borbón-Nuñez, Jorge N. Díaz de León, Ernesto García Mendoza, Iván Estrada, Yanis Toledano-Magaña, Hugo Tiznado, et al. 2020. "Green Synthesis of Silver Nanoparticles Using *Lysiloma acapulcensis* Exhibit High-Antimicrobial Activity." *Scientific Reports* 10 (1): 12805. doi:10.1038/S41598-020-69606-7.

Ghandehari, Sara, Masoud Homayouni Tabrizi, Pouran Ardalan, Ali Neamati, and Reyhaneh Shali. 2019. "Green Synthesis of Silver Nanoparticles Using *Rubia tinctorum* Extract and Evaluation the Anti-Cancer Properties in Vitro." *IET Nanobiotechnology* 13 (3): 269–74. doi:10.1049/IET-NBT.2018.5190.

Ghosh, Ashis. 2003. "Herbal Folk Remedies of Bankura and Medinipur Districts, West Bengal." *Indian Journal of Traditional Knowledge* 2 (4): 393–96. https://www.semanticscholar.org/paper/Herbal-folk-remedies-of-Bankura-and-Medinipur-West-Ghosh/f0d3c3fbc20e3f64b74fe20a012d4211413990e5.

Gopi, Sreeraj, and Augustine Amalraj. 2016. "Introduction of Nanotechnology in Herbal Drugs and Nutraceutical: A Review." *Journal of Nanomedicine & Biotherapeutic Discovery* 6 (2): 1–8. doi:10.4172/2155-983X.1000143.

Irshad, Muhammad Atif, Rab Nawaz, Muhammad Zia ur Rehman, Muhammad Imran, Jamil Ahmad, Sajjad Ahmad, Aqil Inam, Abdul Razzaq, Muhammad Rizwan, and Shafaqat Ali. 2020. "Synthesis and Characterization of Titanium Dioxide Nanoparticles by Chemical and Green Methods and Their Antifungal Activities against Wheat Rust." *Chemosphere* 258 (November): 127352. doi:10.1016/j.chemosphere.2020.127352.

Khan, Shakeel Ahmad, Sammia Shahid, Basma Shahid, Urooj Fatima, and Saddam Akber Abbasi. 2020. "Green Synthesis of MnO Nanoparticles Using *Abutilon indicum* Leaf Extract for Biological, Photocatalytic, and Adsorption Activities." *Biomolecules* 10 (5): 785. doi:10.3390/biom10050785.

Khattak, Aishma, Bashir Ahmad, Abdur Rauf, Sami Bawazeer, Umar Farooq, Jawad Ali, Seema Patel, Eman Ramadan El-Sharkawy, Rabia Ikram, and Huang Linfang. 2019. "Green Synthesis, Characterisation and Biological Evaluation of Plant-based Silver Nanoparticles Using *Quercus semecarpifolia* Smith Aqueous Leaf Extract." *IET Nanobiotechnology* 13 (1): 36–41. doi:10.1049/iet-nbt.2018.5063.

Kumar, Gourav, Manik Ghosh, and Dev Mani Pandey. 2019. "Method Development for Optimised Green Synthesis of Gold Nanoparticles from *Millettia pinnata* and Their Activity in Non-small Cell Lung Cancer Cell Lines." *IET Nanobiotechnology* 13 (6): 626–33. doi:10.1049/iet-nbt.2018.5410.

Maghimaa, M., and Sulaiman Ali Alharbi. 2020. "Green Synthesis of Silver Nanoparticles from *Curcuma longa* L. and Coating on the Cotton Fabrics for Antimicrobial Applications and Wound Healing Activity." *Journal of Photochemistry and Photobiology B: Biology* 204 (March): 111806. doi:10.1016/j.jphotobiol.2020.111806.

Majoumouo, Michele S., Jyoti R. Sharma, Nicole R. S. Sibuyi, Marius B. Tincho, Fabrice F. Boyom, and Mervin Meyer. 2020. "Synthesis of Biogenic Gold Nanoparticles from *Terminalia mantaly* Extracts and the Evaluation of Their In Vitro Cytotoxic Effects in Cancer Cells." *Molecules* 25 (19): 4469. doi:10.3390/molecules25194469.

Manosalva, Nixson, Gonzalo Tortella, María Cristina Diez, Heidi Schalchli, Amedea B. Seabra, Nelson Durán, and Olga Rubilar. 2019. "Green Synthesis of Silver Nanoparticles: Effect of Synthesis Reaction Parameters on Antimicrobial Activity." *World Journal of Microbiology and Biotechnology* 35 (6): 88. doi:10.1007/s11274-019-2664-3.

Mothana, Ramzi A. A., and Ulrike Lindequist. 2005. "Antimicrobial Activity of Some Medicinal Plants of the Island Soqotra." *Journal of Ethnopharmacology* 96 (1–2): 177–81. doi:10.1016/j.jep.2004.09.006.

Moustafa, Nagy E., and Abdulaziz Ali Alomari. 2019. "Green Synthesis and Bactericidal Activities of Isotropic and Anisotropic Spherical Gold Nanoparticles Produced Using *Peganum harmala* L. Leaf and Seed Extracts." *Biotechnology and Applied Biochemistry* 66 (4): 664–72. doi:10.1002/bab.1782.

Nascimento, Gislene G. F., Juliana Locatelli, Paulo C. Freitas, and Giuliana L. Silva. 2000. "Antibacterial Activity of Plant Extracts and Phytochemicals on Antibiotic-Resistant Bacteria." *Brazilian Journal of Microbiology* 31 (4): 247–56. doi:10.1590/S1517-83822000000400003.

Ogunyemi, Solabomi Olaitan, Yasmine Abdallah, Muchen Zhang, Hatem Fouad, Xianxian Hong, Ezzeldin Ibrahim, Md. Mahidul Islam Masum, Afsana Hossain, Jianchu Mo, and Bin Li. 2019. "Green Synthesis of Zinc Oxide Nanoparticles Using Different Plant Extracts and Their Antibacterial Activity against *Xanthomonas oryzae* pv. oryzae." *Artificial Cells, Nanomedicine, and Biotechnology* 47 (1): 341–52. doi:10.1080/21691401.2018.1557671.

Okigbo, R. N., C. L. Anuagasi, and J. E. Amadi. 2009. "Advances in Selected Medicinal and Aromatic Plants Indigenous to Africa." *Journal of Medicinal Plants Research*. 3. http://www.academicjournals.org/JMPR.

Ragab, Gehad A., and Khalil M. Saad-Allah. 2020. "Green Synthesis of Sulfur Nanoparticles Using *Ocimum basilicum* Leaves and Its Prospective Effect on Manganese-Stressed *Helianthus annuus* (L.) Seedlings." *Ecotoxicology and Environmental Safety* 191 (March): 110242. doi:10.1016/j.ecoenv.2020.110242.

Rajput, Shubhra, Dinesh Kumar, and Veena Agrawal. 2020. "Green Synthesis of Silver Nanoparticles Using Indian Belladonna Extract and Their Potential Antioxidant, Anti-Inflammatory, Anticancer and Larvicidal Activities." *Plant Cell Reports* 39 (7): 921–39. doi:10.1007/s00299-020-02539-7.

Sandhiya, V., and U. Ubaidulla. 2020. "A Review on Herbal Drug Loaded into Pharmaceutical Carrier Techniques and Its Evaluation Process." *Future Journal of Pharmaceutical Sciences* 6 (51): 1–16. doi:10.1186/S43094-020-00050-0.

Sati, Satish C., Gurpreet Kour, Ankit S. Bartwal, and Maneesha D. Sati. 2020. "Biosynthesis of Metal Nanoparticles from Leaves of *Ficus palmata* and Evaluation of Their Anti-Inflammatory and Anti-Diabetic Activities." *Biochemistry* 59 (33): 3019–25. doi:10.1021/acs.biochem.0c00388.

Selim, Yasser A., Maha A. Azb, Islam Ragab, and Mohamed H. M. Abd El-Azim. 2020. "Green Synthesis of Zinc Oxide Nanoparticles Using Aqueous Extract of *Deverra tortuosa* and Their Cytotoxic Activities." *Scientific Reports* 10 (1): 3445. doi:10.1038/s41598-020-60541-1.

Shah, Muzamil, Sabir Nawaz, Hasnain Jan, Noor Uddin, Ashaq Ali, Sumaira Anjum, Nathalie Giglioli-Guivarc'h, Christophe Hano, and Bilal Haider Abbasi. 2020. "Synthesis of Bio-Mediated Silver Nanoparticles from *Silybum marianum* and Their Biological and Clinical Activities." *Materials Science and Engineering: C* 112 (July): 110889. doi:10.1016/j.msec.2020.110889.

Sharma, Vikrant, Sulochana Kaushik, Pooja Pandit, Divya Dhull, Jaya Parkash Yadav, and Samander Kaushik. 2018. "Green Synthesis of Silver Nanoparticles from Medicinal Plants and Evaluation of Their Antiviral Potential against Chikungunya Virus." *Applied Microbiology and Biotechnology* 103 (2): 881–91. doi:10.1007/S00253-018-9488-1.

Shobana, Sampath, Sunderam Veena, S. S. M. Sameer, K. Swarnalakshmi, and L. A. Vishal. 2020. "Green Synthesis of Silver Nanoparticles Using *Artocarpus hirsutus* Seed Extract and Its Antibacterial Activity." *Current Pharmaceutical Biotechnology* 21 (10): 980–89. doi:10.2174/1389201021666200107115849.

Sofowora, Abayomi, Eyitope Ogunbodede, and Adedeji Onayade. 2013. "The Role and Place of Medicinal Plants in the Strategies for Disease Prevention." *African Journal of Traditional, Complementary, and Alternative Medicines* 10 (5): 210–29. http://www.ncbi.nlm.nih.gov/pubmed/24311829.

Subhapriya, S., and P. Gomathipriya. 2018. "Green Synthesis of Titanium Dioxide (TiO_2) Nanoparticles by *Trigonella foenum-graecum* Extract and Its Antimicrobial Properties." *Microbial Pathogenesis* 116 (March): 215–20. doi:10.1016/j.micpath.2018.01.027.

Wu, Shuang, Shanmugam Rajeshkumar, Malini Madasamy, and Vanaja Mahendran. 2020. "Green Synthesis of Copper Nanoparticles Using *Cissus vitiginea* and Its Antioxidant and Antibacterial Activity against Urinary Tract Infection Pathogens." *Artificial Cells, Nanomedicine, and Biotechnology* 48 (1): 1153–58. doi:10.1080/21691401.2020.1817053.

Yuan, Haidan, Qianqian Ma, Li Ye, and Guangchun Piao. 2016. "The Traditional Medicine and Modern Medicine from Natural Products." *Molecules (Basel, Switzerland)* 21 (5). doi:10.3390/molecules21050559.

Zhang, Dan, Xin-lei Ma, Yan Gu, He Huang, and Guang-wei Zhang. 2020. "Green Synthesis of Metallic Nanoparticles and Their Potential Applications to Treat Cancer." *Frontiers in Chemistry* 8 (October): 799. doi:10.3389/FCHEM.2020.00799.

Zhang, Ke, Xiaodong Liu, Samson Oliver Abraham Samuel Ravi, Arunkumar Ramachandran, Ibrahim Abdel Aziz Ibrahim, Anmar M. Nassir, and Jiapei Yao. 2019a. "Synthesis of Silver Nanoparticles (AgNPs) from Leaf Extract of *Salvia miltiorrhiza* and Its Anticancer Potential in Human Prostate Cancer LNCaP Cell Lines." *Artificial Cells, Nanomedicine, and Biotechnology* 47 (1): 2846–54. doi:10.1080/21691401.2019.1638792.

Zhang, Zhiping, Guangda Xin, Guangyu Zhou, Qianyu Li, Vishnu Priya Veeraraghavan, Surapaneni Krishna Mohan, Dayu Wang, and Feng Liu. 2019b. "Green Synthesis of Silver Nanoparticles from *Alpinia officinarum* Mitigates Cisplatin-Induced Nephrotoxicity via Down-Regulating Apoptotic Pathway in Rats." *Artificial Cells, Nanomedicine, and Biotechnology* 47 (1): 3212–21. doi:10.1080/21691401.2019.1645158.

9 Green and Chemical Synthesis of Nanoparticles from Pteridophytes and Their Biological Application

M. Johnson, A. Vinoth, Sakthi S. Vasantha and B. Shivananthini

CONTENTS

9.1 INTRODUCTION

Pteridophytes (lycophytes, ferns and fern allies) are seedless vascular plants that include about 13,600 known species globally. As per the new Pteridophyte Phylogeny Group I (PPG I 2016) classification, pteridophytes are classified into two phylogenetically distinct groups: the lycophytes (less than 1500 species) and the ferns (some 10,500 species). According to Cao et al., "In the botanical kingdom, ferns represent an important phylogenetic bridge between lower and higher plants. Because of their unique evolutionary history and biology, they produce a distinct set of secondary metabolites, many of which are not found in other plants" (Cao et al., 2017).

Phytochemical studies on ferns have revealed and confirmed the presence of alkaloids (Dong et al., 2012), flavonoids (Xia et al., 2014), polyphenols (Socolsky et al., 2012), terpenoids (Socolsky et al., 2007) and steroids (Ho et al., 2012). Pharmacological and ethnopharmacological studies have revealed the versatile application of ferns viz., cytotoxicity (Johnson et al., 2020a), hepatoprotective activity (Wills and Asha 2006), antihyperglycemic activity (Zheng et al., 2011), leishmanicidal activity (Socolsky et al., 2016), trypanocidal activity (Morais-Braga et al., 2013), antioxidant (Johnson et al., 2020b), antinociceptive and anti-inflammatory activity (Johnson et al., 2017), immunomodulatory activity (Wu et al., 2005) and chemopreventive effects (Wills and Asha 2009).

Fierascu et al. (2021) summarized the synthesis and characterization techniques employed for the nanoparticles synthesized from ferns. They tabulated source, shape, size, methods, characterization of phytosynthesized nanoparticles of ferns and information regarding various applications of phytosynthesized nanoparticles of ferns. The researchers explained the application of the class Polypodiopsida in nanotechnology and they supplemented current bottlenecks and future perspectives. The present review is intended to serve as a broad introduction to the role of nanoparticles of pteridophytes in biological application and will focus on various nanoparticle synthesis procedures, factors influencing nanoparticle synthesis and characterization of the nanoparticles.

DOI: 10.1201/9781003231745-9

9.2 SILVER NANOPARTICLES OF FERNS

Silver nanoparticles (AgNPs) have received enormous attention of researchers due to their extraordinary defense against a wide range of micro-organisms and also due to the appearance of drug resistance against commonly used antibiotics (Sharma et al., 2009). In addition, AgNPs are useful in inks, adhesives, paint industries, electronic devices, pastes etc. (Park et al., 2008). AgNPs have shown wide-ranging applications in biolabelling, antibiotics (Sarkar et al., 2011), antibacterial (Mollick et al., 2012), antifouling and antiparasitic effects (Srikar et al., 2016) and drug delivery etc. AgNPs displayed an extended role as a catalyst for the reduction of dye like methylene blue (Edison and Sethuraman, 2012). AgNPs enhanced the rate of radiation therapy, are used in electron spin resonance (ESR) dosimetry (Guidelli et al., 2012) and as glyconano sensors for diseases diagnosis (Schofield et al., 2006).

Chemical reduction (Prow et al., 2011), gamma ray radiation (Chen et al., 2007), microemulsion (Zhang et al., 2006), electrochemical (Reicha et al., 2012), laser ablation (Abid et al., 2002), autoclave (Yang and Pan, 2012), microwave (Khan et al., 2011) and photochemical reduction (Alarcon et al., 2012) are the generally employed physico-chemical methods for nanoparticle synthesis. Although chemical synthesis is generally a rapid process, it necessitates the use of toxic chemicals such as $NaBH_4$, N_2H_4, NH_2OH and ethanol. Moreover, control of the growth, stability and aggregation of particles is difficult through chemical synthesis; for stabilizing the nanoparticle size, use of capping agents is necessary (Amooaghaie et al., 2015). Scientists, therefore, have devised alternative methods for AgNP synthesis using micro-organisms, plant extracts (Song and Kim, 2009) and natural polymers (Huang and Yang, 2004) as reducing and capping agents (Philip, 2010). The biological activity of AgNPs depends on various factors including surface chemistry, size, size distribution, shape, particle morphology, particle composition, coating/capping, agglomeration, dissolution rate, particle reactivity in a solution, efficiency of silver ion release and cell type (Jain et al., 2011). The biogeneration of nanoparticles using plant extracts as reducers might be considered as an effective beneficial approach for the pharmaceutical industry as the method is not expensive, not cumbersome and ecofriendly (Kedi et al., 2020).

A number of bio-inspired pathways were explored by scientists over the years using various sources viz., bacteria, mushroom, bark, peel extract and enzymes as reducing agents (Ghosh et al., 2012). For the last two decades, many researchers focused their attention to the synthesis of AgNPs using natural products as the source (Johnson et al., 2017). The published reviews focused on the synthesis and several characterization techniques for the analysis of AgNPs synthesized from plant and microbial sources, but not from pteridophytes.

The use of pteridophytes for the green synthesis of AgNPs is limited to the members of the genera *Acrostichum* (Anju et al., 2018), *Actinopteris* (Koteswaramma et al., 2017), *Adiantum* (Nalwadea et al., 2013; Sant et al., 2013; Britto et al., 2014; Santhoshkumar and Nagarajan 2014; Kalita et al., 2016; Anju et al., 2018; Omidi et al., 2018; Chatterjee et al., 2019; Zamani et al., 2020), *Pteris* (Britto et al., 2014; Baskaran et al., 2016), *Azolla* (Korbekandi et al., 2014; Jha and Prasad, 2016), *Christella* (Xavier et al., 2016; Anju et al., 2018), *Cibotium* (Wang et al., 2017), *Cyathea* (Johnson et al., 2018), *Cyclosorous* (Paul et al., 2015), *Diplazium* (Paul et al., 2015), *Gleichenia* (Femi-Adepoju et al., 2019), *Microsorum* (Kedi et al., 2020), *Nephrolepis* (Bhor et al., 2014; Chrislyn et al., 2016; Xavier et al., 2016), *Odontosoria* (Johnson et al., 2020a), *Phlepodium* (Johnson et al., 2020b) and *Selaginella* (Narayani and Johnson, 2015; Riya, 2018).

9.3 GOLD NANOPARTICLES OF FERNS

Earlier research studies are being reported on the synthesis of green Gold Nanoparticles (AuNPs) mostly by the use of enzymes and micro-organisms (Ghosh et al., 2012). The main disadvantage of the microbes' mediated synthesis is the need of a bioreactor to manipulate the reaction parameters i.e., pH, temperature and incubation time for the synthesis of nanoparticles (Krumov et al.,

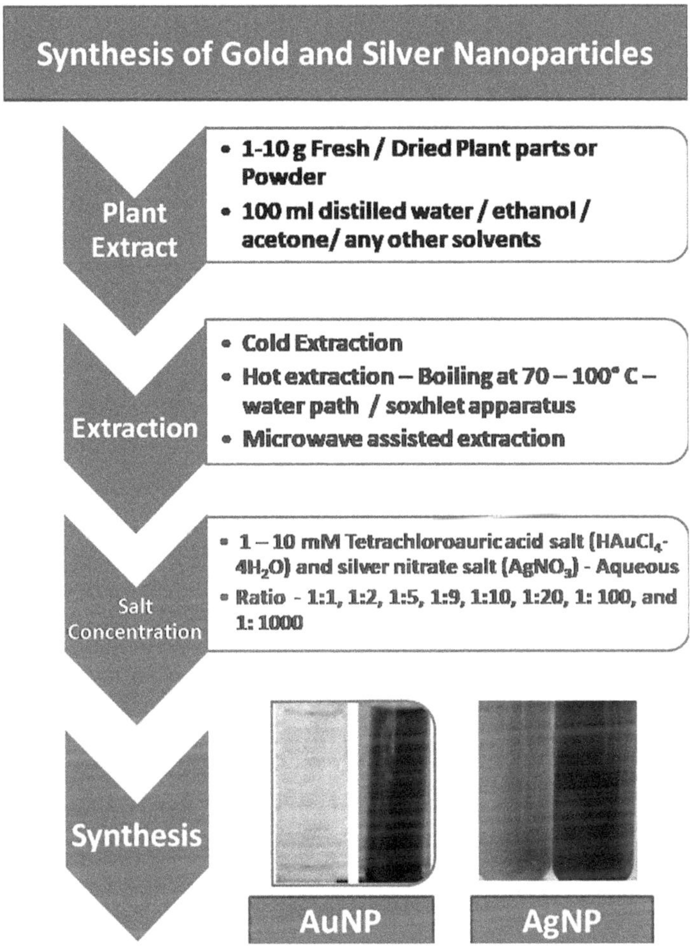

FIGURE 9.1 Synthesis of silver and gold nanoparticles.

2009). Plant phytochemicals such as polyphenols, flavonoids, triterpenes, tannins, glycosides, vitamins, proteins and steroids are capable of reducing the gold ions (Au^{3+} or Au^{+}) to neutral GNPs (Au^{0}) and ensure the size uniformity and crystalline nature of the nanoparticles (Shankar et al., 2004). Recent reports confirmed gold nanoparticle synthesis from *Azolla microphylla* (Kunjiappan et al., 2015), *Salvinia molesta* (Tasneem et al., 2016), *Cibotium barometz* (Wang et al., 2017), *Marsilea quadrifolia* (Balashanmugam et al., 2018) and *Adiantum capillus-veneris* (Rautray and Rajananthini, 2020)

9.4 SYNTHESIS OF AgNPs AND AuNPs

The flow chart (Figure 9.1) explains the synthesis of silver and gold nanoparticles from aqueous, ethanolic and other solvent extracts of ferns.

9.5 CHARACTERIZATION OF AuNPs AND AgNPs

Due to advancements in the instrumentation, various analytical techniques are available to characterize noble metal nanoparticles. Figure 9.2 explains the different available spectroscopic and microscopic techniques employed for the characterization of nanoparticles. The available reports

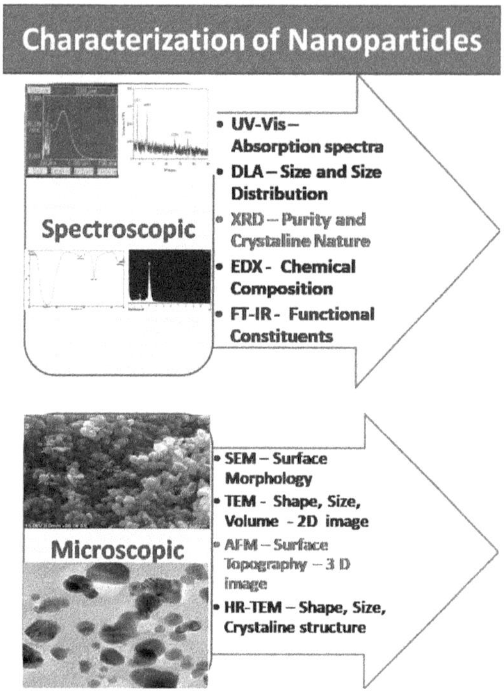

FIGURE 9.2 Characterization of nanoparticles.

suggest that the appearance of straw yellow to slight brownish-yellow colour is an indicator of AgNP synthesis. The surface plasmon resonance (SPR) peak of the synthesized AgNPs was viewed in the range of 380–450 nm (Johnson et al., 2017). The available literature reported the SPR peak of the synthesized AgNPs ranged between 400 nm and 460 nm.

UltraViolet-Visible (UV-Vis) spectral analyses have been used to confirm nanoparticle formation and reveal the size-stability of synthesized AgNPs by exhibiting a red shift in the SPR peak with the increase in size of nanoparticles and a blue shift for a decrease in size (Srikar et al., 2016).

The characterization of gold nanoparticles begins when a colour change from colourless to dark pink/blue confirms AuNP formation. In addition, a spectroscopic technique is employed to analyse the composition, structure and crystal phase of AuNPs. Poinern (2014) noted the absorption spectra of AuNPs is within the range of 500–550 nm. Shukla and Iravani (2017) suggested broadening the SPR bandwidth, which is illustrated by a red shift, can be used as an index of their state of aggregation, dispersity, size and shape. The available literature on AuNPs of ferns also confirms that the absorption spectra of AuNPs of ferns range from 520 to 573 nm.

The purity and crystalline nature of AuNPs and AgNPs are confirmed through X-ray diffraction (XRD), which shows a rough idea of the particle size, determined by the Debye-Scherrer equation. The Joint Committee on Powder Diffraction Standards (JCPDS) is employed as a standard pattern to confirm the gold crystalline nature. The JCPDS data card number 04-0783 was employed to confirm the silver crystalline nature. The available literature suggest that lattice planes (111), (200), (220) and (311) with lattice constant $\alpha = 4.086$ Å confirm the AgNPs, lattice planes (111), (200), (220) and (311) with lattice constant $\alpha = 4.078$ Å verify the AuNPs, lattice planes 12.71° (111), 26.69° (320), 34.38° (332) and 40.50° (521) authenticate the presence of ruthenium nanoparticles (RuNPs) and lattice planes (220), (222), (104), (115) and (440) (JCPDS File No.-01-076-1470) and JCPDS-05-0661 validate the iron nanoparticles (FeNPs) and copper oxide (CuO) nanoparticles. The authors confirmed the silver, gold, copper (Cu), ruthenium (Ru) and iron (Fe) nanoparticles formation from the ferns using the lattice planes proposed by the JCPDS.

The chemical and metal compositions of AgNPs, CuNPs, RuNPs and AuNPs are confirmed by energy-dispersive X-ray spectroscopy (EDX) (Shah et al., 2012). The available literature on nanoparticles synthesized from fern sources also confirmed the metal composition by EDX. The percentage of silver composition ranges from 2.36 to 87.86% (3 KeV) and gold ranges from 41.85 to 48.23% (2 KeV).

Fourier-transform infrared spectroscopy (FT-IR) is employed to determine the functional atoms or groups bound to the surface of nanoparticles. Due to advancement in the microscopy, the morphology of nanoparticles is well characterized. Different types of electron microscopy are employed for the characterization of nanoparticles such as electron microscopy (SEM), transmission electron microscopy (TEM), high-resolution transmission electron microscopy (HRTEM) and atomic force microscopy (AFM) (Azharuddin et al., 2019). To determine the surface morphology and dispersion of the nanoparticles, SEM is used. Generally, TEM is adopted to measure the size, volume and shape of the nanoparticles and to observe the two-dimensional (2D) image of three-dimensional (3D) view of the synthesized nanoparticles (Quester et al., 2013). In addition, TEM provides second-hand information about the number of material layers and broad evidence of uptake and localization, composition, polymer tethering and physical properties. HR-TEM is employed to determine the exact shape, size and crystalline structure of the synthesized nanoparticles (Khanna et al., 2019). The available literature on fern nanoparticles validated the application of electron microscopy on the characterization of silver, gold, copper, ruthenium and iron nanoparticles.

Figures 9.3, 9.4 and 9.5 explain the biological potential of phytosynthesized nanoparticles of ferns.

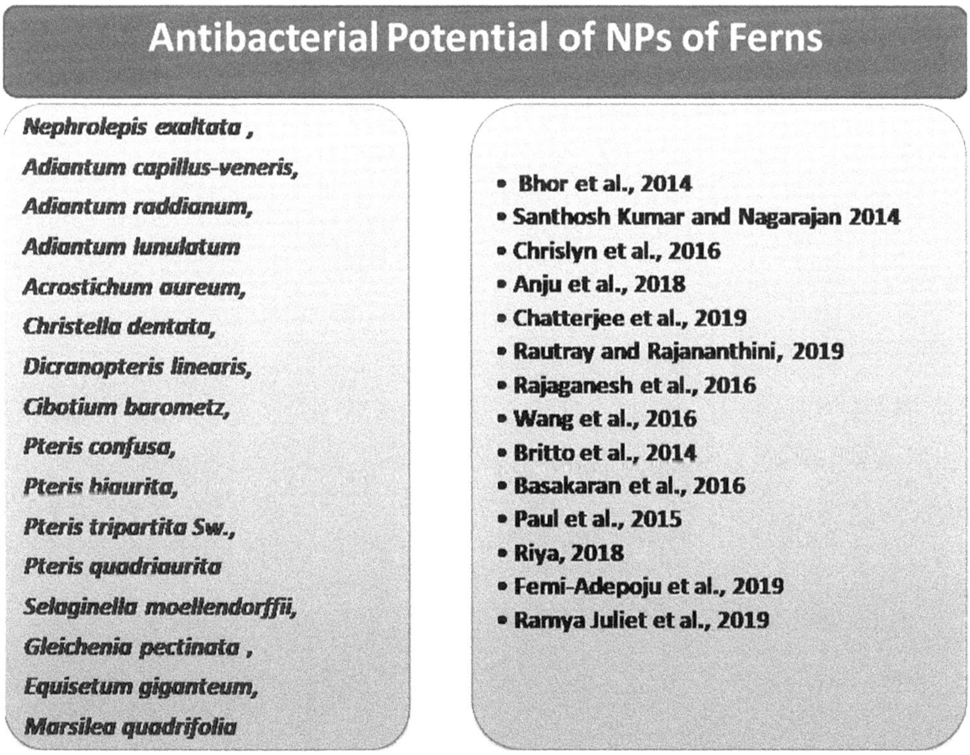

FIGURE 9.3 Antibacterial potential of phytosynthesized nanoparticles of ferns.

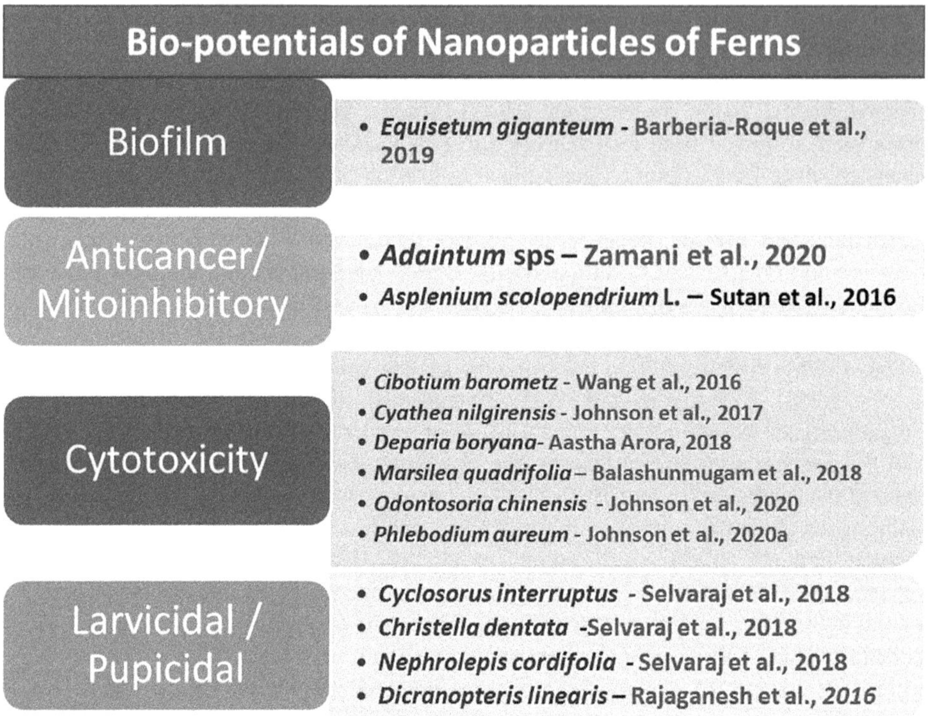

FIGURE 9.4 Biopotentials of phytosynthesized nanoparticles of ferns.

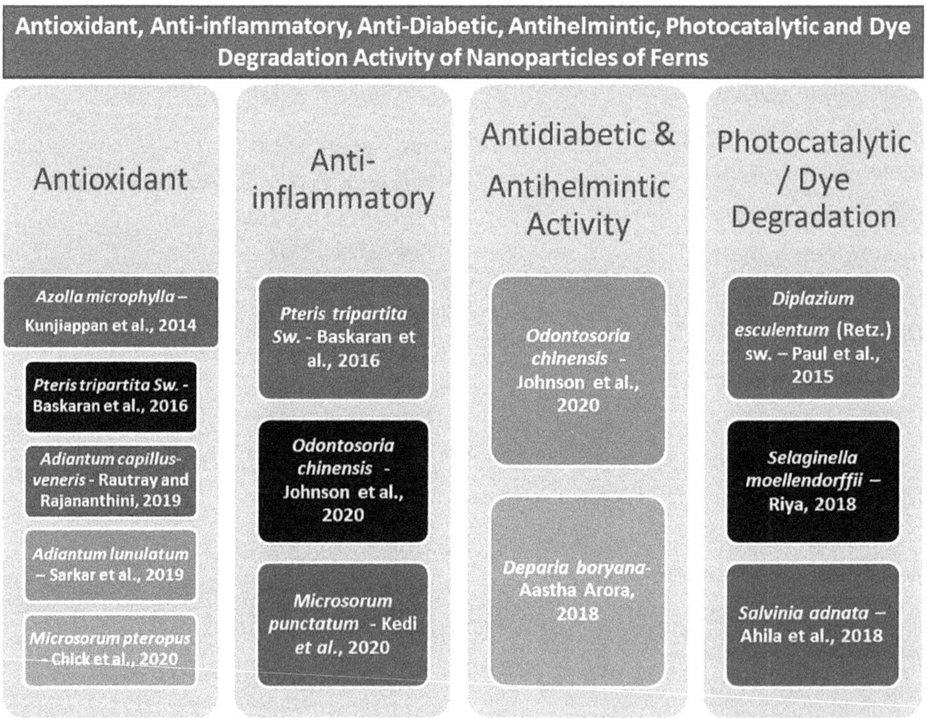

FIGURE 9.5 Antioxidant, anti-inflammatory, anti-diabetic, antihelmintic, photocatalytic and dye degradation activity of nanoparticles of ferns.

9.6 CONCLUSION

Green nanomaterial synthesis is rapidly growing and the outcomes of studies are regularly published as research articles. The recent research outcomes confirmed the use of plant extracts is an alternative method to reduce toxic potential and synthesize an ecofriendly nanomaterial that is cost-effective. This chapter summarized the various methods of synthesis, characterization of nanoparticles and the biological application of ferns in nanotechnology. Angiosperms, seaweed and seagrasses are focused on in much of the current research works, but ferns are not being explored to their real potential in this field (only 59 nanoparticles synthesized from 38 ferns). With reference to the phytochemical composition and biopotential of ferns, the number of nanomaterial synthesized is very low. This bottleneck could be overcome by interdisciplinary research, orientation programmes and workshops on green nanobiotechnology that would create awareness about the importance of ferns among nonbiologists towards this unexplored resource. This would facilitate more study on the green synthesis of nanoparticles and their applications.

REFERENCES

Abid, Jean-Pierre, A. W. Wark, Pierre-Francois Brevet, and H. H. Girault. "Preparation of silver nanoparticles in solution from a silver salt by laser irradiation." *Chemical Communications* 7, (2002): 792–793.

Alarcon, Emilio I., Klas Udekwu, Mårten Skog, Natalia L. Pacioni, Kevin G. Stamplecoskie, María González-Béjar, Naresh Polisetti et al. "The biocompatibility and antibacterial properties of collagen-stabilized, photochemically prepared silver nanoparticles." *Biomaterials* 33, no. 19 (2012): 4947–4956.

Amooaghaie, Rayhaneh, Mohammad Reza Saeri, and Morteza Azizi. "Synthesis, characterization and biocompatibility of silver nanoparticles synthesized from *Nigella sativa* leaf extract in comparison with chemical silver nanoparticles." *Ecotoxicology and Environmental Safety* 120 (2015): 400–408.

Anju, Dev, Shyam Kumar, Prashob Peter Kj, and Thomas Antony. "Antimicrobial potential of biosynthesized silver nanoparticle against human pathogen *Pseudomonas aeruginosa* and *Klebsiella pneumoniae*." *Life Science* 6, no. 1 (2018): 1–4.

Azharuddin, Mohammad, Geyunjian H. Zhu, Debapratim Das, Erdogan Ozgur, Lokman Uzun, Anthony P. F. Turner, and Hirak K. Patra. "A repertoire of biomedical applications of noble metal nanoparticles." *Chemical Communications* 55, no. 49 (2019): 6964–6996.

Balashanmugam, P., K. Mosachristas, and E. Kowsalya. "In vitro cytotoxicity and antioxidant evaluation of biogenic synthesized gold nanoparticles from *Marselia quadrifolia* on lung and ovarian cancer cells." *International Journal of Applied Pharmaceutics* 10, no. 5 (2018): 153–158.

Baskaran, Xavierravi, Antony Varuvel Geo Vigila, Thangaraj Parimelazhagan, Doulathabad Muralidhara-Rao, and Shouzhou Zhang. "Biosynthesis, characterization, and evaluation of bioactivities of leaf extract-mediated biocompatible silver nanoparticles from an early tracheophyte, *Pteris tripartita* Sw." *International Journal of Nanomedicine* 11 (2016): 5789.

Bhor, G., S. Maskare, S. Hinge, L. Singh, and A. Nalwade. "Synthesis of silver nanoparticles using leaflet extract of *Nephrolepi sexaltata* L. and evaluation antibacterial activity against human and plant pathogenic bacteria." *Asian Journal of Pharmaceutical Technology and Innovation* 2, no. 7 (2014): 6.

Britto, J. D., D. H. S. Gracelin, and P. B. J. R. Kumar. "Antibacterial activity of silver nanoparticles synthesized from a few medicinal ferns." *International Journal of Pharmaceutical Research and Development* 6 (2014): 25–29.

Cao, Hui, Tsun-Thai Chai, Xin Wang, Maria Flaviana B. Morais-Braga, Jing-Hua Yang, Fai-Chu Wong, Ruibing Wang et al. "Phytochemicals from fern species: Potential for medicine applications." *Phytochemistry Reviews* 16, no. 3 (2017): 379–440.

Chatterjee, A., S. Khatua, K. Acharya, and J. Sarkar. "A green approach for the synthesis of antimicrobial bio-surfactant silver nanoparticles by using a fern." *Digest Journal of Nanomaterials and Biostructure* 14 (2019): 479–490.

Chen, Peng, Linyong Song, Yankuan Liu, and Yue-E. Fang. "Synthesis of silver nanoparticles by γ-ray irradiation in acetic water solution containing chitosan." *Radiation Physics and Chemistry* 76, no. 7 (2007): 1165–1168.

Chrislyn G., S. Prachi, S. Sangeeta, and S. Priya. "Green synthesis and characterization of silver nanoparticles." *International Journal of Advance Research* 4, no. 8 (2016): 1563–1568.

Dong, Liao-Bin, Jing Yang, Juan He, Huai-Rong Luo, Xing-De Wu, Xu Deng, Li-Yan Peng, Xiao Cheng, and Qin-Shi Zhao. "Lycopalhine A, a novel sterically congested Lycopodium alkaloid with an unprecedented skeleton from *Palhinhaea cernua*." *Chemical Communications* 48, no. 72 (2012): 9038–9040.

Edison, T. Jebakumar Immanuel, and M. G. Sethuraman. "Instant green synthesis of silver nanoparticles using *Terminalia chebula* fruit extract and evaluation of their catalytic activity on reduction of methylene blue." *Process Biochemistry* 47, no. 9 (2012): 1351–1357.

Femi-Adepoju, Abiola Grace, Adewumi Oluwasogo Dada, Kabir Opeyemi Otun, Adeyinka Olufemi Adepoju, and Ojo Paul Fatoba. "Green synthesis of silver nanoparticles using terrestrial fern (*Gleichenia pectinata* (Willd.) C. Presl.): Characterization and antimicrobial studies." *Heliyon* 5, no. 4 (2019): e01543.

Fierascu, Irina, Radu Claudiu Fierascu, Camelia Ungureanu, Oana Alexandra Draghiceanu, and Liliana Cristina Soare. "Application of polypodiopsida class in nanotechnology: Potential towards development of more effective bioactive solutions." *Antioxidants* 10, no. 5 (2021): 748.

Ghosh, Sougata, Sumersing Patil, Mehul Ahire, Rohini Kitture, Sangeeta Kale, Karishma Pardesi, Swaranjit S. Cameotra et al. "Synthesis of silver nanoparticles using *Dioscorea Bulbifera tuber* extract and evaluation of its synergistic potential in combination with antimicrobial agents." *International Journal of Nanomedicine* 7, (2012): 483.

Guidelli, Eder Jose, Ana Paula Ramos, Maria Elisabete D. Zaniquelli, Patricia Nicolucci, and Oswaldo Baffa. "Synthesis of silver nanoparticles using DL-alanine for ESR dosimetry applications." *Radiation Physics and Chemistry* 81, no. 3 (2012): 301–307.

Ho, Raimana, Jean-Pierre Girault, Phila Raharivelomanana, and René Lafont. "E-and Z-isomers of new phytoecdysteroid conjugates from French Polynesian *Microsorum membranifolium* (Polypodiaceae) fronds." *Molecules* 17, no. 10 (2012): 11598–11606.

Huang, Haizhen, and Xiurong Yang. "Synthesis of polysaccharide-stabilized gold and silver nanoparticles: A green method." *Carbohydrate Research* 339, no. 15 (2004): 2627–2631.

Jain S., A. Bhargava, S. Majumdar, and J. Panwar. "Extracellular biosynthesis and characterization of silver nanoparticles using *Aspergillus flavus* NJP08: A mechanism prospective." *Nanoscale* 3 (2011), 635–641.

Jha, A. K., and K. Prasad. "Aquatic fern (*Azolla* sp.) assisted synthesis of gold nanoparticles." *International Journal of Nanoscience* 15, no. 01n02 (2016): 1650008.

Johnson, M., P. Ramakrishnan, S. Perumal, and T. Shibila. "Anti-inflammatory activity of selected pteridophytes from Western Ghats." *International Journal of Complementary Alternative Medicine* 9, no. 4 (2017): 00307.

Johnson, M., A. Santhanam, S. Thangaiah, and J. Narayanan. "Green synthesis of silver nanoparticles using *Cyathea nilgirensis* holttum and their cytotoxic and phytotoxic potentials." *Particulate Science and Technology* 36, no. 5 (2018): 578–582.

Johnson M., T. Shibila, S. Amutha, I. R. A. Menezes, J. C. M. D. Costa., N. F. L. Sampaio., and H. D. M. Coutinho. "Synthesis of silver nanoparticles using *Odontosoria chinensis* (L.) J. Sm. and evaluation of their biological potentials." *Pharmaceuticals* 13, no. 66 (2020a).

Johnson M., T. Shibila, S. Amutha, and G. Vidyarani. "Evaluation of cytotoxic effect of silver nanoparticles (AgNPs) synthesized from *Phlepodium aureum* (L.) J. Smith extracts." *Anticancer Agents in Medicinal Chemistry* 21, no. 18 (2020b): 2603–2609.

Kalita, Sanjeeb, Raghuram Kandimalla, Kaustav Kalyan Sharma, Amal Chandra Kataki, Manab Deka, and Jibon Kotoky. "Amoxicillin functionalized gold nanoparticles reverts MRSA resistance." *Materials Science and Engineering: C* 61, (2016): 720–727.

Kedi, Philippe Belle Ebanda, Chick Christian Nanga, Awawou Paboudam Gbambie, Vandi Deli, Francois Eya'ane Meva, Hamza Elsayed Ahmed Mohamed, Agnes Antoinette Ntoumba, and Moise Henri Julien Nko'o. "Biosynthesis of silver nanoparticles from *Microsorum punctatum* (L.) copel fronds extract and an in-vitro anti-inflammation study." *Journal of Nanotechnology Research* 2, no. 2 (2020): 25–41.

Khan, Aslam, Ahmed Mohamed El-Toni, Salman Alrokayan, Mohamad Alsalhi, Mansour Alhoshan, and Abdullah S. Aldwayyan. "Microwave-assisted synthesis of silver nanoparticles using poly-N-isopropylacrylamide/acrylic acid microgel particles." *Colloids and Surfaces A: Physicochemical and Engineering Aspects* 377, no. 1–3 (2011): 356–360.

Khanna, Prerna, Amrit Kaur, and Dinesh Goyal. "Algae-based metallic nanoparticles: Synthesis, characterization and applications." *Journal of Microbiological Methods* 163 (2019): 105656.

Korbekandi, Hassan, Mohammad Reza Chitsazi, Gholamreza Asghari, Rahim Bahri Najafi, Akbar Badii, and Siavash Iravani. "Green biosynthesis of silver nanoparticles using *Azolla pinnata* whole plant hydroalcoholic extract." *Green Processing and Synthesis* 3, no. 5 (2014): 365–373.

Koteswaramma, B., J. Kamakshamma, and S. Varalaksmi. "Biological synthesis of silver nanopartical from liquid extract of *Actiniopteris radiata* and evaluation of their antimicrobial activity." *International Journal of Pharmaceutical Biology Science* 8 (2017): 121–125.

Krumov, Nikolay, Iris Perner-Nochta, Stephanie Oder, Velitchka Gotcheva, Angel Angelov, and Clemens Posten. "Production of inorganic nanoparticles by microorganisms." *Chemical Engineering & Technology: Industrial Chemistry-Plant Equipment-Process Engineering-Biotechnology* 32, no. 7 (2009): 1026–1035.

Kunjiappan, Selvaraj, Chiranjib Bhattacharjee, and Ranjana Chowdhury. "In vitro antioxidant and hepato-protective potential of *Azolla microphylla* phytochemically synthesized gold nanoparticles on acet-aminophen–induced hepatocyte damage in *Cyprinus carpio* L." *In Vitro Cellular & Developmental Biology-Animal* 51, no. 6 (2015): 630–643.

Mollick, M. M. R., B. Bhowmick, D. Maity, D. Mondal, M. K. Bain, K. K. Bankura et al. "Green synthesis of silver nanoparticles using *Paederia foetida* L. leaf extract and assessment of their antimicrobial activi-ties," *International Journal of Green Nanotechnology* 4 (2012): 230–239.

Morais-Braga, Maria F. B., Teógenes M. Souza, Karla K. A. Santos, Gláucia M. M. Guedes, Jacqueline C. Andrade, Celeste Vega, Miriam Rolón, José G. M. Costa, Antonio A. F. Saraiva, and Henrique D. M. Coutinho. "Phenol composition, cytotoxic and anti-kinetoplastidae activities of *Lygodium venustum* SW. (Lygodiaceae)." *Experimental Parasitology* 134, no. 2 (2013): 178–182.

Nalwade, A. R., M. N. Badhe, C. B. Pawale, and S. B. Hinge. "Rapid biosynthesis of silver nanoparticles using fern leaflet extract and evaluation of their antibacterial activity." *International Journal of Biological Technology* 4, no. 2 (2013): 12–18.

Narayani, M., and M. Johnson, "Novel synthesis of silver nanopeptides using protein extracts of *Selaginella intermedia* (Bl.) String." In: V. K. Gupta, G. D. Singh, S. Singh, and A. Kaul (eds.), *Medicinal Plants: Phytochemistry, Pharmacology and Therapeutics*, pp. 489–497. Daya Publishing House, New Delhi, 2015.

Omidi, Sariyeh, Sajjad Sedaghat, Kambiz Tahvildari, Pirouz Derakhshi, and Fereshte Motiee. "Biosynthesis of silver nanoparticles with *Adiantum capillus-veneris* L. leaf extract in the batch process and assess-ment of antibacterial activity." *Green Chemistry Letters and Reviews* 11, no. 4 (2018): 544–551.

Park, K., D. Seo, and J. Lee. "Conductivity of silver paste prepared from nanoparticles." Colloids and Surfaces A. 313–314 (2008): 351–354. http://dx.doi.org/10.1016/j.colsurfa.2007.04.147.

Paul, Bappi, Bishal Bhuyan, Debraj Dhar Purkayastha, and Siddhartha Sankar Dhar. "Green synthesis of silver nanoparticles using dried biomass of *Diplazium esculentum* (retz.) sw. and studies of their photo-catalytic and anticoagulative activities." *Journal of Molecular Liquids* 212 (2015): 813–817.

Philip, Daizy. "Honey mediated green synthesis of silver nanoparticles." *Spectrochimica Acta Part A: Molecular and Biomolecular Spectroscopy* 75, no. 3 (2010): 1078–1081.

Poinern, Gerrard Eddy Jai. "The Nano World." In *A Laboratory Course in Nanoscience and Nanotechnology*, pp. 32–45. CRC Press, Boca Raton, FL, Taylor & Francis, 2014.

PPG I. "A community-derived classification for extant lycophytes and ferns." *Journal of Systematics and Evolution* 54, no. 6 (2016): 563–603, 414–423.

Prow, T. W., J. E. Grice, L. L. Lin, R. Faye, M. Butler, W. Becker, E. M. T. Wurm, C. Yoong, T. A. Robertson, H. P. Soyer, and M. S. Roberts. "Nanoparticles and microparticles for skin drug delivery." *Advanced Drug Delivery Reviews* 63, no. 6 (2011): 470–491.http://dx.doi.org/10.1016/j.addr.2011.01.012.

Quester, Katrin, M. Avalos-Borja, and E. Castro-Longoria. "Biosynthesis and microscopic study of metallic nanoparticles." *Micron* 54 (2013): 1–27.

Rautray, Satabdi, and A. S. Rajananthini. "Therapeutic potential of green, synthesized gold nanoparticles." *BioPharm International* 33 (2020): 30–38.

Reicha, Fikry M., Afaf Sarhan, Maysa I. Abdel-Hamid, and Ibrahim M. El-Sherbiny. "Preparation of silver nanoparticles in the presence of chitosan by electrochemical method." *Carbohydrate Polymers* 89, no. 1 (2012): 236–244.

Riya G. Green Synthesis of Silver Nanoparticles Using Aqueous Extract of *Selaginella moellendorffii*, a Himalayan Pteridophyte and Its Catalytic Activities. M. Sc. Thesis submitted to School of Chemistry and Biochemistry Thapar Institute of Engineering and Technology, Patiala (Punjab), India (2018).

Sant, Duhita G., Tejal R. Gujarathi, Shrikant R. Harne, Sougata Ghosh, Rohini Kitture, Sangeeta Kale, Balu A. Chopade, and Karishma R. Pardesi. "*Adiantum philippense* L. frond assisted rapid green synthe-sis of gold and silver nanoparticles." *Journal of Nanoparticles* (2013), Article ID 182320, 1–9. DOI: 10.1155/2013/182320.

Santhoshkumar, S., and N. Nagarajan. "Biological synthesis of silver nanoparticles of *Adiantum capillus-veneris* L. and their evaluation of antibacterial activity against human pathogenic bacteria." *International Journal of Pharmaceutical Sciences and Research* 5 no. 12 (2014), 5511–5518.

Sarkar, J., S. Saha, D. Chattopadhyay, S. Patra, and K. Acharya, "Mycosynthesis of silver nanoparticles and investigation of their antimicrobial activity." *Journal of NanoScience, NanoEngineering & Applications* 1 (2011): 17–26.

Schofield, Claire L., Alan H. Haines, Robert A. Field, and David A. Russell. "Silver and gold glyconanoparticles for colorimetric bioassays." *Langmuir* 22, no. 15 (2006): 6707–6711.
Shah, Ritu, Goldie Oza, Sunil Pandey, and Madhuri Sharon. "Biogenic fabrication of gold nanoparticles using *Halomonas salina*." *Journal of Microbiology and Biotechnology Research* 2, no. 4 (2012): 485–492.
Shankar, S. Shiv, Akhilesh Rai, Absar Ahmad, and Murali Sastry. "Rapid synthesis of Au, Ag, and bimetallic Au core–Ag shell nanoparticles using Neem leaf broth." *Journal of Colloid and Interface Science* 275, no. 2 (2004): 496–502.
Sharma, V. K., R. A. Yngard, and Y. Lin. "Preparation of highly positively charged silver nanoballs and their stability." *Advances in Colloid and Interface Science* 145 (2009): 83–96.
Shukla, Ashutosh Kumar, and Siavash Iravani. "Metallic nanoparticles: Green synthesis and spectroscopic characterization." *Environmental Chemistry Letters* 15, no. 2 (2017): 223–231.
Socolsky, Cecilia, Yoshinori Asakawa, and Alicia Bardón. "Diterpenoid glycosides from the bitter fern Gleichenia quadripartita." *Journal of Natural Products* 70, no. 12 (2007): 1837–1845.
Socolsky, Cecilia, Laura Domínguez, Yoshinori Asakawa, and Alicia Bardón. "Unusual terpenylated acylphloroglucinols from *Dryopteris wallichiana*." *Phytochemistry* 80 (2012): 115–122.
Socolsky, Cecilia, Efrain Salamanca, Alberto Giménez, Susana A. Borkosky, and Alicia Bardón. "Prenylated acylphloroglucinols with leishmanicidal activity from the fern *Elaphoglossum lindbergii*." *Journal of Natural Products* 79, no. 1 (2016): 98–105.
Song, Jae Yong, and Beom Soo Kim. "Rapid biological synthesis of silver nanoparticles using plant leaf extracts." *Bioprocess and Biosystems Engineering* 32, no. 1 (2009): 79–84.
Srikar, S. K., D. D. Giri, D. B. Pal, P. K. Mishra, and S. N. Upadhyay. "Green synthesis of silver nanoparticles: A review," *Green and Sustainable Chemistry* 6 (2016): 34–56.
Şuţan, Nicoleta Anca, Irina Fierăscu, Radu Claudiu Fierăscu, Denisa Ştefania Manolescu, and Liliana Cristina Soare. "Comparative analytical characterization and in vitro cytogenotoxic activity evaluation of *Asplenium scolopendrium* L. leaves and rhizome extracts prior to and after Ag nanoparticles phytosynthesis." *Industrial Crops and Products* 83 (2016): 379–386.
Tasneem, A., J. Anuradha, and S. A. Abbasi. "Utilization of the pernicious aquatic weed salvinia (*Salvinia molesta* D S Mitchell) in generating gold nanoparticles." *Indian Journal of Biotechnology* 15 (2016): 382–391.
Wang, Dandan, Josua Markus, Chao Wang, Yeon-Ju Kim, Ramya Mathiyalagan, Verónica Castro Aceituno, Sungeun Ahn, and Deok Chun Yang. "Green synthesis of gold and silver nanoparticles using aqueous extract of *Cibotium barometz* root." *Artificial Cells, Nanomedicine, and Biotechnology* 45, no. 8 (2017): 1548–1555.
Wills, P. J., and V. V. Asha. "Chemopreventive action of *Lygodium flexuosum* extract in human hepatoma PLC/PRF/5 and Hep 3B cells." *Journal of Ethnopharmacology* 122, no. 2 (2009): 294–303.
Wills, P. J., and V. V. Asha. "Protective effect of *Lygodium flexuosum* (L.) Sw. extract against carbon tetrachloride-induced acute liver injury in rats." *Journal of Ethnopharmacology* 108, no. 3 (2006): 320–326.
Wu, Ming-Jiuan, Ching-Yi Weng, Lisu Wang, and Tzi-Wei Lian. "Immunomodulatory mechanism of the aqueous extract of sword brake fern (*Pteris ensiformis* Burm.)." *Journal of Ethnopharmacology* 98, no. 1–2 (2005): 73–81.
Xavier, G. S. A., P. Selvaraj, and Nida John. "Impact of phytoecdysone fractions of the ferns *Cyclosorous interruptus*, *Christella dentata* and *Nephrolepis cordifolia* on the biology of *Spodoptera litura* (Fab.)." *Journal of Biopesticides* 9, no. 2 (2016): 125.
Xia, Xian, Jianguo Cao, Yunxia Zheng, Quanxi Wang, and Jianbo Xiao. "Flavonoid concentrations and bioactivity of flavonoid extracts from 19 species of ferns from China." *Industrial Crops and Products* 58 (2014): 91–98.
Yang, Jisheng, and Jin Pan. "Hydrothermal synthesis of silver nanoparticles by sodium alginate and their applications in surface-enhanced Raman scattering and catalysis." *Acta Materialia* 60, no. 12 (2012): 4753–4758.
Zamani, Sabiha, Danish Idrees, Babita Jha, and Anal K. Jha. "Green synthesis of metal nanoparticles from *Adiantum frond*: Comparative analysis on cancer cell lines." *Nanoscience & Nanotechnology-Asia* 10, no. 6 (2020): 806–816.
Zhang, Wanzhong, Xueliang Qiao, and Jianguo Chen. "Synthesis and characterization of silver nanoparticles in AOT microemulsion system." *Chemical Physics* 330, no. 3 (2006): 495–500.
Zheng, Xiao-ke, Yu-jie Li, Li Zhang, Wei-sheng Feng, and Xin Zhang. "Antihyperglycemic activity of *Selaginella tamariscina* (Beauv.) spring." *Journal of Ethnopharmacology* 133, no. 2 (2011): 531–537.

10 Silver Nanoparticles for the *In Vitro* Production of Phytochemicals

R. Sreelekshmi, S. Muthukrishnan, S. Bhagya, T. Parimelazhagan and E.A. Siril

CONTENTS

10.1 INTRODUCTION

The nano size range gives unique features to nanoparticles (NPs), namely high surface energy, large surface area per unit mass and quantum confinement (Nel et al. 2006). These features improve their biochemical reactivity, competitive binding activity and agricultural applications (Duhan et al. 2017) by enhancing yield, compound production and nutritional value (Lala 2021; Rivero-Montejo et al. 2021).

Biosynthesis of nanoparticles provides an eco-friendly, cost-effective and most efficient approach with many advantages over chemical and physical synthesis (Prasad 2014). This paves the way for modifications in protocols to access biological applications without any toxicity. The best route of biosynthesis is a green protocol, which is nontoxic and has safety advantages (Prasad 2014).

Among nanomaterials, silver (Ag), copper oxide (CuO), titanium dioxide (TiO_2), zinc oxide (ZnO) and silicon dioxide (SiO_2) are employed in the agriculture field (Parisi et al. 2015; Duhan et al. 2017). However, silver nanoparticles (AgNPs) are the most widely used in the field of plant tissue culture (Tiquia-Arashiro and Rodrigues 2016). Silver nitrate ($AgNO_3$) is used in tissue-culture experiments to enhance plant growth, somatic embryogenesis or metabolite production (Kim et al. 2017). Nano silver plays a significant function on plants' metabolic regulation by liberating silver ions. NPs generate reactive oxygen species (ROS) in plants (Marslin et al. 2017). Correspondingly, the plants respond to this by activating enzymatic and nonenzymatic antioxidant defence mechanisms to scavenge excess ROS and plant antioxidant machinery (Sewelam et al. 2016).

NPs can elicit phytochemical production in plant biotechnology systems (Hatami et al. 2016). AgNPs are widely used, as easy transport of AgNPs facilitates regulation of plant metabolism and the heavy metal-based NPs are taken up through the plant's ionic transporter system (Hauck et al. 2008).

The potential effects of AgNPs on the *in vitro* culture of plants contribute to efficient plant cell tissue and organ cultures. AgNPs used *in vitro* elicitation works recorded a high impact on various metabolites in diverse culture systems (Lala 2021). Microscopic investigations are used to study the accumulation and assimilation of AgNPs in plants under *in vitro* conditions (Cvjetko et al. 2017). This chapter discusses the effects and advantages of AgNPs on the production of secondary

DOI: 10.1201/9781003231745-10

metabolites in various *in vitro* cultures, AgNPs as elicitors for metabolite production and future prospectives of AgNPs elicitation.

10.2 WHY SILVER NANOPARTICLES?

AgNPs are important in nanotechnology due to their immense potential on antimicrobial catalytic property and surface-enhanced Raman scattering (Aziz et al. 2015, 2016). The silver-based crystalline nanomaterial production creates the nanostructural, compositional, and specific morphological characteristics. The high surface area property of AgNPs helps in higher interaction with the active site; improving the release of silver ions (Ag^{2+}) becomes a key to their specific functions (Dhrutika et al. 2013). The preparation of AgNPs by the green approach, however, utilizes the advantages of nanotechnology in an eco-friendly manner. Biosynthesis of AgNPs prefers normal environmental conditions for the reduction process, using microbial or phytoextracts as reducing agents. Thus, it is safe and cost-effective. The biological system reduces Ag^{2+} to Ag^0 (Malik et al. 2014). Furthermore, AgNPs can work alone or in combination with other nanoparticles like gold (AuNPs) or silica, thereby improving its application in all aspects of applied sciences (Fazal et al. 2016).

10.3 ADVANTAGE OF SILVER NANOPARTICLES IN PLANT TISSUE CULTURE

In plant tissue culture, NP applications have been explored in micropropagation, organogenesis, somatic embryogenesis and somaclonal variation. In other areas like seed germination, genetic manipulation, culture disinfection, and plant protection, the effectiveness of nanoparticles is explored (Wang et al. 2016). A milestone achievement was made while realizing the antimicrobial activity of AgNPs in *in vitro* culture systems (Yogeswari et al. 2012). Owing to its broad range of antimicrobial efficiency, the researchers eliminated *in vitro* contaminants/plant disease caused by fungal pathogens in plant-tissue culture by the application of AgNPs to the culture medium (Ismail et al. 2017). Mahna et al. (2013) reported that AgNPs (100mg L^{-1}) are 100% effective in surface sterilization of *Arabidopsis* seeds, *Solanum lycopersicum* cotyledons, and *Solanum tuberosum* leaves without showing toxicity of explants. The antimicrobial activity of AgNPs explains that the rapid release of silver ions interacts with the essential functional group (-SH) of enzymes and inactivates microbial activity (Kim et al. 2017). Moreover, it can damage the microbial cytoplasm membrane or inhibit the DNA replication, leading to apoptosis (Aziz et al. 2015).

AgNPs at different morphologies indicate applicability in diverse aspects. Spherical (8 ± 2 nm) and triangular (47 ± 7 nm) AgNPs exhibited the highest and lowest degrees of antimicrobial activity, respectively. Decahedral (45 ± 5 nm) particles caused root growth enhancement, while spherical (8 ± 2 nm) particles increased production of anthocyanin in *Arabidopsis* seedlings (Syu et al. 2014). The contamination during large-scale production of *Vanilla planifolia* via a temporary immersion system was reversed after AgNPs at different concentrations ranging from 25 to 200 mg L^{-1} were effectively applied to Murashige and Skoog (MS) liquid medium (Spinoso-Castillo et al. 2017). The shoot regeneration potential of AgNPs was reported on an MS medium alone or with added cytokinins and auxins. Aghdaei et al. (2012) reported that AgNPs at 10 mg L^{-1} in conjunction with 2.5 mg L^{-1} benzyl adenine (BA) and 0.1 mg L^{-1} indole acetic acid (IAA) resulted in the enhancement of the explant response (%), shoot number and plant survival in *Tecomella undulata*. The delayed explant senescence of AgNP treatments was also proved by generating the down-regulation of ethylene biosynthesis genes. The down-regulation of the ethylene biosynthetic gene in micropropagated plantlets of *Dianthus chinensis* under AgNPs (100 mg L^{-1}) treatment was also observed by our research group (Sreelekshmi et al. 2021). The vase life/water uptake efficiency of the cut flower *Gerbera* was reported under AgNP treatment, where reduced vascular blockage and enhanced turgidity were observed (Solgi et al. 2009).

In *Solanum nigrum*, AgNPs displayed friable white, green or yellowish calli after 30 days of incubation in contrast to compact calli of the control treatment. The MS medium treated with

8 mg L^{-1} AgNPs, 5 mg L^{-1} BA and 3 mg L^{-1} α-naphthaleneacetic acid (NAA) enriched the callus biomass to 4.67 g/explant. Still, higher concentrations adversely affected callus growth and morphology (Ewais et al. 2015). Fazal et al. (2016) proved AgNPs (30 mg L^{-1}) or 1 mg L^{-1} AgNP and 2 mg L^{-1} AuNPs or 2 mg L^{-1} AgNP with 1 mg L^{-1} AuNPs in combination with 2 mg L^{-1} NAA significantly enhanced callusing (100%) in contrast to the control, where 2 mg L^{-1} NAA was used. Sarmast and Salehi (2016) reported the effect of AgNPs during co-cultivation in the *Agrobacterium*-mediated gene transformation study of tobacco and *T. undulata* by eliminating the bacteria from the culture medium. Further research on AgNP-mediated gene delivery in plants warranted further exploration of this tool in genetic manipulation.

The research of AgNPs in plant-tissue culture signifies the positive effect on seed germination, shoot proliferation, callus growth, alteration of antioxidant enzymes and inhibition of ethylene gene expression. In contrast, adverse effects were observed at higher concentrations. However, the actual mechanism behind each parameter deserves explorations. Recently, several studies have initiated on applications of AgNPs in the *in vitro* production of metabolites. In this direction, further investigations are needed to evolve a clear understanding of the mechanism and to determine the applicability of this tool in phytochemical production.

10.4 SILVER NANOPARTICLES AS ELICITORS

Some applied research in the nanotechnology field demonstrates the importance of NPs as secondary metabolite elicitors in plant-tissue culture systems. The mechanism behind the elicitation of AgNPs is their interference with signal pathways of different metabolisms, finally regulating secondary metabolism in plants. The possible mechanism of NPs regulating the phytochemical production of plants involves the initial elevation of the level of ROS and cytoplasmic calcium ions (Ca^{2+}) followed by up-regulation of mitogen-activated protein kinase (MAPK) related to plant stress responses (Marslin et al. 2017). The receptors present in the plasma membrane triggered a Ca^{2+} burst and ROS induction. Likewise, the signalling pathway of proteins and Ca^{2+} levels were up-regulated in the proteomic analysis of AgNP-treated plants (Sosan et al. 2016). ROS themselves act as signalling molecules, where they directly or indirectly modulate other messengers like salicylic acid (SA) (Baxter et al. 2014), jasmonic acid (JA) (Wu and Ge 2004), nitric oxides (Lindermayr and Durner 2015), and brassinosteroids (Xia et al. 2009).

Antioxidant evidence (Mirzajani et al. 2014) postulates that increased accumulation of superoxide dismutase (SOD), ascorbate peroxidase (APX), and glutathione-S-transferase under AgNPs treated *Oryza sativa*. The effect of NPs on *Pisum sativum* seedlings showed a significant effect on SOD and APX, while reducing glutathione reductase and dehydroascorbate reductase (Tripathi et al. 2017). During the interactive phase, the plasma membrane-bounded receptors were recognized by the AgNPs, which led to a Ca^{2+} burst and ROS induction. In the recognition by NP-specific proteins or calcium-binding proteins, NP either mimics the signalling molecule or Ca^{2+} in the plant cytoplasm (Khan et al. 2017). It has been postulated that under AgNP exposure, the plant cell may utilize these MAPK cascades (Kohan-Baghkheirati and Geisler-Lee 2015), which may lead to enhanced production of metabolites and establish the relationship between MAPK phosphorylation and the transcriptional reprogramming of various metabolic pathways.

10.5 AgNP-MEDIATED ENHANCEMENT ON *In Vitro* METABOLITE PRODUCTION

Supplementation of NPs to the *in vitro* culture medium may not only act as a nutrient source but also an elicitor to enhance metabolite production. So far, several reports have been published that evaluate the influence of AgNPs on secondary metabolite production in various culture systems (Chung et al. 2018). Recently, several reports were published on the *in vitro* production of metabolites by elicitation using AgNPs (see Table 10.1). Generally, the enhanced or *de novo* synthesis of

TABLE 10.1
Elicitation of AgNPs in Different *In Vitro* Culture Systems

Plant Species	Culture System	Metabolite(s)	Reference
Caralluma tuberculata	Callus culture	Total phenolic and flavonoid	Ali et al. (2019)
Prunella vulgaris	Callus culture	Total phenolic and flavonoid	Fazal et al. (2016)
Echinacea purpurea	Cell suspension culture	Cichoric acid	Ramezannezhad et al. (2019)
Aloe vera	Cell suspension culture	Aloin	Raei et al. (2014)
Corylusa vellana	Cell suspension culture	Taxol	Jamshidi et al. (2016)
Capsicum frutescens	Cell suspension culture	Capsaicin	Bhat and Bhat 2016
Artemisia annua	Hairy root culture	Artemisinin	Zhang et al. (2013)
Vanilla planifolia	Adventitious shoot culture	Phenolic	Spinoso-Castillo et al. (2017)
Momordica charantia	Cell suspension culture	Total phenolic and flavonoid	Chung et al. (2018)

secondary metabolites by plants is a response to biotic and abiotic stress or a defence mechanism. Plant response against stress occurs both at the organismic as well as the cellular level. The stress signal is first perceived by cell membrane receptors and activates large and complex signalling cascades including the generation of secondary signal molecules. An overview of elicitation mechanism is depicted in a schematic representation in Figure 10.1.

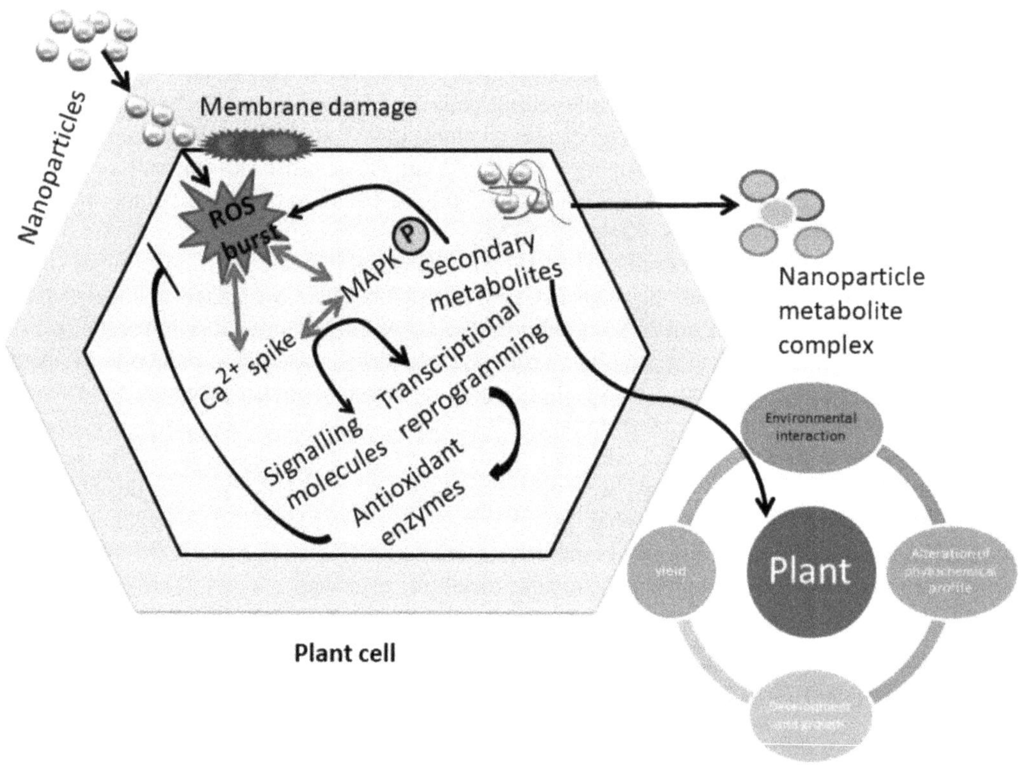

FIGURE 10.1 Schematic representation of the probable mechanism of secondary metabolite production directed by silver nanoparticles.

Ali et al. (2019) investigated the elicitation potential of AgNPs (90 μg L^{-1}) for increasing the amount of total flavonoid content (TFC), phenolic content (TPC), phenylalanine ammonia lyase (PAL) and antioxidant activity in callus cultures of *Caralluma tuberculate*. The highest value of TFC (1.8 mg), TPC (3.0 mg), and PAL (5.8 U/mg) were displayed in the callus culture at 90 μg/l AgNPs, added alone in the MS medium. Moreover, higher concentrations of AgNPs significantly decreased the production of secondary metabolites. The enhancement of essential oil content in *Calendula officinalis* was reported by the elicitation of AgNPs at 3 mg L^{-1} in callus induction on the MS medium. When *Phaseolus vulgaris* callus cultures were treated with different concentrations of AgNPs, percentage of callus induction and biomass were maximum at 50 mg mL^{-1} treatments (Mustafa et al. 2017). Fazal et al. (2016) established the combined effect of AgNPs and AuNPs on the enhanced production of metabolites and showed maximum accumulation of total phenolic and flavonoids in *Prunella vulgaris*.

Ramezannezhad et al. (2019) investigated the elicitation potential of AgNPs on enhancing the content of the polyphenolic compound cichoric acid in a cell suspension culture of *Echinacea purpurea*. The production of cichoric acid from the leaf-derived cell suspension showed a two-fold increment when the cultures were supplemented with 2 mg L^{-1} AgNPs for 72 hours of incubation. A notable increment in cichoric acid accumulation was observed in the cell suspension culture from root-derived callus after 48 hours of treatment with 2 mg L^{-1} AgNPs. The report indicates that the duration of AgNP elicitation is specific to different tissue-derived cultures. The production of saponin was greatly increased in the MS medium amended with 0.4 mM, whereas maximum flavonoid or anthocyanin accumulation was recorded at 1.2 mM AgNPs. Raei et al. (2014) reported that the addition of AgNPs (0.625 mg L^{-1}) to an *Aloe vera* suspension culture increased the content of anthraquinone after 48 hours. The NP elicitation on the medicinally important aloin compound was enhanced at 48 hours of treatment and thereafter it decreased gradually. In a *Corylus avellana* cell suspension culture, the exposure of AgNPs (5 mg L^{-1}) gave maximum production of taxol. However, the increased concentration (10 mg L^{-1}) reduced taxol production efficiency to 56% in contrast to the control (Jamshidi et al. 2016). The elicited production (two-fold) of capsaicin via cell-suspension cultures of *Capsicum frutescens* was reported under *in vitro* treatment of 3 mg L^{-1} AgNPs (Bhat and Bhat 2016). In addition to capsaicin production, significant callus growth was also reported in these cultures.

A substantial enhancement of metabolite production was demonstrated through hairy root cultures. Zhang et al. (2013) established elicitation by AgNPs on artemisinin production in hairy root cultures of *Artemisia annua*. The incorporation of 900 mg L^{-1} AgNPs in medium gave maximum artemisinin production (3.9-fold) after three days of exposure. The total enhancement was associated with lipid peroxidation, H_2O_2 production and catalase activity. Similarly, the enhanced artemisinin content along with the down-regulation of squalene synthase and artemisinic aldehyde reductase genes were noticed in cobalt nanoparticle (CoNP)-treated cell suspension cultures of *A. annua* (Ghasemi et al. 2015). The authors concluded that the AgNPs exposed for 12, 24 and 48 hours showed increased atropine production via hairy root cultures. Among these various elicitors, AgNPs were the most successful for the enhanced accumulation of atropine in hairy root cultures.

Spinoso-Castillo et al. (2017) reported the accumulation of the total phenolic content in adventitious shoots of *Vanilla planifolia* depended on the relative level of AgNPs in the MS medium. The enhancement on flavonoid synthesis was obtained on an MS medium containing 25 or 50 mg L^{-1} AgNPs. Most recently, a significant increase of total saponin content by AgNP (250 μM) elicitation in adventitious shoot cultures of *Dianthus chinensis* was also recorded in our research (unpublished data).

All the investigations mentioned in this section offer hope of AgNPs as promising and successful bioactive elicitors in the plant cell, tissue and organ cultures. Further studies are needed to explain the genetic regulations and elicitation potential of various nanomaterials on metabolite production and their mechanisms on related metabolic pathways.

10.6 FUTURE PROSPECTS

The mechanisms of action of nanoscale and colloidal silver nanoparticles in humans and the environment have yet to be fully explained, even though AgNPs have a long history of use in various by-products. Most of the recent research revealed several unsolved problems. Controlled, synthesized nanoparticles can be used for various applications in the pharmaceutical industry, cosmetics, agriculture, environment, food and secondary metabolite enhancement, etc. This review suggests that the application of AgNPs can be a novel system to upgrade the quality, quantity and enhancement of bioactive compounds. Moreover, AgNPs act as a novel elicitor in secondary metabolite and improve the primary metabolite pathway in plants. These responses are based on AgNP size, route of application, plant host cells and their secondary metabolite pathways.

Elicitation is the most practical strategy for increasing the production of desirable secondary metabolites, including flavonoids, carotenoids, alkaloids, volatile oils and taxol. The use of AgNPs as elicitors has unlocked a new area of research that could have an impact on the agricultural and pharmaceutical industries through alternate and improved production of bioactive compounds. In recent years, the application of AgNPs as elicitors has focused on improving the enhancement of secondary metabolites through tissue culture techniques. Additionally, researchers focused their studies on discovering the optimal concentration of AgNPs without any toxic impact and the mechanisms involved in the secondary metabolism of plants resulting from nanoparticle elicitation through tissue-culture techniques. The accomplished review on AgNPs on cell culture suggests that the literature is still too limited to establish the effect of secondary metabolism pathways and the role of nanomaterials on the development of specific tissue systems. Recent research shows that AgNPs can be chosen as powerful elicitors to enhance the metabolism or biological activity leading to production of bioactive compounds without phytotoxic effects. Still, the application of AgNPs as elicitors needs accurate investigation on the sequential effects on the environment and human health.

REFERENCES

Aghdaei, M., M. K. Sarmast, and H. Salehi. "Effects of silver nanoparticles on *Tecomella undulata* (Roxb.) Seem. micropropagation." *Advances in Horticultural Science* 26 (2012): 21–24.

Ali, Amir, Sher Mohammad, Mubarak Ali Khan, Naveed Iqbal Raja, Mohammad Arif, Atif Kamil, and Zia-ur-Rehman Mashwani. "Silver nanoparticles elicited in vitro callus cultures for accumulation of biomass and secondary metabolites in *Caralluma tuberculata*." *Artificial Cells, Nanomedicine, and Biotechnology* 47, no. 1 (2019): 715–724.

Aziz, Nafe, Mohd Faraz, Rishikesh Pandey, Mohd Shakir, Tasneem Fatma, Ajit Varma, Ishan Barman, and Ram Prasad. "Facile algae-derived route to biogenic silver nanoparticles: Synthesis, antibacterial, and photocatalytic properties." *Langmuir* 31, no. 42 (2015): 11605–11612.

Aziz, Nafe, Rishikesh Pandey, Ishan Barman, and Ram Prasad. "Leveraging the attributes of Mucorhiemalis-derived silver nanoparticles for a synergistic broad-spectrum antimicrobial platform." *Frontiers in Microbiology* 7 (2016): 1984.

Baxter, Aaron, Ron Mittler, and Nobuhiro Suzuki. "ROS as key players in plant stress signalling." *Journal of Experimental Botany* 65, no. 5 (2014): 1229–1240.

Bhat, Pooja, and Akash Bhat. "Silver nanoparticles for enhancement of accumulation of capsaicin in suspension culture of *Capsicum* sp." *Journal of Experimental Sciences* 7 (2016): 1–6.

Chung, Ill-Min, Kaliyaperumal Rekha, Govindasamy Rajakumar, and Muthu Thiruvengadam. "Elicitation of silver nanoparticles enhanced the secondary metabolites and pharmacological activities in cell suspension cultures of bitter gourd." *3 Biotech* 8, no. 10 (2018): 1–12.

Cvjetko, Petra, Anita Milošić, Ana-Marija Domijan, Ivana Vinković Vrček, Sonja Tolić, Petra Peharec Štefanić, Ilse Letofsky-Papst, Mirta Tkalec, and Biljana Balen. "Toxicity of silver ions and differently coated silver nanoparticles in *Allium cepa* roots." *Ecotoxicology and Environmental Safety* 137 (2017): 18–28.

Dhrutika, P., P. Miral, and R. Krishnamurthy. "Silver nanoparticles biosynthesis and its antimicrobial activity." *CIB Tech Journal of Bio-protocols* 2 (2013): 50–57.

Duhan, Joginder Singh, Ravinder Kumar, Naresh Kumar, Pawan Kaur, Kiran Nehra, and Surekha Duhan. "Nanotechnology: The new perspective in precision agriculture." *Biotechnology Reports* 15 (2017): 11–23.

Ewais, Emad A., Said A. Desouky, and Ezzat H. Elshazly. "Evaluation of callus responses of *Solanum nigrum* L. exposed to biologically synthesized silver nanoparticles." *Nanoscience and Nanotechnology* 5, no. 3 (2015): 45–56.

Ghasemi, Bita, Ramin Hosseini, and Fatemeh Dehghan Nayeri. "Effects of cobalt nanoparticles on artemisinin production and gene expression in *Artemisia annua*." *Turkish Journal of Botany* 39, no. 5 (2015): 769–777.

Hatami, Mehrnaz, Khalil Kariman, and Mansour Ghorbanpour. "Engineered nanomaterial-mediated changes in the metabolism of terrestrial plants." *Science of the Total Environment* 571 (2016): 275–291.

Hauck, Tanya S., Arezou A. Ghazani, and Warren C. W. Chan. "Assessing the effect of surface chemistry on gold nanorod uptake, toxicity, and gene expression in mammalian cells." *Small* 4, no. 1 (2008): 153–159.

Ismail, Massalimov, Ram Prasad, Amr I. M. Ibrahim, and Ahmed I. S. Ahmed. "Modern prospects of nanotechnology in plant pathology." In *Nanotechnology*, pp. 305–317. Springer, Singapore, 2017.

Jamshidi, Mitra, Faezeh Ghanati, Ayatollah Rezaei, and Ebrahim Bemani. "Change of antioxidant enzymes activity of hazel (Corylus avellana L.) cells by AgNPs." *Cytotechnology* 68, no. 3 (2016): 525–530.

Khan, M. N., M. Mobin, Z. K. Abbas, K. A. Almutairi, and Z. H. Siddiqui. "Role of nanomaterials in plants under challenging environments." *Plant Physiology and Biochemistry* 110 (2017): 194–209.

Kim, Doo Hwan, Judy Gopal, and Iyyakkannu Sivanesan. "Nanomaterials in plant tissue culture: The disclosed and undisclosed." *RSC Advances* 7, no. 58 (2017): 36492–36505.

Kohan-Baghkheirati, Eisa, and Jane Geisler-Lee. "Gene expression, protein function and pathways of *Arabidopsis thaliana* responding to silver nanoparticles in comparison to silver ions, cold, salt, drought, and heat." *Nanomaterials* 5, no. 2 (2015): 436–467.

Lala, Sanchaita. "Nanoparticles as elicitors and harvesters of economically important secondary metabolites in higher plants: A review." *IET Nanobiotechnology* 15, no. 1 (2021): 28–57.

Lindermayr, Christian, and Jörg Durner. "Interplay of reactive oxygen species and nitric oxide: nitric oxide coordinates reactive oxygen species homeostasis." *Plant Physiology* 167, no. 4 (2015): 1209–1210.

Mahna, Nasser, Sepideh Zununi Vahed, and Sajjad Khani. "Plant in vitro culture goes nano: nanosilver-mediated decontamination of ex vitro explants." *Journal of Nanomedicine and Nanotechnology* 4, no. 2 (2013): 3.

Malik, Parth, Ravi Shankar, Vibhuti Malik, Nitin Sharma, and Tapan Kumar Mukherjee. "Green chemistry based benign routes for nanoparticle synthesis." *Journal of Nanoparticles* 2014, (2014): 14.

Marslin, Gregory, Caroline J. Sheeba, and Gregory Franklin. "Nanoparticles alter secondary metabolism in plants via ROS burst." *Frontiers in Plant Science* 8 (2017): 832.

Mirzajani, Fateme, Hossein Askari, Sara Hamzelou, Yvonne Schober, Andreas Römpp, Alireza Ghassempour, and Bernhard Spengler. "Proteomics study of silver nanoparticles toxicity on *Oryza sativa* L." *Ecotoxicology and Environmental Safety* 108 (2014): 335–339.

Mustafa, Harir S., Asma G. Oraibi, Kadhim M. Ibrahim, and Nabeel K. Ibrahim. "Influence of silver and copper nanoparticles on physiological characteristics of *Phaseolus vulgaris* L. in vitro and in vivo." *International Journal of Current Microbiology and Applied Science* 6 (2017): 834–843.

Nel, A., T. Xia, L. Madler, and N. Li. "Toxic potential of materials at the nano level." *Science* 311, (2006): 622–627.

Parisi, Claudia, Mauro Vigani, and Emilio Rodríguez-Cerezo. "Agricultural nanotechnologies: what are the current possibilities?" *Nano Today* 10, no. 2 (2015): 124–127.

Prasad, R. "Synthesis of silver nanoparticles in photosynthetic plants." *Journal of Nanoparticles* 2014 (2014): 963961. http://dx.doi.org/10.1155/2014/963961.

Raei, Mona, S. Abdolhamid Angaji, M. Omidi, and M. Khodayari. "Effect of abiotic elicitors on tissue culture of *Aloe vera*." *International Journal of Bioscience* 5, no. 1 (2014): 74–81.

Ramezannezhad, Rezvan, Mahnaz Aghdasi, and Mohammad Fatemi. "Enhanced production of cichoric acid in cell suspension culture of *Echinacea purpurea* by silver nanoparticle elicitation." *Plant Cell, Tissue and Organ Culture (PCTOC)* 139, no. 2 (2019): 261–273.

Rivero-Montejo, Samantha de Jesus, Marcela Vargas-Hernandez, and Irineo Torres-Pacheco. "Nanoparticles as novel elicitors to improve bioactive compounds in plants." *Agriculture* 11, no. 2 (2021): 134.

Sarmast, Mostafa K., and H. Salehi. "Silver nanoparticles: an influential element in plant nanobiotechnology." *Molecular Biotechnology* 58, no. 7 (2016): 441–449.

Sewelam, N., K. Kazan, and P. M. Schenk. "Global plant stress signaling: reactive oxygen species at the crossroad." *Frontiers of Plant Science* 7 (2016): 187.

Solgi, Mousa, Mohsen Kafi, Toktam Sadat Taghavi, and Roohangiz Naderi. "Essential oils and silver nanoparticles (SNP) as novel agents to extend vase-life of gerbera (*Gerbera jamesonii* cv. Dune) flowers." *Postharvest Biology and Technology* 53, no. 3 (2009): 155–158.

Sosan, Arifa, Dimitri Svistunenko, Darya Straltsova, Katsiaryna Tsiurkina, Igor Smolich, Tracy Lawson, Sunitha Subramaniam et al. "Engineered silver nanoparticles are sensed at the plasma membrane and dramatically modify the physiology of *Arabidopsis thaliana* plants." *The Plant Journal* 85, no. 2 (2016): 245–257.

Spinoso-Castillo, J. L., R. A. Chavez-Santoscoy, Nina Bogdanchikova, J. A. Pérez-Sato, V. Morales-Ramos, and J. J. Bello-Bello. "Antimicrobial and hormetic effects of silver nanoparticles on in vitro regeneration of vanilla (*Vanilla planifolia* Jacks. ex Andrews) using a temporary immersion system." *Plant Cell, Tissue and Organ Culture (PCTOC)* 129, no. 2 (2017): 195–207.

Sreelekshmi, R., E. A. Siril, and S. Muthukrishnan. "Role of biogenic silver nanoparticles on hyperhydricity reversion in *Dianthus chinensis* L. an in vitro model culture." *Journal of Plant Growth Regulation* (2021): 1–17.

Syu, You-yu, Jui-Hung Hung, Jui-Chang Chen, and Huey-wen Chuang. "Impacts of size and shape of silver nanoparticles on Arabidopsis plant growth and gene expression." *Plant Physiology and Biochemistry* 83 (2014): 57–64.

Tiquia-Arashiro, S., and D. Rodrigues. "Application of nanoparticles." In Extremophiles: applications in nanotechnology. pp. 163–193, Springer, Cham, 2016.

Tripathi, Durgesh Kumar, Swati Singh, Shweta Singh, Prabhat Kumar Srivastava, Vijay Pratap Singh, Samiksha Singh, Sheo Mohan Prasad et al. "Nitric oxide alleviates silver nanoparticles (AgNps)-induced phytotoxicity in *Pisum sativum* seedlings." *Plant Physiology and Biochemistry* 110 (2017): 167–177.

Wang, Peng, Enzo Lombi, Fang-Jie Zhao, and Peter M. Kopittke. "Nanotechnology: a new opportunity in plant sciences." *Trends in Plant Science* 21, no. 8 (2016): 699–712.

Wu, Jianyong, and Xiuchun Ge. "Oxidative burst, jasmonic acid biosynthesis, and taxol production induced by low-energy ultrasound in *Taxus chinensis* cell suspension cultures." *Biotechnology and Bioengineering* 85, no. 7 (2004): 714–721.

Xia, Xiao-Jian, Yan-Jie Wang, Yan-Hong Zhou, Yuan Tao, Wei-Hua Mao, Kai Shi, Tadao Asami, Zhixiang Chen, and Jing-Quan Yu. "Reactive oxygen species are involved in brassino steroid-induced stress tolerance in cucumber." *Plant Physiology* 150, no. 2 (2009): 801–814.

Yogeswari, Rout, Sikha Behera, Akshya Kumar Ojha, and P. L. Nayak. "Green synthesis of silver nanoparticles using *Ocimum sanctum* (Tulashi) and study of their antibacterial and antifungal activities." *Journal of Microbiology and Antimicrobials* 4, no. 6 (2012): 103–109.

Zhang, Ben, Li Ping Zheng, Wan Yi Li, and Jian Wen Wang. "Stimulation of artemisinin production in *Artemisia annua* hairy roots by Ag-SiO2 core-shell nanoparticles." *Current Nanoscience* 9, no. 3 (2013): 363–370.

11 Nanoparticles Synthesized by Compounds of Marine Macroalgae for Drug Delivery

Priyanka Rathod and Kulanthaiyesu Arunkumar

CONTENTS

DOI: 10.1201/9781003231745-11

11.1 INTRODUCTION

Algae are photosynthetic autotrophic thallophytes. These organisms, which can be either prokaryotic or eukaryotic, mainly grow in aquatic ecosystems and are divided into fresh-water algae and marine-water algae. Fresh-water algae live in ponds, rivers and streams. While marine-water algae live in oceans and seas; some are also present in brackish waters. Marine algal species are highly productive because 70% of the earth is covered with marine-water ecosystem. Microalgae are unicellular algae that occupy both benthic and littoral habitats. In contrast, macroalgae occupy littoral zones of oceans and seas. Macroalgae are also known as seaweeds, a highly used food source for humans in various countries. Algae are also used for various applications such as food additives, in industrial crude oil, biodiesel and ethanol production, in medicine and in the pharmaceutical and nanoscale industries for preparing cosmetics, dyes, agriculture, drugs, etc. (Stengel and Connan, 2015). Based on pigmentation, marine algae are divided into three groups: red, green and brown (Rioux and Turgeon, 2015).

11.2 BIOACTIVE COMPOUNDS OF SEAWEEDS FOR DRUGS

Seaweeds contain pigments, carbohydrates, fats, proteins, polyunsaturated fatty acids and phytochemicals of both food and pharmaceutical value (Plaza et al., 2008).

Functional properties of these marine bioactive compounds have diverse utility in drug delivery for various biological applications such as antioxidants, immunomodulation, anticoagulation and antiulcerogenic, anticancerous and antidiabetic activities (Suleria et al., 2015).

11.3 NANOPARTICLE SYNTHESIS USING BIOACTIVE COMPOUNDS OF SEAWEEDS

The bioactive compounds found in seaweeds include natural polymers (fucoidan, alginate, agar, laminarin, ulvan, porphyrin, carrageenan), natural pigments (carotenoids), polyunsaturated compounds, carbohydrates, proteins, sterols, alkaloids, etc. From these bioactive compounds, various types of nanoparticles (NPs) are synthesized. Synthesis of NPs involves the following steps in a charge-reduction process.

11.3.1 Nanoparticles

NPs are petite colloidal solid particles with remarkable properties. Their range varies from 1 to 100 nm, but sometimes it is extended to include the range of 1–1000 nm (Kang et al., 2008). The particle is made up of three layers (top surface layer, middle shell layer and inner core layer) that functionalize the NPs in different forms (Shin et al., 2016). In the field of nanomedicine and nanodiagnosis, the NP size range is about 5–100 nm. NPs have demonstrated extensive use in biomedical applications because of their unique physiochemical properties (biochemical, magnetic, optical, electronic) at the cellular, atomic and molecular levels due to their surface-to-volume ratio.

11.3.2 Mechanism of Synthesis of NPs

Synthesis of NPs relies on two different approaches: top-down and bottom-up. In the top-down process, heavy bulk materials are minced into very small pieces and grouped into nanoscale objects. This is a microfabrication process. The top-down approach to fabricating NPs includes various techniques — mechanical milling, laser ablation, etching, sputtering and electro-expansion (Khanna et al., 2019). The bottom-up approach differs with the top-down process since it relies on the assembly of small particles by the self-organization of molecules. This technique is based on molecular nanotechnology involving the assembly of molecules and atoms by joining a defined structure (Thakkar et al., 2010). Various techniques use this bottom-up method: the sol-gel process, spinning, laser pyrolysis, chemical vapour deposition, molecular condensation, supercritical fluid extraction, chemical reduction and green synthesis.

11.3.3 Green Synthesis of NPs

Factors involved in NP synthesis are time, temperature, concentration and pH. There are two types of green synthesis: intracellular synthesis and extracellular synthesis. Intracellular synthesis of algae is a dose-dependent method. In this process, three steps are involved (photosynthesis, respiration and nitrogen fixation) in which nicotinamide adenine dinucleotide phosphate (NADPH) or NADPH-dependent reductase molecules act as reducing agents in the metabolic process (Sharma et al., 2016). Extracellular biosynthesis takes place outside the surface of algal cells and depends on exudates of cell metabolism and metabolites such as proteins, pigments, lipids, enzymes, etc. (Khanna et al., 2019) that become attached to the metal ions, thereby reducing the charge to a zero-valent stage. Extracellular synthesis involves a pretreatment process like washing and blending of algal biomass for the purification of algal extracts.

11.3.4 Polyelectrolyte Complexation Method

Polymers of marine bioactive compounds with positive or negative charges (polyelectrolytes) bind with other complex groups – either polymers or proteins (chitosan, carrageenan, alginates, hyaluronic acid, albumin, etc.) – to formulate NPs. This formulation is mainly based on ionic gelation (cross-linkage of molecules in a carbon chain) and coacervation, a phase separation process (Umerska et al., 2014) mainly used for protein drug delivery.

11.4 NP SYNTHESIS USING RED ALGAE BIOACTIVE COMPOUNDS

Bioactive compounds in red algae include carrageenan, porphyran, agar and carotenoids. Modification of these natural polymers form various types of NPs. A biopolymer is combined with a nanomaterial, resulting in a bionanomaterial, which is highly used in various research studies and now in the preparation of various drugs (Jayachandran et al., 2016). The NPs can be produced by hydrophilic conditions that prevent the use of organic solvents and chemicals.

11.4.1 Carrageenan NPs

Carrageenan is an abundant source of red algal polymers with the D-galactose and anhydro-galactose subunits with glycosidic and ester sulphate linkages. Carrageenan can be extracted based on its gelling nature with various alkali pretreatment processes or washings (Sudha et al., 2014). NP synthesis leads to some modification by polyelectrolyte complexation with the self-assembly of compounds or the ionic gelation method with strong ionic character to exhibit binding capacity with other proteins for reaction (Malafaya et al., 2007) and encapsulation of organic and inorganic materials. A review study shows the role of ferrous oxide in magnetic NPs commonly used in nanoscale materials. The core-shell coated organic polymer carrageenan is used to modify the structural chemistry for various applications in diagnosis (Abida et al., 2017). Carrageenan is bound with polyvinyl chlorides, calcium carbonates, chitosan and polyacrylic acid to form various carrageenan-based nanocomposites for advent uses in biological fields (Abida et al., 2017). Magnetic nanomaterials of carrageenan can be used to assess various morphological characters and chemical stability with analytical tools such as scanning electron microscopy (SEM), transmission electron microscopy (TEM), X-ray diffraction (XRD) and Fourier-transform infrared spectroscopy (FTIR).

11.4.2 Agar NPs

Agar-based NPs are derived from the red seaweeds of *Gracilaria* and *Gelidium*. Agar NPs contain two different components, for example, a mix of 30% agaropectin (nongelling agent) branched with 1,3-glycosidic bonded D-galactoseunits and 70% agarose (gelling agent) branched with the D-galactose and 3,6-anhydrogalactose units (Marinho-Soriano and Bourret, 2005). Agar is mainly used in culture media in the form of agar plates or as beads in the electrophoresis process to carry out experiments. Surface modification in galactose subunits of agar NPs was formed by the water emulsion encapsulation method (Wang et al., 1997). A study reported agarose NPs were capsulated with protein and peptides for various drug delivery processes. Agar is used as a capping agent to the silver nanoparticles (AgNPs) that produce composite biofilms by reducing the silver nitrate using the green synthesis method. Agar NPs were formed as solvents in the casting surface method. It was characterized by SEM and XRD (Rhim et al., 2013).

11.4.3 Porphyran NPs

Porphyran NPs are extracted from red algae and formed by the skeletal linkage of 3,6-anhydro-galactose units. Porphyran is similar to the agar compound because it also shows low viscosity. A hot-water-soluble part of the cell wall is a main feature of this algae. Porphyran acts as a capping agent on metal NPs such as gold and silver. It also acts as a reducing agent. Some studies reveal the structure of porphyran based on gelling and emulsifying properties (Bhatia et al., 2010). Green synthesis of gold nanoparticles (AuNPs) uses porphyran loaded with doxorubicidonen for anticancer activity (Venkatpurwar et al., 2011).

11.4.4 Carotenoid NPs

Carotenoid NPs were formed by the solvent-displacement method. The size of the carotenoid NPs were encapsulated in a range below 100 nanometres. Due to modifications, carotenoid extract is capped with AuNPs and AgNPs (Harshada et al., 2016).

11.5 NP SYNTHESIS USING GREEN ALGAE BIOACTIVE COMPOUNDS

Ulvan NPs are synthesized from green algal cell walls. Ulvan NP synthesis occurs through various methods such as ionic gelation and complexation, polyelectrolyte complexation, functionalization with chitosan NPs, etc., due to the modification of ulvan polysaccharides to ulvan NPs. AgNPs synthesised by silver ions with sodium borohydride using Ulvan polymer through an irradiation technique, heating method and chemical reduction posseses more stability and observed as Ulvan provide stabilized to NPs during synthesis (Alessio et al., 2019).

The ulvan NPs act as a stabilizing and emulsifying agent with amphiphilic characteristics. Ulvan-based NPs were prepared by an electrospinning process with a polyelectrolyte oxide solution, polyvinyl alcohol, polycaprolactone solution, etc. (Santhana et al., 2020). Pigments and vitamins derived from green algae modulate their function and bind with the NPs showing NP synthesis.

11.6 NP SYNTHESIS USING BROWN ALGAE BIOACTIVE COMPOUNDS

Bioactive compounds in brown algae are alginates, fucoidan, laminarin, etc. With some modifications of these bioactive compounds, NPs were formed.

11.6.1 Alginate NPs

There are various methods used to modify alginate into alginate NPs: an emulsification technique and solvent evaporation technique, complexation, a layer-by-layer approach and a spray drying method. In the emulsification process, to modify alginate into alginate NPs, emulsified droplets were fabricated on the core cross-linked bridges in a gelation process using calcium carbonate (Paques et al., 2014). Alginate complexation occurs using divalent ions to cross-link other groups of oppositely charged ions by incorporating them. In the layer-by-layer approach, cross-linked bridges of NPs were formed by the electrostatic attraction of oppositely charged particles to form a core shell NP by alginates (Choukaife et al., 2020). Sodium alginates are used as a precursor for metallic NPs such as gold, silver and copper to form alginates capped with metal NPs using a microwave irradiation method for a one-step green synthesis process. Metal NPs were used as a reducing agent.

11.6.2 Fucoidan NPs

Fucoidan is a sulphated polysaccharide derived from brown algae and used for many applications. Like alginates, fucoidan NPs are prepared by different strategies including complexation, conjugation and functionalization of metallic NPs. Negatively charged sulphate groups in fucoidan bind with other positively charged groups for NP synthesis. The biphasic release of fucoidan causes it to bind with chitosan NPs due to the electrostatic interaction (Huang et al., 2016). Functionalization of metallic NPs is required for stabilizing the biological environment during the loading of a drug.

Different types of metallic NPs were present such as iron oxide, silver and gold for functionalization with the sulphated polysaccharides to form bioactive NPs. Fucoidan-coated NP biosynthesis plays an important role that can be analysed by various techniques such as FTIR, 2,2-Diphenyl-1-picrylhydrazyl (DPPH) assay, agar bioassay, etc. Magnetic NPs can also stabilize with the fucoidan to enhance drug delivery in various biomedical applications. Fucoidan magnetic stabilized NPs show a general synergetic effect due to the combination of fucoidan and its nanoparticles. Fucoidan-coated metallic NPs are present inside the core of nanospheres such as liposomes and micelles. Fucoidans can also be designed as surface core polymeric NPs instead of coating the surface metallic NPs (Lira et al., 2011). There are two basic methods used for fabricating the fucoidan surface coated with polymeric micelles: redox radical emulsion polymerization and anionic emulsion polymerization (Bertholon et al., 2006).

11.6.3 Laminarin NPs

Laminarin NPs were formed by the modification of natural polysaccharides with various steps such as ionic gelation, emulsion, conjugation and polyelectrolyte complexation that were commonly performed by adding cationic or anionic polymers to produce these NPs (Jayachandran et al., 2016). The laminarin polysaccharide acted as a modifier to form selenium NPs and it was prepared by a simple oxidation-reduction method to study the cytotoxicity and apoptotic reactions by various characteristic techniques (Dongxiao et al., 2019). Biosynthesis of laminarin NPs used AgNPs (Remya et al., 2018).

11.7 EXTRACTION OF MARINE ALGAE BIOACTIVE COMPOUNDS

Marine algae groups present a high number of bioactive compounds such as polyunsaturated compounds, proteins, fatty acids, sterols, vitamins, polyphenols, trace metal elements and pigments (Figure 11.1). Extraction of these bioactive compounds includes various conventional methodologies such as hydrodistillation, soxhlet extraction, maceration, percolation and infusion, while non-conventional methods include enzyme-assisted extraction, supercritical fluid extraction, microwave assisted extraction and pressure-liquid extraction. All common methods of extraction are carried

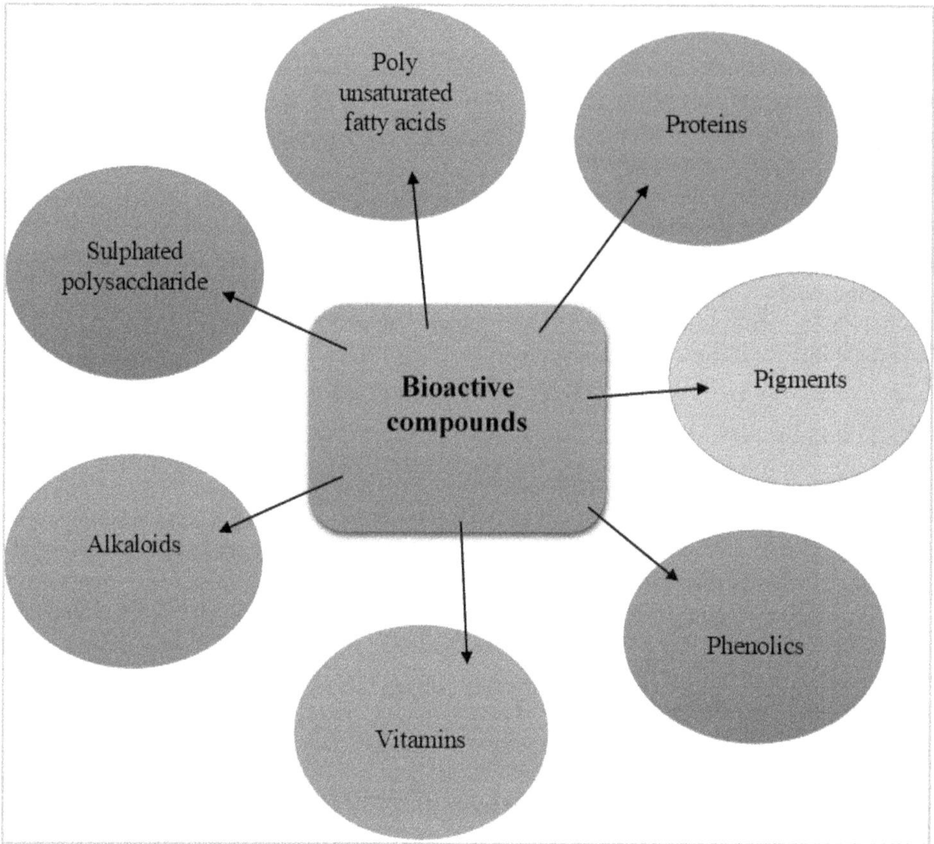

FIGURE 11.1 Bioactive compounds of marine macroalgae.

out for the removal of pigments, chlorophyll and other compounds as a pretreatment process. (Sosa-Hernández et al., 2011).

Many factors influence NP synthesis with algal bioactive compounds, such as pH, concentration, time, etc. Algal extract acts as a stabilizer and reduction agent during the preparation of NPs. Green seaweed (*Caulerpa racemose*) samples were washed repeatedly with Milli-Q purified water, chopped into pieces, dried in the shade without contamination and then ground into powder (Kathiraven et al., 2015). Next, 1 g of powdery extract was placed into a 250 ml conical flask with 100 ml of Milli-Q water at room temperature for 24 hours. This extract was filtered with Whatman No.1 filter paper for any analysis and was stored at room temperature. For biosynthesis of NPs like AgNPs, 10 ml of seaweed-filtered extract was added to 90 ml of 10^3 M aqueous silver nitrate ($AgNO_3$) solution at room temperature (Govindaraju et al., 2009). Colour change can indicate the NP synthesis.

11.8 NP CHARACTERIZATION

NPs are characterized based on their size and morphology and its surface charge distribution can be determined using various microscopic techniques such as atomic force microscopy (AFM), TEM and SEM. Properties such as size distribution, average particle distribution, diameter and the stability of particles can be determined by electron microscopy techniques including FTIR, X-ray photoelectron spectroscopy (XPS), AFM, dynamic light scattering (DLS) and ultraviolet-visible (UV-Vis) spectroscopy. These techniques are useful parameters for understanding and analysing particle size, shape, crystallinity, dimension of pore size and surface area (Bhatia, 2016).

11.8.1 Size

Particle size distribution and morphology can be evaluated with the help of electron microscopy techniques. For the release of various drugs, various tools are used in each application level. Smaller-sized NPs have a large surface area that results in the fast release of drugs. There is maximum stability in smaller-sized particles. The degradation effects of the NP will be greater in the smaller-sized particles in drug therapy.

11.8.2 UV-Vis Spectroscopy

When bioactive compound extracts were added with metal salts, the first significant change noticed in the NP formation process was the solution's change in colour observed by UV-spectral analysis by surface plasma resonance (SPR) peaks (Tu et al., 2012).

11.8.3 Photon-Correlation Spectroscopy (PCS)/Dynamic Light Scattering (DLS)

In determining particle size, current research methods use photon-correlation spectroscopy (PCS), also known as DLS. This is a fast and popular technique used to determine the size of Brownian NPs in colloidal suspensions and submicron ranges. In this method, the particle causes a Doppler shift when exposed to laser light, which shows a change in wavelength of light when the particle is hit by light. This technique evaluates particle-size distribution and the motion of particles in the medium (De Assis et al., 2008).

11.8.4 Scanning Electron Microscopy (SEM)

SEM is used to determine size, shape and surface morphology using the visualization of NPs. A solution of NPs is converted into a dry powder and this powder is mounted on a sample holder and coated with the conductive metals (silver, gold). Later, they are scanned with a focused beam of electrons (Jores et al., 2004).

11.8.5 Transmission Electron Microscopy (TEM)

TEM can provide imaging diffraction and spectroscopic information of a specimen simultaneously or in a sequence with spatial resolution. TEM operates on a different principle from SEM. TEM is a complicated and time-consuming process due to the requirement of ultra-thin specimens for electron transmittance. During TEM, NPs are dispersed and deposited on the film/supporter grid and are later fixed on staining material for support or embedded with plastic material to facilitate handling. Next, NPs were exposed to liquid nitrogen temperature. Later, when a beam of light scatters through the ultra-thin sample, it shows a high resolution image (Molpeceres et al., 2000).

11.8.6 Atomic Force Microscopy (AFM)

AFM is a height resolution technique also known as scanning force microscopy, which scans samples at the submicron level using a probe tip of atomic scale. AFM provides an accurate description of size and size distribution with a high resolution image and is useful for understanding various biological processes (Polakovic et al., 1999).

11.9 MARINE MACROALGAE NPs FOR DRUG DELIVERY

Seaweed-based polysaccharide NPs (Table 11.1) have abundant properties that are used in drug-delivery systems (Saravankumar et al., 2012). Seaweed polysaccharides including alginate, agar, fucoidan, carrageenan and laminarin have been isolated (Figure 11.2).

TABLE 11.1
Sulphated Polysaccharide-Based Nanoparticle Drugs

Sulphated Nanoparticles	Drug/Drug Delivery	References
Alginate NPs	Insulin/Protein	Reis et al., 2006
Alginate NPs	BSA/Protein	Nesamony et al., 2012
Alginate NPs	Doxorubicin/Cancer	Zhang et al., 2016
Alginate NPs	Bupivacaine/Neural	Grillo et al., 2010
Fucoidan NPs	BSA/Protein	Liu et al., 2012
Fucoidan/Chitosan NPs	Curcumin/Cancer	Huang et al., 2011
Laminarin NPs	HLDM/Cancer	Yueming et al., 2019
Carrageenan NPs	Riboflavin	Mahdavinia and Etemadi, 2019
Agarose NPs	Ovalbumin/Protein	Wang and Wu, 1997
Porphyran NPs	Amphotericin-B/Antifungal	Bhatia et al., 2014
Ulvan NPs	Dexamethasone/Wound healing	Alves et al., 2012

NPs can be used as a targeted drug delivery system to improve the bioavailability of a drug (tumour targeting delivery system using NPs, oral delivery of peptides and proteins, gene delivery of NPs, drug delivery into the brain, cancer treatment, etc.). The marine bioactive-modified algae-biosynthesized NPs include AgNPs, AuNPs and some metal oxide NPs such as zinc oxide, copper dioxide and ferric oxide synthesized from marine as well as fresh algae. The emerging growth for developing NPs may also be carried out by various physical and chemical processes, but these are highly toxic and hazardous methods (Chaudhary et al., 2017). The green method is a simple, cost-effective and ecofriendly method. Current research mostly depends on this method and it is also used for drug delivery to treat many diseases. The NPs are coated by polymers that release the drug by controlled diffusion or erosion from the polymeric matrix core, which acts as a barrier because of the ionic interaction (Huang et al., 2016).

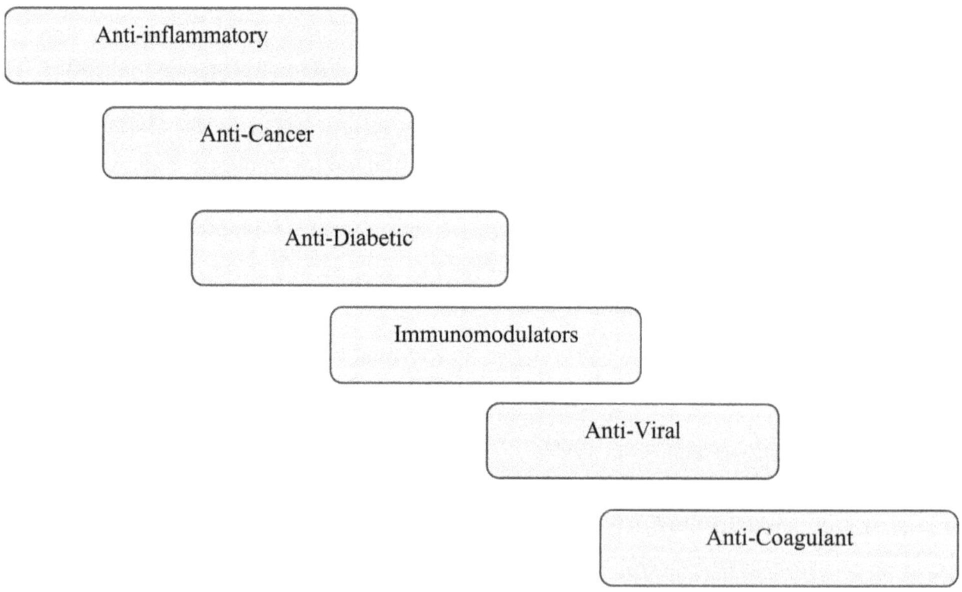

FIGURE 11.2 Potential marine compounds for drugs.

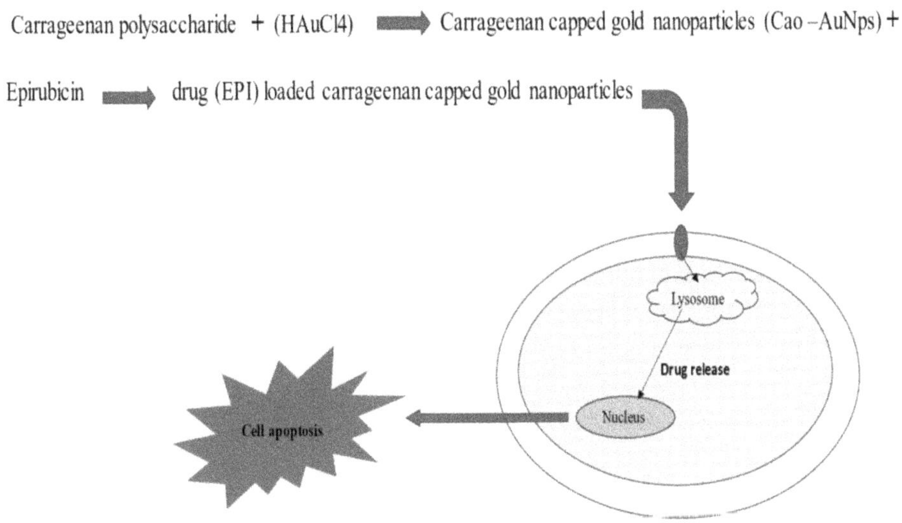

FIGURE 11.3 Sulphated polysaccharide-based nanodrug delivery for cancer treatment.

Active targeting approaches the surface functionalization of drug carriers with ligands that may recognize the specific receptors when binding to the targeted cell of interest during drug delivery. It includes two main factors: specific targeting and time release: (1) Specific targeting of the NPs with the diseased tissue includes the selective binding of drugs with the appropriate size and functionalized antibodies/NPs, which can resolve and reduce the toxicity level and show the specific binding of drugs with the receptors 2. (2) Time release involves the release of a drug in a sequenced order (because the NP is encapsulated with many layers) so that particle binding shows appropriate timing for the drug delivery (Bhatia, 2016).

11.9.1 Red Algae NPs for Drug Delivery

11.9.1.1 Carrageenan NPs

Today, carrageenan is used as stabilizer, a reducing agent for preparing novel nanocarrier particles for emulsifying agents, a form of nanosuspension of polysaccharide compounds with the help of lyophilization process for use in drug delivery and other biomedical applications. Marine-derived chitosan-carrageenan NPs are developed by ionic complexation with the interaction of positive charged groups of amine and negatively charged groups of sulphate. These formulated NPs were extensively used in drug delivery and cytotoxicity effects of mouse (L929) fibroblast cell lines (Grenha et al., 2010). It was prepared using a polyelectrolyte method and chitosan-carrageenan NPs were also developed using a polyelectrolyte method and an ionotropic gelation method for the release of the recombinant human erythropoietin in a controlled process of drug delivery. The efficiency of encapsulation of the recombinant human erythropoietin was approximately 47.97% ± 4.10. This result will be used by various analytical techniques including response surface methodology (RSM) (Bulmer et al., 2012). Another study reveals the interaction of chitosan-carrageenan-triphosphate-binding NPs that were used in mucosal delivery of macromolecules.

11.9.1.2 Agarose NPs

Agarose is a bioactive compound extract from red algae that synthesizes agarose NPs for drug delivery. A study was carried out and agarose hydrogel NPs were developed for the drug delivery of proteins and peptides by using an emulsion-free method in which ovalbumin was used as a protein drug (Wang and Wu, 1997). Agar NPs were loaded with bupropion, an antidepressant drug, and processed for drug delivery with oral administration (Varshosaz et al., 2015).

11.9.1.3 Porphyran NPs

Porphyran bioactive compounds have some important biomedical and pharmaceutical applications. The biological activities of porphyran were modified into porphyran NPs for various applications. Porphyran shows a highly immunomodulating structure. In one study, amphotericin B NPs showed antifungal properties synthesized by polyelectrolyte complexation method using tripolyphosphate (TPP), binding to NPs with porphyran amphotericin B NPs (Bhatia et al., 2014). Biosynthesis of porphyran with AuNPs was used for the treatment of cancer and loaded with the drug doxorubicin (Venkatpurwar et al., 2011).

11.9.2 Green Algae NPs for Drug Delivery

11.9.2.1 Ulvan NPs

Ulvan develops from the cell walls of marine macro green algae. The current generation is mainly dependant on bioactive compounds for various applications. These ulvan-based NPs also show a greater impact on biomedical applications. To study antibacterial activity, Gram-positive bacteria were synthesized by a polyelectrolyte complexation method with a lysozyme showing a ubiquitous nature in which it degrades the bacteria, causing lysis in the cell (Tziveleka et al., 2013). Polyelectrolyte complexation with chitosan molecules constructed the dexamethasone model drug for wound healing and drug delivery (Alves et al., 2012).

11.9.3 Brown Algae NPs for Drug Delivery

11.9.3.1 Alginate NPs

Alginate NPs have been used extensively in drug delivery processes because of their high encapsulation efficiency of the drugs, proteins and peptides. Alginate NPs do not show agglomeration in organs when they deliver a drug (Sarei et al., 2013). Alginates showing stability in the NPs. Particle size and zeta potential were measured for a period of time because storing alginate NPs for months may aggregate their size due to electrostatic interactions. To maintain the stability of alginates, vitamin B2 is added as a stabilizer.

11.9.3.2 Alginate NPs in Protein and Peptide Delivery

Oral delivery of alginate NPs is used for delivering proteins and peptides drugs. Today, diabetes is the leading disease for about 10% of the population. A nanotechnology using alginate NPs for oral insulin delivery was developed via the emulsion method and cross-linkage of calcium ions to play an important role to control particle size. Encapsulation efficiency is higher and it lowers the release of insulin (Reis et al., 2006). In another example, insulin-loaded NPs prepared alginate NPs via an ionotropic pregelation method with the use of calcium ions followed by a polyelectrolyte complex coating with the chitosan. The particle size was influenced by time and the stirring process. An approximate 800 nm particle size is produced. Insulin was determined by the HPLC method. The FTIR method revealed the efficiency of encapsulation of insulin (Sarmento et al., 2007).

11.9.3.3 Alginate NPs for Cancer Drug Delivery

Many experimental studies reported on cancer treatment by using the NPs such as liposome-polymers, dendrimers and inorganic NPs. Today, marine algae polysaccharides can also be used to develop drugs to treat cancer in the form of doxorubicin and 5-fluorouracil. Alginate NPs were either developed with calcium ions and poly-L-Lysin by a gelification method or calcium ions cross-linkage method. The developed NPs have the efficiency to kill the carcinoma cells (Rajaonarivony et al., 1993). Another co precipitation method is used to prepare the alginate calcium carbonate doxorubicin p53 drug. The p53 drug has an influential role of suppressing the gene in the cell cycle and DNA repair. The combination of doxorubicin and p53 may show effective treatment in cancer. This was mainly studied in gene drug delivery purposes (Zhao et al., 2012).

11.9.3.4 Fucoidan NPs for Cancer Drug and Growth Factor Delivery

Fucoidan NPs for cancer drug delivery have a number of research studies that report the ability to remove cancer cells by inducing apoptosis (Kim et al., 2013). Fucoidan-based NPs were prepared by the self-assembly, ionic cross-linkage and ionic gelation methods. Curcumin is an anticancer drug because of its low bioavailability hindered in utilizing drug delivery. The chitosan-fucoidan NPs shows the maximum encapsulation efficiency (Singh and Lillard, 2014). Fucoidan NPs with chitosan were developed by using ionic gelation and polyelectrolyte complex methods. These chitosan-fucoidan NPs were the drivers for stromal cell factor (SDF-1) (Huang et al., 2016). The efficiency of the binding of chitosan-TPP-fucoidan NPs with SDF-1 is reported at 60–68%. This is used in nerve tissue engineering technology.

11.9.3.5 Laminarin NPs for Cancer Drug Delivery

Laminarin-synthesized NPs are used mainly for drug delivery in cancer therapy. A review study reveals a photodynamic therapy against human breast cancer cells by Hematin-Laminarin-Dithiodipropionic Acid-MGK (HLDM), a drug derived from laminarin bioactive compounds. HLDM is an amphiphilic carrier used to load the hydrophobic drug and is combined with protoporphyrin IX as a photosensitizer to fabricate NPs for treating cancer cells (Yueming et al., 2019). The nanofibrous particles were prepared by an electrospinning method from laminarin and a conjugated compound, i.e., poly (3-hydroxybutyrate-co-3-hydroxyvalerate) (PHBV), were synthesized for studying the potential cell growth in cancer therapy (Iliou et al., 2022). Another study reveals conjugated nonviral vector-based polyethyleneimine-laminarin, which carries the small interfering RNA (siRNA) to target the B-cells for targeting the gene delivery in human breast cancer cells (Ren et al., 2015).

11.10 SUMMARY AND CONCLUSION

This review is focused on algal NPs synthesis. The NPs synthesized from marine algal bioactive compounds have diverse uses in various applications, including biomedical applications, gene delivery, drug delivery and diagnosis. Marine algal bioactive compounds are pigments, sulphated polysaccharides (carrageenan, agar, porphyran, fucoidan, laminarin, alginate, ulvan), cellulose, carotenoids, phlorotannin, etc., that were categorized based on the pigmentation of algae. The preparation of bioactive compound-capped NPs depends on various methods such as polyelectroglycation, polyelectrolyte complexation, conjugation, emulsification, gelation and green synthesis. Synthesis of NPs using marine algal extracts/bioactive compounds by pretreatment methods was used for preparing NPs by adding distilled water in different volumes and a precursor metal ion and watching for a colour change in solution, which shows that NPs were formed. Synthesized NP compounds were detected by different characterization techniques such as UV-Vis spectroscopy, TEM, SEM, XRD, FTIR, DPPH, agar bioassay, etc., helping to explain NP synthesis and its use. The colour change of the solution can be characterized by the UV-Vis technique. Size and morphology can be characterized by SEM, TEM offers the ability to study the variable shapes of NPs and XRD depicts the crystallinity of biosynthesized NPs. Bottom-up approach and Top-down approach. Green synthesis/biosynthesis of NPs will follow the bottom-up approach method. The biosynthesis method involves the factors of reaction time, pH, temperature and concentration. Nanoencapsulated bioactive compounds mainly show high lipophilic ability and lower solubility in water, which are effectively used in the drug delivery process by immobilizing active compounds. Medical, food and nutraceutical industries can use these bioactive compounds for various diseases and applications, including anticancer, antidiabetic, anticoagulant activity, neural diseases and antioxidant agents.

11.11 CONCLUSION

It is concluded from this review that biosynthesized NPs from compounds of marine macroalgae are natural source that acts as a drug and is capable of delivering drugs known for curing many diseases. Biosynthesized NPs will mainly help to understand the properties and antiproliferative activity of tumour growth and their anticancerous activity.

REFERENCES

Abida K. K., Saba Ain Us, Nawazish Shamyla, Akhtar Fahad, Rashid Rehana, Mir Sadullah, Nasir Bushra, Iqbal Furqan, Afzal Samina, Pervaiz Fahad, and Murtaza Ghulam. "Carrageenan based bionanocomposites as drug delivery tool with special emphasis on the influence of ferromagnetic nanoparticles." *Oxidative Medicine and Cellular Longevity* 2017, 8158315 (2017): 1–13. https://doi.org/10.1155/2017/8158315

Alessio M., Morelli Andrea, Grassi Lucia, Puppi Dario, Braccini Simona, Maisetta Giuseppantonio, Esin Semih, Batoni Giovanna, Pina Cristina Della, and Chiellini Federica. "Ulvan as novel reducing and stabilizing agent from renewable algal biomass: Application to green synthesis of silver nanoparticles." *Carbohydrate Polymers* 203 (2019): 310–321.

Alves A., Pinho E. D., Neves N. M., Sousa R. A., and Reis R. L. "PDLLA enriched with ulvan particles as a novel 3D porous scaffold targeted for bone engineering." *Journal of Supercritical Fluids* 65 (2012): 32–38.

Bertholon I., Lesieur Sylviane, Labarre Denis, Besnard Madeleine, and Vauthier Christine. "Characterization of dextran–poly(isobutylcyanoacrylate) copolymers obtained by redox radical and anionic emulsion polymerization." *Macromolecules* 39, no. 10 (2006): 3559–3567. doi: 10.1021/ma060338z

Bhatia S., Namdeo A. G., and Nanda S. "Factors effecting the gelling and emulsifying properties of a natural polymer." *Systematic Reviews in Pharmacy* 1 (2010): 86–92.

Bhatia Saurabh, Kumar Vikash, Sharma Kiran, Nagpal Kalpana, and Bera Tanmoy. "Significance of algal polymer in designing amphotericin B nanoparticles." *Scientific World Journal* 2014 (2014). 564573. https://doi.org/10.1155/2014/564573.

Bhatia Saurabh. "Natural polymers vs synthetic polymer." In *Natural Polymer Drug Delivery Systems*, pp. 95–118. Springer, Cham, 2016.

Bulmer Cody, Margaritis Argyrios, and Xenocostas Anargyros. "Encapsulation and controlled release of recombinant human erythropoietin from chitosan-carrageenan nanoparticles." *Current Drug Delivery* 9, no. 5 (2012): 527–537.

Chaudhary D., Kumar R., Kumari A., and Rashmi J. R. "Biosynthesis of nanoparticles by microorganisms and their significance in sustainable agriculture." In V. Kumar, M. Kumar, S. Sharma, and R. Prasad (eds.) *Probiotics in Agroecosystem*, pp. 93–115. Springer, Singapore, 2017. doi: 10.1007/978-981-10-4059-7_5

Choukaife H., Doolaanea A. A., and Alfatama M. "Alginate nanoformulation: Influence of process and selected variables." *Pharmaceuticals (Basel)* 13, no. 11 (2020): 335. doi: 10.3390/ph13110335

De Assis D. N., Mosqueira V. C., Viela J. M., Andrade M. S., and Cardoso V. N. "Release profiles and morphological characterization by atomic force microscopy and photon correlation spectroscopy of 99m Technetium-fluconazole nano capsules." *International Journal of Pharmaceutics* 349 (2008): 152–160.

Dongxiao C., Ma Jun, Liang Tingting, Sun Liqian, Meng Liaqiang, Liang Taigang, and Li Qingshan. "Selenium nanoparticles fabricated in laminarin polysaccharides solutions exert their cytotoxicities in HepG2 cells by inhibiting autophagy and promoting apoptosis." *International Journal of Biological Macromolecules* 137 (2019): 829–835.

Govindaraju K., Kiruthiga V., Ganesh Kumar V., and Singaravelu G. "Extracellular synthesis of silver nanoparticles by a marine alga, *Sargassum wightii* Grevilli and their antibacterial effects." *Journal of Nanoscience Nanotechnology* 9, no. 9 (2009): 5497–5501. doi: 10.1166/jnn.2009.1199.

Grenha A, Gomes M. E., Rodrigues M., Santo V. E., Mano J. F., Neves N. M., and Reis R. L. "Development of new chitosan/carrageenan nanoparticles for drug delivery applications." *Journal of Biomedical Materials Research A* 92, no. 4 (2010): 1265–1272. doi: 10.1002/jbm.a.32466. PMID: 19322874.

Grillo R., de Melo N. F., de Araújo D. R., de Paula E., Rosa A. H., and Fraceto L. F. "Polymeric alginate nanoparticles containing the local anesthetic bupivacaine." *Journal of Drug Targeting* 18, no. 9 (2010 Nov): 688–699. doi: 10.3109/10611861003649738. PMID: 20196632.
Harshada S., Mohite Pallavi, Damale Shailesh, Kulkarni Mohan, and Zinjarde Smita. "Carotenoid stabilized gold and silver nanoparticles derived from the Actinomycete *Gordonia amicalis* HS-11 as effective free radical scavengers." *Enzyme and Microbial Technology* 95 (2016): 164–173.
Huang Y. C., Li R. Y., Chen J. Y., and Chen J. K. "Biphasic release of gentamicin from chitosan/fucoidan nanoparticles for pulmonary delivery." *Carbohydrate Polymers* 138 (2016 Mar 15): 114–122. doi: 10.1016/j.carbpol.2015.11.072. Epub 2015 Nov 28. PMID: 26794744.
Iliou K., Kikionis S., Ioannou E., and Roussis V. "Marine biopolymers as bioactive functional ingredients of electrospun nanofibrous scaffolds for biomedical applications." *Marine Drugs* 20, no. 5 (2022): 314. https://doi.org/10.3390/md20050314
Jayachandran V., Kim Se-Kwon, and Shim Min Suk. "Antimicrobial, antioxidant, and anticancer activities of biosynthesized silver nanoparticles using marine algae *Ecklonia cava*." *Nanomaterials and Nanotechnology* 6, no. 12 (2016): 235.
Jores K., Mehnert W., Drecusler M., Bunyes H., Johan C., and Mader K. "Investigation on the stricter of solid lipid nanoparticles and oil-loaded solid nanoparticles by photon correlation spectroscopy, field-flow fractionation and transmission electron microscopy." *Journal of Controlled Release Society* 17 (2004): 217–227.
Kang Huizhi, Wang Lin, O'Donoghue Meghan, Charles Cao Y., and Tan Weihong. "Nanoparticles for biosensors." In *Optical Biosensors*, pp. 583–621. Elsevier, Hungary, 2008.
Kathiraven T., Sundaramanickam A., Shanmugam N., and Balasubramanian T. "Green synthesis of silver nanoparticles using marine algae *Caulerpa racemosa* and their antibacterial activity against some human pathogens." *Applied Nanoscience* 5, no. 4 (2015): 499–504.
Khanna Prerna, Kaur Amrit, and Goyal Dinesh. "Algae-based metallic nanoparticles: Synthesis, characterization and applications." *Journal of Microbiological Methods* 163 (2019): 105656.
Kim S. H., Kang J. H., and Yang M. P. "Fucoidan directly regulates the chemotaxis of canine peripheral blood polymorphonuclear cells by activating F-actin polymerization." *Veterinary Immunology and Immunopathology*. 151 (2013): 124–131. doi: 10.1016/j.vetimm.2012.11.001
Lira M. C. B., Santos-Magalhães N. S., Nicolas V., Marsaud V., Silva M. P. C., Ponchel G., and Vauthier C. "Cytotoxicity and cellular uptake of newly synthesized fucoidan-coated nanoparticles." *European Journal of Pharmaceutics and Biopharmaceutics* 79, no. 1 (2011): 162–170.
Liu X., Chen X., Li Y., Cui Y., Zhu H., and Zhu W. "Preparation of superparamagnetic sodium alginate nanoparticles for covalent immobilization of *Candida rugosa lipase*." *Journal of Nanoparticle Research* 14, no. 3 (2012): 1–7.
Mahdavinia G. R. and Etemadi H. "Surface modification of iron oxide nanoparticles with κ-carrageenan/carboxymethyl chitosan for effective adsorption of bovine serum albumin." *Arabian Journal of Chemistry* 12, no. 8 (2019): 3692–3703. https://doi.org/10.1016/j.arabjc.2015.12.002
Malafaya P. B., Silva G. A., and Reis R. L. "Natural-origin polymers as carriers and scaffolds for biomolecules and cell delivery in tissue engineering applications." *Advance Drug Delivery Review* 59, no. 4–5 (2007): 207–233.
Marinho-Soriano E. and Bourret E. "Polysaccharides from the red seaweed *Gracilaria dura* (Gracilariales, Rhodophyta)." *Bioresource Technology* 96, no. 3 (2005 Feb): 379–382. doi: 10.1016/j.biortech.2004.04.012. PMID: 15474941.
Molpeceres J., Aberturas M. R., and Guzman M. "Biodegradable nanoparticles as a delivery system for cyclosporine: preparation and characterization." *Journal of Microencapsulation* 17 (2000): 599–614.
Nesamony J., Singh P. R., Nada S. E., Shah Z. A., and Kolling W. M. "Calcium alginate nanoparticles synthesized through a novel interfacial cross-linking method as a potential protein drug delivery system." *Journal of Pharmaceutical Science* 101 (2012): 2177–2184. https://doi.org/10.1002/jps.23104
Paques Jerome P., van der Linden Erik, van Rijn Cees J. M., and Sagis Leonard M. C. "Preparation methods of alginate nanoparticles." *Advances in Colloid and Interface Science* 209 (2014): 163–171.
Plaza Merichel, Cifuentes Alejandro, and Ibáñez Elena. "In the search of new functional food ingredients from algae." *Trends in Food Science & Technology* 19, no. 1 (2008): 31–39.
Polakovic M., Gorner T., Gref R., and Dellacherie E. "Lidocaine loaded biodegradable nanospheres. II. Modelling of drug release." *Journal of Controlled Release* 60 (1999): 169–177.
Prasad, R. "Synthesis of silver nanoparticles in photosynthetic plants." *Journal of Nanoparticles* 2014 (2014): 963961. 8 pages. http://dx.doi.org/10.1155/2014/963961.
Rajaonarivony M., Vauthier C., Couvrraze G., Puisieux F., and Couvreur P. "Development of a new drug carrier made from alginate." *Journal of Pharmaceutical Science* 82 (1993): 912–917. https://doi.org/10.1002/jps.2600820909

Reis Catarina P., Neufeld R. J., Ribeiro António J., and Veiga Francisco. "Design of insulin-loaded alginate nanoparticles: Influence of the calcium ion on polymer gel matrix properties." *Chemical Industry and Chemical Engineering Quarterly* 12, no. 1 (2006): 47–52.

Remya R. R., Radhika Rajasree S. R., and Suman T. Y. "Laminarin based AgNPs using brown seaweed *Turbinaria ornata* and its induction of apoptosis in human retinoblastoma Y79 cancer cell lines." *Materials Research Express* 5, no. 3 (2018). doi: 10.1088/2053-1591/aab2d8

Ren W. X., Han J., Uhm S., Jang Y. J., Kang C., Kim J. H., and Kim J. S. Recent development of biotin conjugation in biological imaging, sensing, and target delivery. *Chemical Communications (Camb.)* 51 (2015): 10403–10418.

Rhim J. W., Wang L. F., and Hong S. I. "Preparation and characterization of agar/silver nanoparticles composite films with antimicrobial activity." *Food Hydrocolloids* 33, no. 2 (2013): 327–335.

Rioux Laurie-Eve and Turgeon, Sylvie L. "Seaweed carbohydrates." In *Seaweed Sustainability*, pp. 141–192. Academic Press, 2015.

Santhana V., Thangaraju D., Tanaka A., Inami W., JayaKumar S., and Matsuda S. "Development of hybrid TiO_2/paint sludge extracted microbe composite for enhanced photocatalytic dye degradation." *Journal of Inorganic and Organometallic Polymers and Materials* 30(2020): 2805–2813. https://doi.org/10.1007/s10904-020-01448-2

Saravankumar G., Jo D.-G., and Park J. H. "Polysaccharide-based nanoparticles: A versatile platform for drug delivery and biomedical imaging." *Current Medical Chemistry* 19 (2012): 3212–3229.

Sarei F., Dounighi N. M., Zolfagharian H., Khaki P., and Bidhendi S. M. "Alginate nanoparticles as a promising adjuvant and vaccine delivery system." *Indian Journal of Pharmaceutical Science* 75 (2013): 442–449.

Sarmento B., Ribeiro A. J., Veiga F., Ferreira D. C., and Neufeld R. J. "Insulin-loaded nanoparticles are prepared by alginate ionotropic pre-gelation followed by chitosan polyelectrolyte complexation." *Journal of Nanoscience and Nanotechnology* 7, no. 8 (2007): 2833–2841.

Sharma Aishwarye, Sharma Shruti, Sharma Kuldeep, Chetri Siva P. K., Vashishtha Amit, Singh Pushpa, Kumar Ravindra, Rathi Brijesh, and Agrawal Veena. "Algae as crucial organisms in advancing nanotechnology: A systematic review." *Journal of Applied Phycology* 28, no. 3 (2016): 1759–1774.

Shin J., Gwak J., Kamarajan P., Fenno J., Rickard A., and Kapila Y. "Biomedical applications of nisin." *Journal of Applied Microbiology* 120 (2016): 1449–1465. https://doi.org/10.1111/jam.13033

Singh R. and Lillard J. W. "Targeted delivery of curcumin to improve therapeutic outcome in breast cancer." *Cancer Research* 74 (2014): 4476.

Sosa-Hernández J. E., Escobedo-Avellaneda Z., Iqbal H. M. N., and Welti-Chanes J. "State-of-the-art extraction methodologies for bioactive compounds from algal biome to meet bio-economy challenges and opportunities." *Molecules* 23, no. 11 (2018 Nov 12): 2953. doi: 10.3390/molecules23112953. PMID: 30424551; PMCID: PMC6278541.

Stengel D. B. and Connan S. "Marine algae: A source of biomass for biotechnological applications." *Methods in Molecular Biology* 1308 (2015): 1–37. doi: 10.1007/978-1-4939-2684-8_1. PMID: 26108496.

Sudha P. N., Gomathi T., Vinodhini P. A., and Nasreen K. "Marine carbohydrates of wastewater treatment." *Advances in Food and Nutrition Research* 73 (2014): 103–143. doi: 10.1016/B978-0-12-800268-1.00007-X. PMID: 25300545.

Suleria H. A., Osborne S., Masci P., and Gobe G. "Marine-based nutraceuticals: An innovative trend in the food and supplement industries." *Marine Drugs* 13, no. 10 (2015): 6336–6351. Published 2015 Oct 14. doi: 10.3390/md13106336

Thakkar K. N., Mhatre S. S., and Parikh, R. Y. "Biological synthesis of metallic nanoparticles." *Nanomedicine* 6 (2010): 257–262. doi: 10.1016/j.nano.2009.07.002

Tu M. H., Sun T., and Grattan K. T. V. "Optimization of gold-nanoparticle-based optical fibre surface plasmon resonance (SPR)-based sensors." *Sensors and Actuators B: Chemical* 164, no. 1 (2012): 43–53. https://doi.org/10.1016/j.snb.2012.01.060.

Tzouvelekis L. S., Miriagou V., Kotsakis S. D., Spyridopoulou K., Athanasiou E., Karagouni E., Tzelepi E., and Daikos G. L. "KPC-producing, multidrug-resistant *Klebsiella pneumoniae* sequence type 258 as a typical opportunistic pathogen." *Antimicrobial Agents Chemother* 57 (2013): 5144–5146.

Umerska A., Corrigan O. I., and Tajber L. "Intermolecular interactions between salmon calcitonin, hyaluronate, and chitosan and their impact on the process of formation and properties of peptide-loaded nanoparticles." *International Journal of Pharmaceutics* 477 (2014): 102–112.

Varshosaz J., Zaki Mohammad Reza, Minaiyan Mohsen, and Banoozadeh Jaafar. "Preparation, optimization, and screening of the effect of processing variables on agar nanospheres loaded with bupropion Hcl by a D-optimal design." *Biomedical Research International* (2015): 571816.

Venkatpurwar V., Shiras A., and Pokharkar V. "Porphyran capped gold nanoparticles as a novel carrier for delivery of anticancer drug: in vitro cytotoxicity study." *International Journal of Pharmaceutics* 409, no. 1–2 (2011): 314–320. doi: 10.1016/j.ijpharm.2011.02.054. Epub 2011 Mar 2. PMID: 21376108.

Wang N. and Wu X. S. "Preparation and characterization of agarose hydrogel nanoparticles for protein and peptide drug delivery." *Pharmaceutical Development and Technology* 2, no. 2 (1997): 135–142.

Wang Y. J., Pan M. H., Cheng A. L., Lin L. I., Ho Y. S., Hsieh C. Y., and Lin J. K. "Stability of curcumin in buffer solutions and characterization of its degradation products." *Journal of Pharmaceutical and Biomedical Analysis* 15, no. 12 (1997): 1867–1876.

Yueming Yu, Wang Bingjie, Guo Chunjing, Zhao Feng, and Chen Daquan "Protoporphyrin IX-loaded laminarin nanoparticles for anticancer treatment, their cellular behaviour, ROS detection, and animal studies." *Nanoscale Research Letters* 14 (2019): 316.

Zhang X. F., Liu Zhi-Guo, Shen Wei, and Gurunathan Sangiliyandi. "Silver nanoparticles: Synthesis, characterization, properties, applications, and therapeutic approaches." *International Journal of Molecular Sciences* (2016): 27649147. doi:10.3390/ijms17091534

Zhao D., Zhuo R. X., and Cheng S. X. "Alginate modified nanostructured calcium carbonate with enhanced delivery efficiency for gene and drug delivery." *Molecular BioSystems* 8, no. 3 (2012): 753–759. doi: 10.1039/c1mb05337j. Epub 2011 Dec 12. PMID: 22159070.

12 Role of Nanotechnology in Diagnosis and Treatment of Diabetes

Eveline M. Anto, Anaga Nair and Jayamurthy Purushothaman

CONTENTS

12.1 INTRODUCTION

Diabetes is a chronic metabolic disease affecting 463 million people globally and is estimated to reach 700 million by 2045 (Saeedi et al., 2019). The current treatment approach focuses on disease management by insulin therapy for type 1 diabetes and oral hypoglycaemic drugs for type 2 diabetes. Diabetic patients need to regularly monitor their blood glucose levels using a glucometer to detect episodes of hypoglycaemia or hyperglycaemia that may otherwise be missed. Current treatment strategies for diabetes are inadequate in terms of the invasive nature of blood glucose monitoring, lack of consistent glucose monitoring when a patient is engaged in some way such as sleeping or driving, multiple insulin injections leading to lower patient compliance, low oral bioavailability and adverse side effects of antihyperglycaemic drugs. Hence, there is a need to develop better diagnostic and treatment strategies for effective disease management to avoid long-term microvascular complications, such as, neuropathy, retinopathy and nephropathy. Nanotechnology-based approaches are gaining momentum in terms of efficient glucose monitoring and better drug delivery compared to conventional methods. This chapter will discuss the various emerging nanotechnology approaches for invasive and noninvasive continuous blood glucose monitoring and nanotechnology-based oral delivery systems for insulin for better management of diabetes (Figure 12.1).

DOI: 10.1201/9781003231745-12

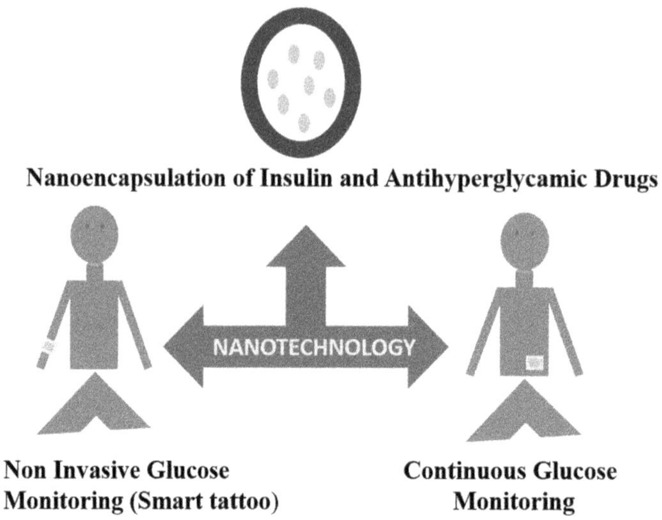

FIGURE 12.1 Role of nanotechnology in the management of diabetes.

12.2 NANOTECHNOLOGY-BASED SENSORS FOR GLUCOSE MONITORING

A sensor can be defined as a device that detects and quantifies a physical, chemical and biological component and converts it into readable data, whereas biosensors specifically detect analytes such as glucose in a biological environment (Kruss et al., 2013; McGrath and Scanaill, 2013). Glucose sensing is the gold standard for monitoring glucose levels in diabetic patients. There are two main approaches for glucose monitoring: electrochemical and fluorescent/optical methods. Both these sensors and their nanotechnology-based improvements will be discussed here.

12.2.1 Electrochemical-Based Enzymatic Sensors

Electrochemical-based glucose biosensors are mainly classified into enzymatic and nonenzymatic categories. Enzymatic glucose sensors have been classified into three generations based on the modes of electron transfer. A brief overview regarding the same is discussed here.

12.2.1.1 First-Generation Sensors

Clark and Lyons (1962) can be deemed as the pioneers of the enzymatic electrochemical glucose sensor. Their design involves the immobilization of glucose oxidase (GOx) enzymes on an oxygen electrode and a subsequent glucose measurement by detecting oxygen consumption. Updike and Hicks (1967) carried forward this work and developed the first generation of enzymatic glucose sensors by developing an oxygen electrode coated by a polymerized gel membrane containing immobilized GOx for the detection of glucose. The reaction involves the conversion of glucose by GO into gluconic acid and hydrogen peroxide (H_2O_2). The oxidation of H_2O_2 then generates a current that is proportional to the amount of glucose present. Conventional glucometers follow the above principle. The first-generation sensors are limited by their oxygen dependence, electroactive interference due to the requirement of a high potential for hydrogen peroxide, reduced activity over time and requirement of frequent recalibration (Veiseh et al., 2015).

12.2.1.2 Second-Generation Sensors

Few of the limitations of the first-generation sensors were overcome by the second generation of glucose sensors. Due to a thick protein layer around the Flavin adenine nucleotide (FAD) redox centre, direct electron transfer from GOx to conventional electrodes is not possible.

Second-generation glucose sensors utilise redox mediators other than oxygen to shuttle electrons from the FAD to the electrode surface. The redox mediators used are ferrocene derivatives, quinones, ferricyanide, transition metal complexes and phenoxazine compounds (Wang, 2008). Electroactive interference is also reduced as measurements can be done at a lower potential. Despite these advantages, second-generation sensors are not that prevalent in *in vivo* glucose sensors due to the toxicity associated with the mediator. Also, interference by electroactive compounds is only minimized but not completely prevented, this may interfere with the reading (Wang, 2008).

12.2.1.3 Third-Generation Sensors

The limitations of the second-generation sensors gave rise to the development of a third generation of sensors that are mediatorless, where electrons from glucose to electrodes are directly transferred via enzymes. Third-generation glucose sensors have the advantage of being more biocompatible, stable and requiring lower operating potential, which decreases the electroactive species interference, but are limited by lower detection over a smaller linear range as compared to first- and second-generation sensors (Si et al., 2013).

12.2.1.4 Nanomaterials Incorporated in Enzymatic Sensors

One of the prerequisites for an efficient glucose sensor is the rapid electron transfer that can be achieved by decreasing the distance between the enzyme and electrode. Nanomaterials are finding applications in sensor development due to their higher surface-area-to-volume ratio and electrocatalysis, which make them ideal for fabrication strategies for improved rates of electron transfer compared to their bulk analogs (Soares et al., 2018). The most used nanomaterials are discussed below.

12.2.1.4.1 Carbon Nanotubes

The widespread applicability of carbon nanotubes (CNTs) is attributed to their high surface-volume ratio, mechanical strength, high chemical stability (which implies no reactivity towards the analyte), no biofouling and high electrical conductivity (Zhu et al., 2012). CNTs are either single-walled or multiwalled cylindrical tubes of allotropes of carbon with a diameter of approximately 1 nm (Barone and Strano, 2009). The hollow tubular structure of CNTs enables the adsorption of enzymes and also rapid electron transfer from immobilized GOx to the electrode surface, improving overall efficiency (Sireesha et al., 2018). In a CNT-based nanosensor, CNT solution is coated on the electrode surface or preheated electrodes are rubbed onto filter paper supporting the nanotubes (Balasubramanian and Burghard, 2006). Recently, Wang et al. (2019) developed a nanocomposite $ZnFe_2O_4$CNT glucose sensor that detects glucose with high sensitivity through the colorimetric approach using tetramethylbenzidine (TMB) substrate.

12.2.1.4.2 Metal Nanoparticles

Another commonly used nanomaterial in enzymatic sensors is a metal nanoparticle (MNP). These include gold, platinum, silver, copper, iron, zinc and alloys. The characteristics that make them suitable candidates for nanosensors are a high surface-to-volume ratio, electrocatalytic efficiency, chemical reactivity, easy synthesis and characterisation. Out of these, gold nanoparticles have wider application due to their relative inertness, electrical conductivity and rapid immobilization through thiols due to a stronger Au-S bond formation (Davis and Higson, 2014).

12.2.1.4.3 Magnetic Nanoparticles

Magnetic nanoparticles are also gaining prominence for use in sensors due to their ease of assembly on electrodes by applying a magnetic field. This makes them relatively economical. Onay et al. (2018) developed a glucose nanosensor based on Fe_3O_4 chitosan gold magnetic nanoparticles immobilized on graphite electrodes, which showcased rapid amperometric response and prolonged storage when tested in blood samples.

Although nanomaterials have improved the overall performance of the enzymatic sensors, the inherent characteristic of the enzyme, such as decrease in GOx activity over time and lack of consistency in enzyme activity, may hamper the efficiency of the nanosensor over time (Veiseh et al., 2015). GOx activity also depends on optimal temperature, pH and humidity and also enzyme immobilization techniques are complicated (Si et al., 2013). Another limitation is the short life of these sensors, which is because of a phenomenon known as biofouling. However, the biofouling-associated tissue damage can be overcome to a large extent by the incorporation of nanomaterials (Vaddiraju et al., 2010).

12.2.2 Electrochemical-Based Nonenzymatic Glucose Sensors

Due to the previously mentioned limitations, researchers worked towards the development of nonenzymatic sensors. They are superior in terms of their direct oxidation ability, longer lifespan and lower fabrication cost (Deshmukh et al., 2020). Activity of nonenzymatic sensors is propelled by a catalyst that initiates the glucose oxidation reaction. Nanoparticles are incorporated in nonenzymatic sensors due to higher surface area-to-volume ratio and increased number of surface atoms, which enhances the catalytic activity (Si et al., 2013). Transition metals such as platinum, palladium, ruthenium, cobalt, gold, manganese, iron, their alloys and oxides are the most commonly used electrocatalysts due to their rapid electron-accepting ability (Si et al., 2013).

12.2.2.1 Nanocomposites

The high cost of noble metals such as gold, platinum and silver enhances the expense of the nonenzymatic sensor. Recent developments in nonenzymatic sensors involve the use of bimetallics or nanocomposites with better efficiency than single metal particles due to their synergistic effect, relatively lower toxicity, better sensitivity and selectivity (Wang et al., 2013). Recently, Vinoth et al. (2020) have synthesized a polyvinylpyrrolidone-conjugated bimetallic platinum-palladium nanosuperlattice-based nonenzymatic glucose sensor having higher sensitivity, resistance to electroactive species, lower cost, higher selectivity, good recovery from biological samples, such as serum and urine, which further contributes to its practical applicability.

12.3 FLUORESCENT GLUCOSE SENSORS

Most of the electrochemical nanosensors are implantable and require invasive monitoring that may limit their shelf life and also may induce toxicity in the host tissue. Although a lot of work has been done to overcome the flaws of these existing sensors, researchers have worked towards developing sensors that involve noninvasive monitoring. One such approach is a fluorescence-based sensor that involves implantation of glucose-specific fluorescent nanosensors into the skin (also known as a "smart tattoo"). Excitation by near-infrared light detects the fluorescence from the surface of skin (Pickup et al., 2008). The basic design of a fluorescent glucose sensor comprises a glucose receptor, a donor fluorophore and an acceptor fluorophore (Klonoff, 2012). A fluorophore is a molecule that becomes excited by absorbing energy at a certain wavelength and comes back to ground state by re-emitting the energy at a different wavelength. In some cases, the energy from an excited fluorophore may be transferred to another acceptor molecule in close proximity, which may then emit lower energy in the form of light — this phenomenon is known as Forster (or fluorescence) energy transfer (FRET) (Kaláb and Soderholm, 2010). In the absence of glucose, the donor fluorophore and acceptor remain in proximity and energy from the donor is transferred to the acceptor due to increased FRET, leading to overall low fluorescence. However, when glucose binds to the receptor, structural changes occur in the sensor that force the donor and acceptor farther apart, and as a result, energy from the donor cannot be passed onto the acceptor, causing increased fluorescence (Klonoff, 2012). One of the advantages of these sensors over enzyme-based sensors is that the reactions taking place are reversible, which prevents by-product formation, thereby increasing the shelf life of the sensor

(Clark et al., 1999). Most of the fluorescent glucose nanosensor technologies are based on FRET and comprise a glucose receptor, a signal transducer and a detector (Chen et al., 2018). Semiconductor quantum dots, carbon nanomaterials (graphene/graphene oxide and CNTs) are fluorescent nanomaterials commonly incorporated in fluorescent nanosensors.

12.3.1 Semiconductor Quantum Dots

The semiconductor quantum dots (QDs) are semiconductor nanocrystals (diameter 2–100 nm) made up of cadmium selenide, cadmium sulphide, indium arsenide, indium phosphide, lead selenide, zinc sulphide, etc., containing around 10,000 atoms (Thomas et al., 2016). Their excellent fluorescent properties are size-influenced and include a broad absorption/excitation spectrum, narrow emission peak that is tunable and resistance to photobleaching, making them ideal candidates for use in sensors (Klonoff, 2012). However, the QDs lack a glucose-recognition domain and hence should be coupled with moieties with glucose recognition sites for incorporation in glucose sensors (Cash and Clark, 2010). The size-dependent change and tunability of absorption/emission properties can be attributed to the QD confinement effect because QDs of varying sizes emit light of different wavelengths (Liu et al., 2005). A recent study by Hu et al. (2018) demonstrated a novel fluorescent glucose sensor with improved sensitivity, low cost, portability, small sample size and less time of detection based on a microfluidic chip comprising GOx entrapped in quantum dot aerogel.

12.3.2 Carbon Nanomaterials

Carbon dots, CNTs, carbon nanodiamonds, graphene QDs and graphene/graphene oxide are the different allotropes of the fluorescent carbon nanomaterial family. CNTs have already been discussed. In addition to electrochemical properties, the unique electronic structure of CNTs also possesses optical properties that enable its incorporation in fluorescent-based glucose sensors. CNTs have a unique electronic structure that contributes to its optical properties such as photophysical properties, excellent fluorescence and photostability (Taylor-Smith, 2021). Due to the electronic band gap between valence and the conduction band, semiconducting single-walled carbon nanotubes (SWCNTs) are fluorescent in the near infrared range (NIR, 900–1600 nm) (O'Connell et al., 2002). Carbon dots are nanoparticles with diameters less than 10 nm and have emerged as potential candidates for sensing applications due to their excellent photoluminescence, remarkable water solubility, photostability, biocompatibility, high quantum yield, low toxicity, economical and simple methods for synthesis and easy surface modification (Sharma and Das, 2019). Graphene and its oxides (graphene oxide and reduced graphene oxide) are gaining importance in sensor design because of their outstanding properties such as high electrical conductivity, optical transparency, fluorescence quenching, large surface area, flexibility and the oxygenated functional group conferred hydrophilicity (Krishnan et al., 2019).The hydrophilic nature facilitates hybrid formation between graphene and other nanoparticles including MNPs and their oxides, nanoclusters, QDs and CNTs, which improve sensor function (Badhulika et al., 2015).

12.4 NANOTECHNOLOGY-BASED APPROACHES FOR ORAL DELIVERY OF INSULIN

The subcutaneous route of insulin administration is painful as it may require multiple injections leading to localized tissue necrosis and infection, subsequently resulting in patient noncompliance—especially in children. Therefore, studies have focused on alternative routes such as the oral route for insulin administration. But insulin, a peptide, cannot be directly orally ingested as it encounters various chemical, enzymatic and absorption barriers in the gut such as acidic environment, enzymes such as peptidases and mucus layers, which will impede its absorption into systemic circulation (Pickup et al., 2008). Encapsulation of insulin with nanocarriers can improve its oral delivery

TABLE 12.1
Various Nanoencapsulation Systems

Nanoencapsulation System	Description	Reference
Nanocapsule	An active component or core trapped in a shell or matrix.	Rambaran, 2020
Nanosphere	Core uniformly dispersed in the shell or matrix.	Jain and Thareja, 2019
Micro/Nanoemulsion	Transparent dispersions consisting of water, oil, surfactant and co-surfactant with oil droplets dispersed in water or vice versa. Microemulsions are more stable thermo-dynamically than nanoemulsions and the former also require a higher surfactant ratio than the latter.	Ramos et al., 2018 Uppal et al., 2018
Liposome	Active component such as drugs are enclosed within a lipid bilayer composed of phospholipids in the form of spherical vesicles.	Ramos et al., 2018
Niosomes	Self-assembly of nonionic surfactants and cholesterol into bilayer vesicles around an aqueous centre. Hydrophilic heads directed toward the aqueous phase and vice-versa for hydrophobic tails, which enables it to deliver both hydrophobic and hydrophilic drugs.	Kazi et al., 2010
Dendrimers	Nanosized, monodisperse structures consisting of symmetrical branched units surrounding a small molecule or linear core.	Abbasi et al., 2014
Micelles	Aggregates of amphiphilic co-polymers that self-assemble on reaching a concentration the same as the critical micelle concentration with an outer hydrophilic layer and an inner hydrophobic centre.	Kesharwani et al., 2018

(Chen et al., 2011). Nanoencapsulation is the process of trapping nanoparticles of an active component (drug, neutraceutical) known as the core within a secondary structure, namely matrix or shell, forming nanocapsules (Rambaran, 2020). In addition to nanocapsules, the encapsulation structures can have various other forms as described in Table 12.1.

Encapsulation by nanocarriers will not only protect drugs like insulin from the gut environment but also aid in its targeted delivery and sustained release (Lopes et al., 2014). The nanocarriers can be categorized as natural or synthetic depending on the origin of the nanomaterial used.

12.4.1 Naturally Derived Nanocarriers for Oral Insulin Delivery

Chitosan, a naturally abundant linear polymer found in the exoskeleton of crustaceans and insects, is made up of β1-4 linked D-glucosamine and N-acetyl D-glucosamine units linked via glycosidic units (Huang et al., 2004). Chitosan's attributes that make it an ideal candidate for insulin delivery are biocompatibility, low toxicity, polycationic behaviour, facile modification strategies for conjugation with appropriate target ligands, biodegradation permeation enhancement through the paracellular pathway and mucoadhesion to the mucosal surface (due to strong ionic interactions between chitosan amino groups which are positively charged and negatively charged epithelial cell membrane), which mediates a reversible opening of tight junctions between epithelial cells (Yeh et al., 2011; Kumar et al., 2016; Wong et al., 2017). One of the recent works on chitosan-coated insulin delivery was reported by Lee et al. (2020) who developed a colon-targeted oral insulin delivery system by preparing a nanocomplex by mixing insulin and amino clay through spontaneous co-assembly followed by double-coating with glycol-chitosan and Eudragit S®100. This double-coated nanocarrier was found to significantly lower initial blood glucose levels in diabetic rats by 38.9% within eight hours. This can be attributed to its pH dependent release, greater than 90% entrapment efficiency and conformational stability against enzyme degradation. Alginate, another naturally available polysaccharide abundant in brown seaweeds namely *Laminaria* sp., *Ascophyllum* sp., *Macrocystis* sp., has been used as a nanocarrier because of its biocompatibility and low toxicity (Ching et al., 2017). The presence of carboxyl groups on alginates renders pH sensitivity and solubility in neutral and alkaline conditions making them preferable for drug delivery (Hariyadi and Islam, 2020). Recently, Zhou et al. (2020) have developed a bioadhesive nanoparticle system with a glucose-responsive switch based on amphipathic 2-nitoimidazole-L-cysteine-alginate (NI-CYS-ALG) conjugates for oral delivery of insulin. Dextran, a water soluble exocellular bacterial polysaccharide comprising branched linear α-D (1-6) glucopyranose units, is a remarkable nanocarrier due to its biodegradability, biocompatibility, hydrophilicity and presence of hydroxyl groups that enable it to undergo many chemical modifications (Hu and Luo, 2018). Jamwal et al. (2019) demonstrated a new glucose-responsive oral insulin delivery system comprising GOx immobilized on acryloyl crosslinked dextran dialdehyde (ACDD) nanoparticles.

12.4.2 Synthetic Polymer-Based Nanocarriers for Oral Delivery of Insulin

Synthetic polymers include polylactides (PLA), polyglycolides (PGA), poly-ϵ-Caprolactone (PCA), polyvinyl alcohol (PVA) and polylactide co-glycolide (PLGA), which is FDA-approved for drug encapsulation studies (Mansoor et al., 2019). Malathi et al. (2015) synthesized a novel PLGA nanoparticle-based oral insulin delivery system consisting of d-α-tocopherol poly(ethylene glycol) 1000 succinate (TPGS)-emulsified poly(ethylene glycol) (PEG)-capped poly(lactic-co-glycolic acid) (PLGA) nanoparticles. The insulin-loaded nanocarrier was prepared by a water-oil-water emulsion solvent evaporation method, which was reported to have an encapsulation efficiency of around 78%. This was found to lower serum glucose levels and had regenerative effects on damaged livers, pancreas and kidneys in streptozotocin-induced diabetic rats. PVA is a biodegradable, biocompatible, nontoxic, thermally stable polymer with high mechanical strength, able to blend with natural polymers and has an ease of preparation (Mansoor et al., 2019). Li et al. (2021) have recently synthesized a biocompatible, pH sensitive semi-interpenetrating polymer poly(vinyl alcohol)/poly (hydroxy propyl methacrylate-*co*-methacrylic acid) (PVA/P(HPMA-*co*-MAA)-based hydrogel loaded with insulin for its oral delivery. Oral administration of this nanocarrier hydrogel in streptozotocin-induced rats was found to significantly lower blood glucose levels.

12.5 CONCLUSION

Nanotechnology-based therapeutic interventions for the management of diabetes has evolved exponentially. The ability of nanotechnology-based approaches in invasive or noninvasive glucose monitoring as well as strategies for effective oral delivery of insulin by nanocarriers have all contributed towards better patient compliance and enhanced efficacy in disease management. But despite the beneficial attributes of nanoparticles, uncertainty looms around their potential toxicity. This aspect needs to be resolved before bringing nanotechnology-based therapies into the clinical setup. This benchmark is still a distant reality that needs to be achieved. Nanotechnology is still in the nascent stage and it needs to be explored deeply for the development of better strategies for disease treatment and management.

REFERENCES

Abbasi, Elham, Sedigheh Fekri Aval, Abolfazl Akbarzadeh, Morteza Milani, Hamid Tayefi Nasrabadi, Sang Woo Joo, Younes Hanifehpour, Kazem Nejati-Koshki, and Roghiyeh Pashaei-Asl. "Dendrimers: Synthesis, applications, and properties." *Nanoscale Research Letters* 9, no. 1 (2014): 1–10.

Badhulika, Sushmee, Trupti Terse-Thakoor, Claudia Maria Chaves Villarreal, and Ashok Mulchandani. "Graphene hybrids: Synthesis strategies and applications in sensors and sensitized solar cells." *Frontiers in Chemistry* 3 (2015): 38.

Balasubramanian, Kannan, and Marko Burghard. "Biosensors based on carbon nanotubes." *Analytical and Bioanalytical Chemistry* 385, no. 3 (2006): 452–468.

Barone, Paul W., and Michael S. Strano. "Single walled carbon nanotubes as reporters for the optical detection of glucose." *Journal of Diabetes Science and Technology* 3, no. 2 (2009): 242–252.

Cash, Kevin J., and Heather A. Clark. "Nanosensors and nanomaterials for monitoring glucose in diabetes." *Trends in Molecular Medicine* 16, no. 12 (2010): 584–593.

Chen, Longyi, Eugene Hwang, and Jin Zhang. "Fluorescent nanobiosensors for sensing glucose." *Sensors* 18, no. 5 (2018): 1440.

Chen, Mei-Chin, Kiran Sonaje, Ko-Jie Chen, and Hsing-Wen Sung. "A review of the prospects for polymeric nanoparticle platforms in oral insulin delivery." *Biomaterials* 32, no. 36 (2011): 9826–9838.

Ching, Su Hung, Nidhi Bansal, and Bhesh Bhandari. "Alginate gel particles–A review of production techniques and physical properties." *Critical Reviews in Food Science and Nutrition* 57, no. 6 (2017): 1133–1152.

Clark, Haley R., Timothy A. Barbari, and Govind Rao. "Modeling the response time of an in vivo glucose affinity sensor." *Biotechnology Progress* 15, no. 2 (1999): 259–266.

Clark Jr, Leland C., and Champ Lyons. "Electrode systems for continuous monitoring in cardiovascular surgery." *Annals of the New York Academy of Sciences* 102, no. 1 (1962): 29–45.

Davis, F., and S. P. J. Higson. "Electrochemical nanosensors for blood glucose analysis." In *Nanosensors for Chemical and Biological Applications*, pp. 28–53. Woodhead Publishing, Sawston, United Kingdom, 2014.

Deshmukh, Megha A., Byeong-Cheol Kang, and Tae-Jun Ha. "Non-enzymatic electrochemical glucose sensors based on polyaniline/reduced-graphene-oxide nanocomposites functionalized with silver nanoparticles." *Journal of Materials Chemistry C* 8, no. 15 (2020): 5112–5123.

Hariyadi, Dewi Melani, and Nazrul Islam. "Current status of alginate in drug delivery." *Advances in Pharmacological and Pharmaceutical Sciences* 2020 (2020).

Huang, Min, Eugene Khor, and Lee-Yong Lim. "Uptake and cytotoxicity of chitosan molecules and nanoparticles: Effects of molecular weight and degree of deacetylation." *Pharmaceutical Research* 21, no. 2 (2004): 344–353.

Hu, Qiaobin, and Yangchao Luo. "Recent advances of polysaccharide-based nanoparticles for oral insulin delivery." *International Journal of Biological Macromolecules* 120 (2018): 775–782.

Hu, Tao, Yi Ye, Ke Chen, Feifei Long, Wen Sang, Yunlong Zhou, Dongke Sun, and Zhonghua Ni. "Visual detection of glucose based on quantum dots aerogel in microfluidic chips." *Analytical Methods* 10, no. 48 (2018): 5749–5754.

Jain, Akhlesh K., and Suresh Thareja. "In vitro and in vivo characterization of pharmaceutical nanocarriers used for drug delivery." *Artificial Cells, Nanomedicine, and Biotechnology* 47, no. 1 (2019): 524–539.

Jamwal, Shivani, Bhagat Ram, Sunita Ranote, Rohini Dharela, and Ghanshyam S. Chauhan. "New glucose oxidase-immobilized stimuli-responsive dextran nanoparticles for insulin delivery." *International Journal of Biological Macromolecules* 123 (2019): 968–978.

Kaláb, Petr, and Jon Soderholm. "The design of Förster (fluorescence) resonance energy transfer (FRET)-based molecular sensors for Ran GTPase." *Methods* 51, no. 2 (2010): 220–232.

Kazi, Karim Masud, Asim Sattwa Mandal, Nikhil Biswas, Arijit Guha, Sugata Chatterjee, Mamata Behera, and Ketousetuo Kuotsu. "Niosome: A future of targeted drug delivery systems." *Journal of Advanced Pharmaceutical Technology & Research* 1, no. 4 (2010): 374.

Kesharwani, Prashant, Bapi Gorain, Siew Yeng Low, Siew Ann Tan, Emily Chai Siaw Ling, Yin Khai Lim, Chuan Ming Chin et al. "Nanotechnology based approaches for anti-diabetic drugs delivery." *Diabetes Research and Clinical Practice* 136 (2018): 52–77.

Klonoff, David C. "Overview of fluorescence glucose sensing: A technology with a bright future." *Journal of Diabetes Science and Technology* 6, no. 6 (2012): 1242–1250.

Krishnan, Siva Kumar, Eric Singh, Pragya Singh, Meyya Meyyappan, and Hari Singh Nalwa. "A review on graphene-based nanocomposites for electrochemical and fluorescent biosensors." *RSC Advances* 9, no. 16 (2019): 8778–8881.

Kruss, Sebastian, Andrew J. Hilmer, Jingqing Zhang, Nigel F. Reuel, Bin Mu, and Michael S. Strano. "Carbon nanotubes as optical biomedical sensors." *Advanced Drug Delivery Reviews* 65, no. 15 (2013): 1933–1950.

Kumar, Ashwini, Archana Vimal, and Awanish Kumar. "Why Chitosan? From properties to perspective of mucosal drug delivery." *International Journal of Biological Macromolecules* 91 (2016): 615–622.

Lee, Sang Hoon, Seung-Yun Back, Jae Geun Song, and Hyo-Kyung Han. "Enhanced oral delivery of insulin via the colon-targeted nanocomposite system of organoclay/glycol chitosan/Eudragit® S100." *Journal of Nanobiotechnology* 18, no. 1 (2020): 1–10.

Li, Shunying, Tingting Qin, Tingting Chen, Jun Wang, and Qingbing Zeng. "Poly (vinyl alcohol)/poly (hydroxypropyl methacrylate-co-methacrylic acid) as pH-sensitive semi-IPN hydrogels for oral insulin delivery: Preparation and characterization." *Iranian Polymer Journal* 30, no. 4 (2021): 343–353.

Liu, Tiancai, Bisen Liu, Haili Zhang, and Ying Wang. "The fluorescence bioassay platforms on quantum dots nanoparticles." *Journal of Fluorescence* 15, no. 5 (2005): 729–733.

Lopes, Marlene, Barbara A. Abrahim, Raquel Seiça, Francisco Veiga, Carlos R. Rodrigues, and Antonio J. Ribeiro. "Intestinal uptake of insulin nanoparticles: Facts or myths?" *Current Pharmaceutical Biotechnology* 15, no. 7 (2014): 629–638.

Malathi, Sampath, Perumal Nandhakumar, Velayudham Pandiyan, Thomas J. Webster, and Sengottuvelan Balasubramanian. "Novel PLGA-based nanoparticles for the oral delivery of insulin." *International Journal of Nanomedicine* 10 (2015): 2207.

Mansoor, Shazia, Pierre P. D. Kondiah, Yahya E. Choonara, and Viness Pillay. "Polymer-based nanoparticle strategies for insulin delivery." *Polymers* 11, no. 9 (2019): 1380.

McGrath, Michael J., and Cliodhna Ní Scanaill. "Sensing and sensor fundamentals." In *Sensor Technologies*, pp. 15–50. Apress, Berkeley, CA, 2013.

O'Connell, Michael J., Sergei M. Bachilo, Chad B. Huffman, Valerie C. Moore, Michael S. Strano, Erik H. Haroz, Kristy L. Rialon et al. "Band gap fluorescence from individual single-walled carbon nanotubes." *Science* 297, no. 5581 (2002): 593–596.

Onay, Aykut, Üzeyir Dogan, Hakan Ciftci, Demet Cetin, Zekiye Suludere, and Ugur Tamer. "Amperometric glucose sensor based on the glucose oxidase enzyme immobilized on graphite rod electrode modified with Fe_3O_4-CS-Au magnetic nanoparticles." *Ionics* 24, no. 12 (2018): 4015–4022.

Pickup, John C., Zheng-Liang Zhi, Faaizah Khan, Tania Saxl, and David J. S. Birch. "Nanomedicine and its potential in diabetes research and practice." *Diabetes/Metabolism Research and Reviews* 24, no. 8 (2008): 604–610.

Rambaran, Theresa F. "Nano polyphenols: A review of their encapsulation and anti-diabetic effects." *SN Applied Sciences* 2, no. 8 (2020): 1–26.

Ramos, Matheus Aparecido Dos Santos, Patrícia Bento Da Silva, Larissa Spósito, Luciani Gaspar De Toledo, Bruna Vidal Bonifacio, Camila Fernanda Rodero, Karen Cristina Dos Santos, Marlus Chorilli, and Tais Maria Bauab. "Nanotechnology-based drug delivery systems for control of microbial biofilms: a review." *International Journal of Nanomedicine* 13 (2018): 1179.

Saeedi, Pouya, Inga Petersohn, Paraskevi Salpea, Belma Malanda, Suvi Karuranga, Nigel Unwin, Stephen Colagiuri et al. "Global and regional diabetes prevalence estimates for 2019 and projections for 2030 and 2045: Results from the international diabetes federation diabetes atlas." *Diabetes Research and Clinical Practice* 157 (2019): 107843.

Sharma, Anirudh, and Joydeep Das. "Small molecules derived carbon dots: Synthesis and applications in sensing, catalysis, imaging, and biomedicine." *Journal of Nanobiotechnology* 17, no. 1 (2019): 1–24.

Si, Peng, Youju Huang, Taihong Wang, and Jianmin Ma. "Nanomaterials for electrochemical non-enzymatic glucose biosensors." *RSC Advances* 3, no. 11 (2013): 3487–3502.

Sireesha, Merum, Veluru Jagadeesh Babu, A. Sandeep Kranthi Kiran, and Seeram Ramakrishna. "A review on carbon nanotubes in biosensor devices and their applications in medicine." *Nanocomposites* 4, no. 2 (2018): 36–57.

Soares, Sara, João Sousa, Alberto Pais, and Carla Vitorino. "Nanomedicine: Principles, properties, and regulatory issues." *Frontiers in Chemistry* 6 (2018): 360.

Taylor-Smith, Kerry. "An Introduction to Carbon Nanotube Optics." AZoM. 7 August 2019. https://www.azom.com/article.aspx?ArticleID=18320 (Accessed June 20, 2021).

Thomas, Andreas, Lutz Heinemann, Araceli Ramírez, and Alfred Zehe. "Options for the development of noninvasive glucose monitoring: Is nanotechnology an option to break the boundaries?" *Journal of Diabetes Science and Technology* 10, no. 3 (2016): 782–789.

Updike, Stuart J., and George P. Hicks. "The enzyme electrode." *Nature* 214, no. 5092 (1967): 986–988.

Uppal, Siddharth, Kishan S. Italiya, Deepak Chitkara, and Anupama Mittal. "Nanoparticulate-based drug delivery systems for small molecule anti-diabetic drugs: An emerging paradigm for effective therapy." *Acta Biomaterialia* 81 (2018): 20–42.

Vaddiraju, Santhisagar, Ioannis Tomazos, Diane J. Burgess, Faquir C. Jain, and Fotios Papadimitrakopoulos. "Emerging synergy between nanotechnology and implantable biosensors: a review." *Biosensors and Bioelectronics* 25, no. 7 (2010): 1553–1565.

Veiseh, Omid, Benjamin C. Tang, Kathryn A. Whitehead, Daniel G. Anderson, and Robert Langer. "Managing diabetes with nanomedicine: Challenges and opportunities." *Nature Reviews Drug Discovery* 14, no. 1 (2015): 45–57.

Vinoth, Victor, Nalenthiran Pugazhenthiran, Ramalinga Viswanathan Mangalaraja, Asad Syed, Najat Marraiki, Héctor Valdés, and Sambandam Anandan. "Development of an electrochemical enzyme-free glucose sensor based on self-assembled Pt–Pd bimetallic nanosuperlattices." *Analyst* 145, no. 24 (2020): 7898–7906.

Wang, Joseph. "Electrochemical glucose biosensors." *Chemical Reviews* 108, no. 2 (2008): 814–825.

Wang, Guangfeng, Xiuping He, Lingling Wang, Aixia Gu, Yan Huang, Bin Fang, Baoyou Geng, and Xiaojun Zhang. "Non-enzymatic electrochemical sensing of glucose." *Microchimica Acta* 180, no. 3–4 (2013): 161–186.

Wang, Chengke, Jiangyu Li, Rong Tan, Qingqing Wang, and Zexiang Zhang. "Colorimetric method for glucose detection with enhanced signal intensity using $ZnFe_2O_4$–carbon nanotube–glucose oxidase composite material." *Analyst* 144, no. 5 (2019): 1831–1839.

Wong, Chun Y., Hani Al-Salami, and Crispin R. Dass. "Potential of insulin nanoparticle formulations for oral delivery and diabetes treatment." *Journal of Controlled Release* 264 (2017): 247–275.

Yeh, Tzyy-Harn, Li-Wen Hsu, Michael T. Tseng, Pei-Ling Lee, Kiran Sonjae, Yi-Cheng Ho, and Hsing-Wen Sung. "Mechanism and consequence of chitosan-mediated reversible epithelial tight junction opening." *Biomaterials* 32, no. 26 (2011): 6164–6173.

Zhou, Xia, Hongwei Wu, Ruimin Long, Shibin Wang, Haiwang Huang, Yanhua Xia, Pei Wang et al. "Oral delivery of insulin with intelligent glucose-responsive switch for blood glucose regulation." *Journal of Nanobiotechnology* 18, no. 1 (2020): 1–16.

Zhu, Zhigang, Luis Garcia-Gancedo, Andrew J. Flewitt, Huaqing Xie, Francis Moussy, and William I. Milne. "A critical review of glucose biosensors based on carbon nanomaterials: Carbon nanotubes and graphene." *Sensors* 12, no. 5 (2012): 5996–6022.

13 Nanophytochemicals in Cancer Therapy
Update on Preclinical Evidence

Madhumitha Kedhari Sundaram, Asha Inbanathan and Arif Hussain

CONTENTS

13.1 INTRODUCTION

Notwithstanding the significant achievements in cancer therapeutics, 9 million people succumbed to the devastating disease in 2018 (Bray *et al.*, 2018). It is well documented that altered genetic and epigenetic processes can trigger and sustain carcinogenesis via various hallmarks (Hanahan and Weinberg, 2011; Sundaram, Silas and Hussain, 2021b).

Phytochemicals offer a great opportunity to design effective cancer therapeutics to counter the rising cancer epidemic. Phytochemicals can be used to target the redox state or signalling pathways via epigenetic alterations and bring about anticancer effects (Sundaram, Unni, *et al.*, 2019; Sundaram, Khan, *et al.*, 2020; Sundaram, Silas and Hussain, 2020a).

Phytochemicals have some inherent limitations due to their solubility and availability. These limitations can be overcome by formulating phytochemicals into nanoparticle drug delivery systems such as nanoparticles, micelles and liposomes. Novel drug delivery systems increase solubility and absorption of phytochemicals and therefore permit the use of lower dosages in a more targeted delivery (Yadav, Ghune, and Jain, 2011).

This chapter catalogues *in vitro* and *in vivo* studies in which nanocarriers have been used to deliver phytochemicals to cancer cells, as nanophytochemicals, to assess their anticarcinogenic effect.

13.2 ANTICARCINOGENIC EFFECTS OF VARIOUS PHYTOCHEMICAL-NANOPARTICLE FORMULATIONS ON DIFFERENT CANCERS

Genistein is a soy iso-flavonoid that occurs in fava beans and soy products (Huang *et al.*, 2019). Genistein is an effective anticancer agent that can modulate various signalling pathways and epigenetic pathways that regulate proliferation, cell cycle, apoptosis and migration as demonstrated in cervical cancer (Sundaram *et al.*, 2018; Sundaram, Unni *et al.*, 2019). Genistein nanoparticles were seen to exert antiproliferative, proapoptotic and antimigratory effects in various cancer cells such as ovarian, breast, leukaemia, lung and prostate cancer cells (Mittal *et al.*, 2019; Sacko, Thangavel and Shoyele, 2019; Tian *et al.*, 2019; Goorbandi, Mohammadi and Malekzadeh, 2020; Kamel *et al.*,

DOI: 10.1201/9781003231745-13

2019; Shukla *et al.*, 2020). These effects were associated with molecular changes including down-regulation of Akt/HIF1α/VEGF signalling, MCL-1, DNMT3B, VEGF, MMP2 and MMP-9. Genistein nanoparticles also induced apoptosis and autophagy via modulating endogenous antioxidant enzyme levels and increased H_2O_2 production in colon cancer cells (Pool *et al.*, 2018).

Resveratrol is a polyphenol that occurs in large amounts in grapes, peanuts and berries (Amri *et al.*, 2012). Resveratrol nanoparticles were seen to exert antiproliferative, proapoptotic and anti-migratory effects in various cancer cells such as breast, lung, prostate, colon, leukaemia and liver cancer cells (Anwar *et al.*, 2018; El-Far *et al.*, 2018; Nassir *et al.*, 2018; Singh, Lillard and Singh, 2018; Al-Attar and Madihally, 2019; Aldawsari *et al.*, 2020; Balan *et al.*, 2020; Senthil Kumar *et al.*, 2020). These effects were associated with molecular changes including modulation of aromatase, VEGF, NF- κB, ABCB1, ABCC1, ABCG2, apoptotic, cell cycle and inflammation-related genes.

Hesperetin is a flavonoid found mainly in citrus fruits (Erlund, 2004). Hesperetin was seen to have anticancer effects on colon cancer, breast cancer, cervical cancer and prostate cancer (Aranganathan and Nalini, 2013). Hesperetin nanoparticles were seen to exert antiproliferative, pro-apoptotic and antimigratory effects in colon cancer and leukaemia cells as well as in cancer mice models (Krishnan *et al.*, 2017; Mary Lazer *et al.*, 2018; Ersoz *et al.*, 2019; Sheokand, Navik and Bansal, 2019; Trendafilova *et al.*, 2020). These effects were associated with molecular changes including modulation of NF- κB, TNF-α, apoptotic, inflammation and oxidative stress regulatory genes.

Curcumin, is a bioactive polyphenolic compound obtained from *Curcuma longa* (Ngwabebhoh, Erdagi and Yildiz, 2018). Curcumin nanoparticles were seen to exert antiproliferative, proapoptotic and antimigratory effects in various cancer cells such as cervical breast, lung, prostate, colorectal, colon, leukaemia and liver cancer cells (Bagheri, Sanaat and Zarghami, 2018; Justin *et al.*, 2018; Kumari *et al.*, 2018; Ngwabebhoh, Erdagi and Yildiz, 2018; Sauraj *et al.*, 2018; Zhang *et al.*, 2018; Hong *et al.*, 2019; Chen *et al.*, 2020; Hasanpoor *et al.*, 2020). A similar effect was seen in mice models as well (Yan *et al.*, 2016; Li *et al.*, 2018; Seok *et al.*, 2018; Xie *et al.*, 2018). These effects could be attributed to the modulation of hTERT metastasis and apoptotic regulators and signalling pathways like PI3K/Akt and NF-κB. These studies are listed in Table 13.1.

TABLE 13.1
Anticarcinogenic Effects of Phytochemical-Nanoparticle Formulations on Different Cancers (*In Vitro* and *In Vivo*)

Cancer	Nanoparticle	Effect	Reference
		Genistein	
Ovarian	Lipid	Antiproliferative	(Mittal *et al.*, 2019)
Breast	Spermine + lipo-polymeric + heparan sulphate	Apoptosis	(Shukla *et al.*, 2020)
	Spermine + lipo-polymeric + heparan sulphate + doxorubicin	Apoptosis, antiangiogenic, antiproliferative, anti-invasive, antimigratory *In vivo*: Akt/Hif1α/VEGF modulation	
Lung	Lipid-protein shell + retinoic acid	Enhanced uptake, Anti-proliferative	(Kamel *et al.*, 2020)
	Lipid-protein shell + mannitol/ hydroxypropyl β-cyclodextrin/leucine	Antiproliferative *In vivo*: higher effect	
	Tretinoin-lipid–protein	Antiproliferative *In vivo*: apoptosis, modulated VEGF, survivin, caspase-3	
	miRNA-29b + mucin-1 aptamer	Modulated pAKT, p-PI3K, DNMT3B, MCL 1, Antiproliferative, apoptosis	(Sacko, Thangavel, and Shoyele, 2019)

(Continued)

TABLE 13.1 *(Continued)*

Cancer	Nanoparticle	Effect	Reference
Prostate	Liposome	Antiproliferative, antimigratory, modulated Glut-1, COX2, thioredoxin reductase, glucose transport	(Tian *et al.*, 2019)
	Liposome + celecoxib	Apoptosis, antimigratory, ROS, modulated GSH, Glut-1, peroxiredoxin-6 expression, COX2, thioredoxin reductase, caspase-3, glucose transport	
Colon	Polyethylene glycol silica	Apoptosis, autophagy, modulated antioxidant enzymes, H_2O_2	(Pool *et al.*, 2018)
Leukaemia	Ferric oxide carboxymethylated chitosan	Antiproliferative, apoptosis	(Goorbandi, Mohammadi, and Malekzadeh, 2020)
		Resveratrol	
Breast Ehrlich ascites	Monascus yellow pigments	Antiproliferative *In vivo:* tumour regression, modulation of aromatase, VEGF, NF-κB	(El-Far *et al.*, 2018)
Lung	Poly(lactic-co-glycolic acid) Chitosan + poly(lactic-co-glycolic acid)	Scavenges DPPH, apoptosis, modulated Bcl2, Bax, cell cycle, DNA damage	(Aldawsari *et al.*, 2020)
Prostate	Novel planetary ball + folic acid	Apoptosis In resistant cells: targeted FRα, apoptosis, modulated NF-kB p65, COX-2, BCL-2, BCL-XL, BAX, BAK, caspase-3, ABCB1, ABCG2, ABCC1, ABCG2, inhibited ATP pump efflux	(Singh, Lillard and Singh, 2018)
	Docetaxel + folic acid + PCL + PEG	Apoptosis In resistant cells: apoptosis, modulated ABCB1, ABCG2, ABCC1	
	Polymeric NP	Apoptosis, cell cycle arrest, ROS	(Nassir *et al.*, 2018)
Colon	Folic acid+ lipid	Antiproliferative, cell cycle arrest	(Senthil Kumar *et al.*, 2020)
	Ferulic acid + chitosan + folic acid + lipid	Apoptosis, cell cycle arrest, modulated cyclin D1 cyclin E, CDK4, CDK2, CDK-6, Bcl-2, p53, caspases 3 and 9	
Leukaemia	Transferrin + liposomes + siRNA	Antiproliferative, apoptosis	(Al-Attar and Madihally, 2019)
Melanoma	Chitosan	Antiproliferative, ROS, apoptosis and necrosis	(Balan *et al.*, 2020)
Liver	Lactobionic/folate + maltodextrin + micelles + sulfasalazine	Antiproliferative *In vivo:* inhibition of angiogenesis, apoptosis	(Anwar *et al.*, 2018)
		Hesperetin	
Breast	Nanocrystalline solid dispersions	*In vivo:* tumour regression; decreased alanine transaminase, aminotransferase and alkaline phosphatase	(Sheokand, Navik, and Bansal, 2019)
Colon	Chitosan Chitosan + folate	Apoptosis, modulated Bax, Bad	(Mary Lazer *et al.*, 2018)
Leukaemia	Magnesium- or silver + silica + morin	Antiproliferative	(Trendafilova *et al.*, 2020)
Brain	Poly(lactic-co-glycolic acid)	Antiproliferative; apoptosis, oxidative damage	(Ersoz *et al.*, 2019)
Liver	Polyethylene glycol + gold	*In vivo*: antiproliferative, anti-inflammatory, modulated NF-κB and glycoconjugate	(Krishnan *et al.*, 2017)

(Continued)

TABLE 13.1 *(Continued)*
Anticarcinogenic Effects of Phytochemical-Nanoparticle Formulations on Different Cancers (*In Vitro* and *In Vivo*)

Cancer	Nanoparticle	Effect	Reference
		Curcumin	
Cervical	Paramagnetic + iron oxide	Apoptosis, modulated Caspase 3	(Justin *et al.*, 2018)
	Polyethylene glycol	Anti-proliferative	(Zhang *et al.*, 2018)
	Methotrexate	*In vivo:* reduction in tumour growth	(Xie *et al.*, 2018)
Breast	Mesoporous silica + folic acid or hyaluronan	Inhibition of tumour growth	(Li *et al.*, 2018)
	Serum albumin + PDL1 peptide	Antiproliferative, down-regulation of PDL1	(Hasanpoor *et al.*, 2020)
	Coumarin Pickering emulsions	Antiproliferative	(Ngwabebhoh, Erdagi, and Yildiz, 2018)
	Zinc oxide	Antiproliferative	(Kundu *et al.*, 2019)
	Gold	Antiproliferative	(Malekmohammadi *et al.*, 2018)
	Chitosan + halloysite + gold	Antiproliferative, showed pH-controlled drug delivery	(Rao *et al.*, 2018)
	Doxorubicin	Increase in lysosomal pH, inhibited P-gp	(Motevalli *et al.*, 2019)
	Mesoporous silica + hyaluronan or polyethyleneimine-folic acid	*In vivo:* inhibited tumour growth, specifically targets tumour site, enhanced drug delivery	(Li *et al.*, 2018)
	Zinc oxide - phenylboronic acid	*In vivo:* antiproliferative	(Kundu *et al.*, 2019)
Lung	Chitosan	Increased solubility, efficacy, selective cytotoxicity	(Duse *et al.*, 2018)
	U11-doxorubicin	Antitumour effect *In vivo:* tumour inhibition, decreased side effects	(Hong *et al.*, 2019)
Prostate	Poly(lactic-co-glycolic acid)	Decrease in cell viability, apoptosis	(Azandeh *et al.*, 2017)
	Gold	Decrease in cell viability	(Nambiar *et al.*, 2018)
	Cabazitaxel	Decrease in cell viability, modulation of COX-2, p53, NF-κB, phospho-Akt, PI3K, RTK	(Chen *et al.*, 2020)
	Docataxel	*In vivo:* decrease in cell viability, no obvious side effects	(Yan *et al.*, 2016)
	Cabazitaxel	Decrease in cell viability	(Chen *et al.*, 2020)
Colon	Chrysin + poly(lactic-co-glycolic acid)	Antiproliferative, apoptosis, down-regulation of telomerase	(Bagheri, Sanaat, and Zarghami, 2018)
	Selenium	Modulation of NFκB, metastasis-associated proteins, autophagy-associated proteins Bcl-2, Bax, cyt c *In vivo:* decreased tumour burden, increased the survival time	(Kumari *et al.*, 2018)
	Xylan	Efficient drug delivery, apoptosis, redox state	(Sauraj *et al.*, 2018)
	Hyaluronic acid + zein	Cyto- and hemo-compatibility, antiproliferative, apoptosis *In vivo:* reduced tumour volume and weight without immunotoxicity	(Seok *et al.*, 2018)

13.3 SUMMATION AND RECOMMENDATIONS

These studies show that specific nanoformulations ensure consistent release and high tissue accumulation, and with the addition of specific ligands, targeted delivery via receptor-mediated endocytosis can lead to an increased anticarcinogenic response (Goorbandi, Mohammadi and Malekzadeh, 2020). It is seen that these formulations are more capable of eliciting a strong antiproliferative response as well as cell-cycle arrest. Further, they can initiate apoptosis through the activation of the caspase cascade as well as by modulating the expression of apoptotic regulatory genes like BCL2, BCL-xl, Bim, Bax and Bad (Senthil Kumar *et al.*, 2020). Several nanophytochemical formulations were able to modulate the signalling pathways such as PI3K and Wnt. This is an important facet of this therapeutic as these pathways govern numerous hallmark features. Various nanoformulations were able to retain and enhance the effect on inflammation and reactive oxygen species (ROS) generation by suppressing the expression of COX-2 and modulating the NF-kB pathway amongst other molecular mechanisms (Pool *et al.*, 2018; Tian *et al.*, 2019; Balan *et al.*, 2020). Lowering inflammation and modifying the redox environment in cancer cells are important steps in mediating a strong anticancer effect. Redox changes are often linked with the onset of apoptosis and further limitation of proliferation (Sundaram, Khan *et al.*, 2020).

Several nanophytochemicals also demonstrated the ability to inhibit the expression of VEGF, a key entity in angiogenesis, and can be expected to therefore hinder this particular hallmark (El-Far *et al.*, 2018; Kamel *et al.*, 2019; Shukla *et al.*, 2020). This is an important hallmark to be altered as metastasis has been earmarked as the major reason for cancer-related mortality. Change in energy metabolism has been identified as an important enabler in carcinogenesis (Hanahan and Weinberg, 2011).

Epigenetic alterations are key to initiate a wider network of gene expression changes. Previous studies have showcased the role of phytochemicals as epigenetic modulators (Sundaram *et al.*, 2017, 2018; Sundaram, Hussain *et al.*, 2019; Sundaram, Preetha *et al.*, 2021). Some studies have also recorded epigenetic alterations, such as expression and activity of DNA methyltransferases (DNMT) and histone modifiers, by these nanoparticles (Al-Attar and Madihally, 2019; Sacko, Thangavel and Shoyele, 2019).

Several studies have showcased a synergistic increase in the effect of a nanoformulation that has both a phytochemical as well as a known chemotherapeutic agent such as doxorubicin (Singh, Lillard and Singh, 2018). Additionally, nanoformulations have also shown promise in mitigating drug resistance (Singh, Lillard and Singh, 2018). This further substantiates their use and is a direction that needs to be thoroughly studied. Some nanoformulations were found to give encouraging results when tested on various cancers. It is recommended that such experiments be taken up for more formulations to better study the efficacy of these agents against a wider range of target cancers.

Several studies have looked at the effects of their formulations both in *in vitro* cell lines as well as in *in vivo* animals (Anwar *et al.*, 2018; El-Far *et al.*, 2018). These studies also helped demonstrate that the issues of poor solubility, tolerance and altered pharmacokinetics did not occur. The beneficial role of nanophytoformulations as theorized in terms of availability, low dose requirement and targeted delivery also appear to be adequately met by examining results of *in vivo* animal studies. Further, work in expanding the number and scope of *in vivo* studies is recommended. Several studies have evaluated the effect on cell lines but have not elaborated on the molecular basis of these observations. The molecular changes mediated by the nanophytochemicals reviewed here are depicted in Figure 13.1. Further clinical studies can then be more reliably conducted to test the anticancer effects of nanophytochemicals *in vivo*. The various advantages and benefits of phytochemical-based nanoparticle formulations as evidenced in *in vitro* and *in vivo* animal studies are depicted in Figure 13.2.

It is interesting to note that Ayurveda, one of the traditional medicine divisions of India, has often utilized metal compounds such as copper, gold, silver, zinc and lead in nanoscale, often referred to as *bhasmam,* in formulations intended to treat different illnesses (Pal, Sahu and Haldar, 2014). It has been noted that the size of these particles in the Ayurvedic formulations is in the range of 56–57 nm for gold and up to 25–50 nm size range for mercuric oxide (Tambur, Chakra and Rao,

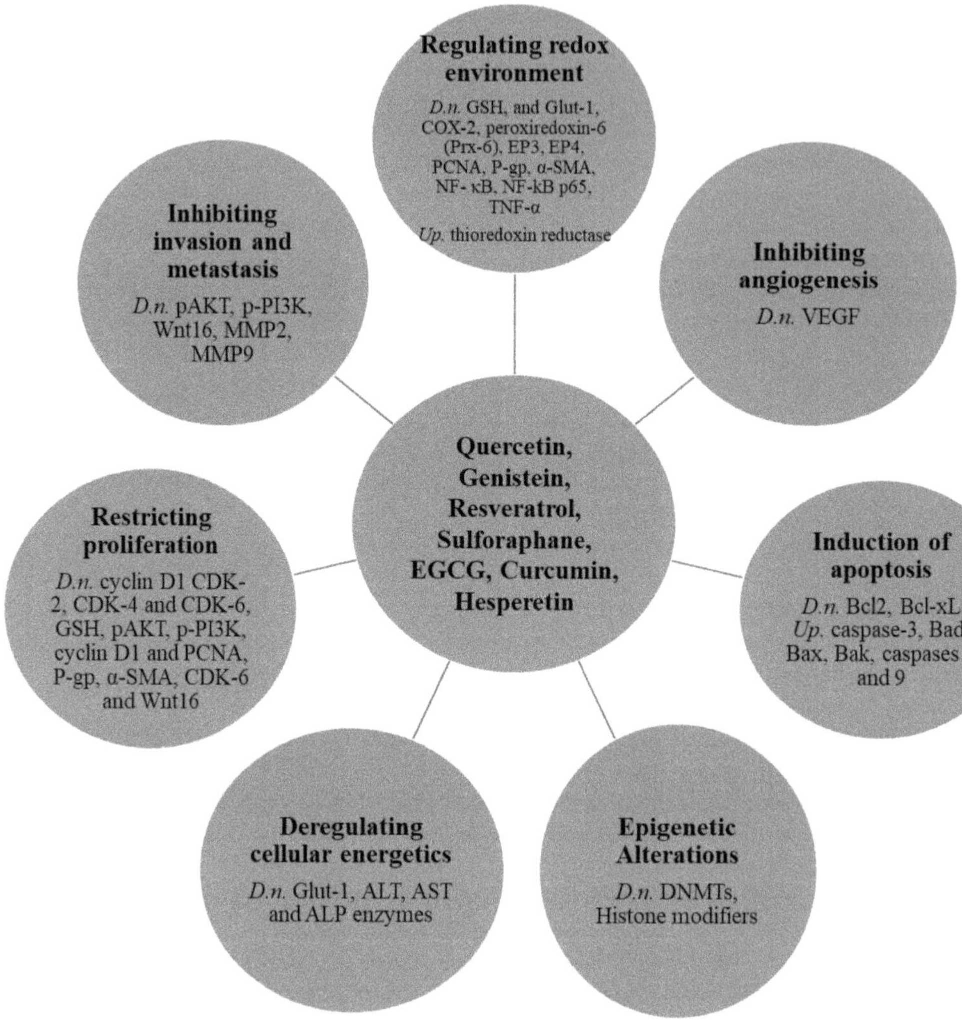

FIGURE 13.1 Effect of phytochemical-based nanoparticle formulations on the various features of cancer as evidenced in *in vitro* and *in vivo* animal studies. *Abbreviations:* Dn, down-regulation; up, up-regulation.

2013). It is indeed worth exploring these traditional formulations to help in the preparation of effective therapeutic moieties.

13.4 CONCLUSION

Phytochemical-loaded nanoparticles provide a superior anticarcinogenic response and are capable of targeting various hallmarks of cancer. The augmented synergistic efficacy can be attributed to the increase in the solubility, bioavailability and tissue-specific targeting. Further, mechanistic studies to increase the understanding and refinement of these nanophytochemicals will pave the way for advancement in cancer therapeutics.

ACKNOWLEDGEMENTS

The authors are grateful to Dr. Jason Fitzsimmons, Academic President, Manipal Academy of Higher Education, Dubai, UAE for his constant support and encouragement. The authors are appreciative and thankful of Ms. Anuradha Lakshminarayanan for her assistance in the gathering of literature.

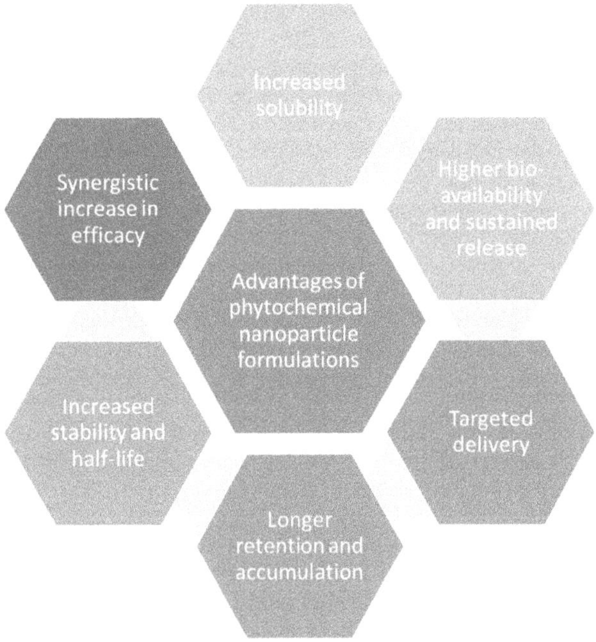

FIGURE 13.2 Advantages and benefits of phytochemical-based nanoparticle formulations as evidenced in *in vitro* and *in vivo* animal studies.

REFERENCES

Al-Attar, Thikrayat, and Sundararajan V. Madihally. "Targeted cancer treatment using a combination of siRNA-liposomes and resveratrol-electrospun fibers in co-cultures." *International Journal of Pharmaceutics* 569 (2019): 118599.

Aldawsari, Hibah M., Nabil A. Alhakamy, Rayees Padder, Mohammad Husain, and Shadab Md. "Preparation and characterization of chitosan coated plga nanoparticles of resveratrol: Improved stability, antioxidant and apoptotic activities in H1299 lung cancer cells." *Coatings* 10, no. 5 (2020): 439.

Amri, A., J. C. Chaumeil, S. Sfar, and C. Charrueau. "Administration of resveratrol: What formulation solutions to bioavailability limitations?" *Journal of Controlled Release* 158, no. 2 (2012): 182–193.

Anwar, Doaa M., Sherine N. Khattab, Maged W. Helmy, Mohamed K. Kamal, Adnan A. Bekhit, Kadria A. Elkhodairy, and Ahmed O. Elzoghby. "Lactobionic/folate dual-targeted amphiphilic maltodextrin-based micelles for targeted codelivery of sulfasalazine and resveratrol to hepatocellular carcinoma." *Bioconjugate Chemistry* 29, no. 9 (2018): 3026–3041.

Aranganathan, Selvaraj, and Namasivayam Nalini. "Retracted: Antiproliferative efficacy of hesperetin (citrus flavanoid) in 1, 2 dimethylhydrazine induced colon cancer." *Phytotherapy Research* 27, no. 7 (2013): 999–1005.

Azandeh, S. S. et al. "Anticancer activity of curcumin-loaded PLGA nanoparticles on PC3 prostate cancer cells." *Iranian Journal of Pharmaceutical Research* 16, no. 3 (2017): 868.

Bagheri, Raana, Zohreh Sanaat, and Nosratollah Zarghami. "Synergistic effect of free and nano-encapsulated chrysin-curcumin on inhibition of hTERT gene expression in SW480 colorectal cancer cell line." *Drug Research* 68, no. 06 (2018): 335–343.

Balan, Poornima, Janani Indrakumar, Padmaja Murali, and Purna Sai Korrapati. "Bi-faceted delivery of phytochemicals through chitosan nanoparticles impregnated nanofibers for cancer therapeutics." *International Journal of Biological Macromolecules* 142 (2020): 201–211.

Bray, Freddie, Jacques Ferlay, Isabelle Soerjomataram, Rebecca L. Siegel, Lindsey A. Torre, and Ahmedin Jemal. "Global cancer statistics 2018: GLOBOCAN estimates of incidence and mortality worldwide for 36 cancers in 185 countries." *CA: A Cancer Journal for Clinicians* 68, no. 6 (2018): 394–424.

Chen, Yougan, Yuanyuan Deng, Chenyao Zhu, and Congming Xiang. "Anti prostate cancer therapy: Aptamer-functionalized, curcumin and cabazitaxel co-delivered, tumor targeted lipid-polymer hybrid nanoparticles." *Biomedicine & Pharmacotherapy* 127 (2020): 110181.

Duse, L. et al. "Preparation and characterization of curcumin loaded chitosan nanoparticles for photodynamic therapy." *Physica Status Solidi (A)* 215, no. 15 (2018): 1700709.

El-Far, Shaymaa W., Maged W. Helmy, Sherine N. Khattab, Adnan A. Bekhit, Ahmed A. Hussein, and Ahmed O. Elzoghby. "Phytosomal bilayer-enveloped casein micelles for codelivery of monascus yellow pigments and resveratrol to breast cancer." *Nanomedicine* 13, no. 5 (2018): 481–499.

Erlund, Iris. "Review of the flavonoids quercetin, hesperetin, and naringenin. Dietary sources, bioactivities, bioavailability, and epidemiology." *Nutrition Research* 24, no. 10 (2004): 851–874.

Ersoz, Melike, Aysegul Erdemir, Dilek Duranoglu, Deniz Uzunoglu, Tulin Arasoglu, Serap Derman, and Banu Mansuroglu. "Comparative evaluation of hesperetin loaded nanoparticles for anticancer activity against C6 glioma cancer cells." *Artificial Cells, Nanomedicine, and Biotechnology* 47, no. 1 (2019): 319–329.

Goorbandi, Ghasemi, R. Mohammad Reza Mohammadi, and Kianoosh Malekzadeh. "Synthesizing efficacious genistein in conjugation with superparamagnetic Fe3O4 decorated with bio-compatible carboxymethylated chitosan against acute leukemia lymphoma." *Biomaterials Research* 24 (2020): 9–9.

Hanahan, Douglas, and Robert A. Weinberg. "Hallmarks of cancer: the next generation." *Cell* 144, no. 5 (2011): 646–674.

Hasanpoor, Zahra, Ali Mostafaie, Iraj Nikokar, and Zuhair Mohammad Hassan. "Curcumin-human serum albumin nanoparticles decorated with PDL1 binding peptide for targeting PDL1-expressing breast cancer cells." *International Journal of Biological Macromolecules* 159 (2020): 137–153.

Hong, Yuan, Shaomin Che, Beina Hui, Yunyi Yang, Xiaoli Wang, Xiaozhi Zhang, Yongqian Qiang, and Hailin Ma. "Lung cancer therapy using doxorubicin and curcumin combination: targeted prodrug based, pH sensitive nanomedicine." *Biomedicine & Pharmacotherapy* 112 (2019): 108614.

Huang, Jian, Zhaoyang Qian, Yuhua Gong, Yanzhou Wang, Yanfang Guan, Yingxin Han, Xin Yi *et al.* "Comprehensive genomic variation profiling of cervical intraepithelial neoplasia and cervical cancer identifies potential targets for cervical cancer early warning." *Journal of Medical Genetics* 56, no. 3 (2019): 186–194.

Justin, C., Antony V. Samrot, Chamarthy Sai Sahithya, Karanam Sai Bhavya, and C. Saipriya. "Preparation, characterization and utilization of coreshell super paramagnetic iron oxide nanoparticles for curcumin delivery." *PLoS One* 13, no. 7 (2018): e0200440.

Kamel, Nayra M., Maged W. Helmy, Elsayeda-Zeinab Abdelfattah, Sherine N. Khattab, Doaa Ragab, Magda W. Samaha, Jia-You Fang, and Ahmed O. Elzoghby. "Inhalable dual-targeted hybrid lipid nanocore–protein shell composites for combined delivery of genistein and all-trans retinoic acid to lung cancer cells." *ACS Biomaterials Science & Engineering* 6, no. 1 (2020): 71–87.

Kamel, Nayra M., Maged W. Helmy, Magda W. Samaha, Doaa Ragab, and Ahmed O. Elzoghby. "Multicompartmental lipid–protein nanohybrids for combined tretinoin/herbal lung cancer therapy." *Nanomedicine* 14, no. 18 (2019): 2461–2479.

Krishnan, Gokuladhas, Jayakumar Subramaniyan, Pramila Chengalvarayan Subramani, Barath Muralidharan, and Devaki Thiruvengadam. "Hesperetin conjugated PEGylated gold nanoparticles exploring the potential role in anti-inflammation and anti-proliferation during diethylnitrosamine-induced hepatocarcinogenesis in rats." *Asian Journal of Pharmaceutical Sciences* 12, no. 5 (2017): 442–455.

Kumari, Manisha, Mahaveer Prasad Purohit, Satyakam Patnaik, Yogeshwer Shukla, Pradeep Kumar, and Kailash Chand Gupta. "Curcumin loaded selenium nanoparticles synergize the anticancer potential of doxorubicin contained in self-assembled, cell receptor targeted nanoparticles." *European Journal of Pharmaceutics and Biopharmaceutics* 130 (2018): 185–199.

Kundu, M., Sadhukhan, P., Ghosh, N., Chatterjee, S., Manna, P., Das, J., & Sil, P. C. (2019). pH-responsive and targeted delivery of curcumin via phenylboronic acid-functionalized ZnO nanoparticles for breast cancer therapy. *Journal of advanced research*, 18, 161–172.

Li, Nana, Zhi Wang, Yongtai Zhang, Kai Zhang, Jianxu Xie, Ying Liu, Wansi Li, and Nianping Feng. "Curcumin-loaded redox-responsive mesoporous silica nanoparticles for targeted breast cancer therapy." *Artificial Cells, Nanomedicine, and Biotechnology* 46, no. sup2 (2018): 921–935.

Malekmohammadi, S. et al. "Immobilization of gold nanoparticles on folate-conjugated dendritic mesoporous silica-coated reduced graphene oxide nanosheets: a new nanoplatform for curcumin pH-controlled and targeted delivery." *Soft Matter. Royal Society of Chemistry* 14, no. 12 (2018): 2400–2410.

Mary Lazer, Lizha, Balaji Sadhasivam, Kanagaraj Palaniyandi, Thangavel Muthuswamy, Ilangovan Ramachandran, Anandan Balakrishnan, Surajit Pathak, Shoba Narayan, and Satish Ramalingam. "Chitosan-based nanoformulation enhances the anticancer efficacy of hesperetin." *International Journal of Biological Macromolecules* 107 (2018): 1988–1998.

Mittal, Pooja, Harsh Vrdhan, Gufran Ajmal, Gunjan Bonde, Ramit Kapoor, and Brahmeshwar Mishra. "Formulation and characterization of genistein-loaded nanostructured lipid carriers: Pharmacokinetic, biodistribution and in vitro cytotoxicity studies." *Current Drug Delivery* 16, no. 3 (2019): 215–225.

Motevalli, S. M. et al. "Co-encapsulation of curcumin and doxorubicin in albumin nanoparticles blocks the adaptive treatment tolerance of cancer cells." *Biophysics Reports* 5, no. 1 (2019): 19–30.

Nassir, Anmar M., Naiyer Shahzad, Ibrahim AA Ibrahim, Iqbal Ahmad, Shadab Md, and Mohammad R. Ain. "Resveratrol-loaded PLGA nanoparticles mediated programmed cell death in prostate cancer cells." *Saudi Pharmaceutical Journal* 26, no. 6 (2018): 876–885.

Nambiar, S. et al. "Synthesis of curcumin-functionalized gold nanoparticles and cytotoxicity studies in human prostate cancer cell line." *Applied Nanoscience* 8, no. 3 (2018): 347–357.

Ngwabebhoh, Fahanwi Asabuwa, Sevinc Ilkar Erdagi, and Ufuk Yildiz. "Pickering emulsions stabilized nano cellulosic-based nanoparticles for coumarin and curcumin nanoencapsulations: In vitro release, anticancer and antimicrobial activities." *Carbohydrate Polymers* 201 (2018): 317–328.

Pal, Dilipkumar, Chandan Kumar Sahu, and Arindam Haldar. "Bhasma: the ancient Indian nanomedicine." *Journal of Advanced Pharmaceutical Technology & Research* 5, no. 1 (2014): 4.

Pool, Héctor, Rocio Campos-Vega, María Guadalupe Herrera-Hernández, Pablo García-Solis, Teresa García-Gasca, Isaac Cornelius Sánchez, Gabriel Luna-Bárcenas, and Haydé Vergara-Castañeda. "Development of genistein-PEGylated silica hybrid nanomaterials with enhanced antioxidant and antiproliferative properties on HT29 human colon cancer cells." *American Journal of Translational research* 10, no. 8 (2018): 2306.

Rao, K. M. et al. "pH and near-infrared active; chitosan-coated halloysite nanotubes loaded with curcumin-Au hybrid nanoparticles for cancer drug delivery." *International Journal of Biological Macromolecules* 112 (2018): 119–125.

Sacko, Koita, Karthik Thangavel, and Sunday A. Shoyele. "Codelivery of genistein and miRNA-29b to A549 cells using aptamer-hybrid nanoparticle bioconjugates." *Nanomaterials* 9, no. 7 (2019): 1052.

Sauraj, Uday Kumar, Vinay Kumar, Ruchir Priyadarshi, P. Gopinath, and Yuvraj Singh Negi. "pH-responsive prodrug nanoparticles based on xylan-curcumin conjugate for the efficient delivery of curcumin in cancer therapy." *Carbohydrate Polymers* 188 (2018): 252–259.

Senthil Kumar, Chinnaiyan, Ramar Thangam, Soloman Agnes Mary, Perumal Ramesh Kannan, Gopinath Arun, and Balaraman Madhan. "Targeted delivery and apoptosis induction of trans-resveratrol-ferulic acid loaded chitosan coated folic acid conjugate solid lipid nanoparticles in colon cancer cells." *Carbohydrate Polymers* 231 (2020): 115682.

Seok, Hae-Yong, N. Sanoj Rejinold, Kamali Manickavasagam Lekshmi, Kondareddy Cherukula, In-Kyu Park, and Yeu-Chun Kim. "CD44 targeting biocompatible and biodegradable hyaluronic acid cross-linked zein nanogels for curcumin delivery to cancer cells: In vitro and in vivo evaluation." *Journal of Controlled Release* 280 (2018): 20–30.

Sheokand, Sneha, Umashanker Navik, and Arvind K. Bansal. "Nanocrystalline solid dispersions (NSD) of hesperetin (HRN) for prevention of 7, 12-dimethylbenz [a] anthracene (DMBA)-induced breast cancer in Sprague-Dawley (SD) rats." *European Journal of Pharmaceutical Sciences* 128 (2019): 240–249.

Shukla, Ravi Prakash, Jayant Dewangan, Sandeep Urandur, Venkatesh Teja Banala, Monika Diwedi, Shweta Sharma, Sristi Agrawal, Srikanta Kumar Rath, Ritu Trivedi, and Prabhat Ranjan Mishra. "Multifunctional hybrid nano constructs facilitate intracellular localization of doxorubicin and genistein to enhance apoptotic and anti-angiogenic efficacy in breast adenocarcinoma." *Biomaterials Science* 8, no. 5 (2020): 1298–1315.

Singh, Santosh Kumar, James W. Lillard Jr, and Rajesh Singh. "Reversal of drug resistance by planetary ball milled (PBM) nanoparticle loaded with resveratrol and docetaxel in prostate cancer." *Cancer Letters* 427 (2018): 49–62.

Sundaram, Madhumitha Kedhari, Arif Hussain, Shafiul Haque, Ritu Raina, and Nazia Afroze. "Quercetin modifies 5′ CpG promoter methylation and reactivates various tumor suppressor genes by modulating epigenetic marks in human cervical cancer cells." *Journal of Cellular Biochemistry* 120, no. 10 (2019): 18357–18369.

Sundaram, Madhumitha Kedhari, Mohammad Zeeshan Ansari, Abdullah Al Mutery, Maryam Ashraf, Reem Nasab, Sheethal Rai, Naushad Rais, and Arif Hussain. "Genistein induces alterations of epigenetic modulatory signatures in human cervical cancer cells." *Anti-cancer Agents in Medicinal Chemistry (Formerly Current Medicinal Chemistry—anti-cancer Agents)* 18, no. 3 (2017): 412–421.

Sundaram, M. K., M. A. Khan, U. Alalami, P. Somvanshi, T. Bhardwaj, S. Pramodh, R. Raina, Z. Shekfeh, S. Haque, and A. Hussain. "Phytochemicals induce apoptosis by modulation of nitric oxide signaling pathway in cervical cancer cells." *European Review for Medical and Pharmacological Sciences* 24 (2020): 11827–11844.

Sundaram, Madhumitha Kedhari, R. Preetha, Shafiul Haque, Naseem Akhter, Saif Khan, Saheem Ahmed, and Arif Hussain. "Dietary isothiocyanates inhibit cancer progression by modulation of epigenome." In *Seminars in Cancer Biology*. Academic Press, 2021.

Sundaram, Madhumitha Kedhari, Ritu Raina, Nazia Afroze, Apurva Dhupkar, Naval Preet Kaur, Aditya Arte, Firdos Alam Khan, and Arif Hussain. "Combinational use of phytochemicals and chemotherapeutic drugs enhance their therapeutic potential on human cervical cancer cells." *International Journal of Cancer Management* 12, no. 6 (2019).

Sundaram, Madhumitha Kedhari, Shefina Silas, and Arif Hussain. "Combinational therapy using chemotherapeutic agents and dietary bioactive compounds: A pragmatic approach to cancer treatment." In Shefina Silas, and Arif Hussain (eds.), *Treating Endocrine and Metabolic Disorders with Herbal Medicines*, pp. 188–214. IGI Global, 2021a.

Sundaram, Madhumitha Kedhari, Shefina Silas, and Arif Hussain. "Epigenetic aberrations in cervical cancer." In Trygve Tollefsbol (ed.), *Epigenetics and Reproductive Health*, pp. 343–370. Academic Press, Elsevier Inc., 2021b.

Sundaram, Madhumitha Kedhari, Sreepoorna Unni, Pallavi Somvanshi, Tulika Bhardwaj, Raju K. Mandal, Arif Hussain, and Shafiul Haque. "Genistein modulates signaling pathways and targets several epigenetic markers in HeLa cells." *Genes* 10, no. 12 (2019): 955.

Tambur, Pavani, Ch Shilpa Chakra, and K. Venkateswara Rao. "A green approach for the synthesis of nano-sized iron oxide, by Indian ayurvedic modified bhasmikaran method." *American Journal of Pharmaceutical Sciences* 1 (2013): 1–7.

Tian, Jingyan, Fengjun Guo, Yingying Chen, Yanqing Li, Bingbing Yu, and Yang Li. "Nanoliposomal formulation encapsulating celecoxib and genistein inhibiting COX-2 pathway and Glut-1 receptors to prevent prostate cancer cell proliferation." *Cancer Letters* 448 (2019): 1–10.

Trendafilova, Ivalina, Judith Mihály, Denitsa Momekova, Ralitsa Chimshirova, Hristina Lazarova, Georgi Momekov, and Margarita Popova. "Antioxidant activity and modified release profiles of morin and hesperetin flavonoids loaded in Mg- or Ag-modified SBA-16 carriers." *Materials Today Communications* 24 (2020): 101198.

Xie, Jiajiang, Zhongxiong Fan, Yang Li, Yinying Zhang, Fei Yu, Guanghao Su, Liya Xie, and Zhenqing Hou. "Design of pH-sensitive methotrexate prodrug-targeted curcumin nanoparticles for efficient dual-drug delivery and combination cancer therapy." *International Journal of Nanomedicine* 13 (2018): 1381.

Yadav, A., M. Ghune, and D. K. Jain. "Nano-medicine based drug delivery system." Journal of Advanced Pharmacy Education & Research 1, no. 4 (2011): 201–213.

Yan, Jieke, Yuzhen Wang, Xufeng Zhang, Shuangde Liu, Chuan Tian, and Hongwei Wang. "Targeted nanomedicine for prostate cancer therapy: docetaxel and curcumin co-encapsulated lipid–polymer hybrid nanoparticles for the enhanced anti-tumor activity in vitro and in vivo." *Drug Delivery* 23, no. 5 (2016): 1757–1762.

Zhang, Huabing, Yubin Zhang, Yilin Chen, Ying Zhang, Yange Wang, Yinying Zhang, Liang Song et al. "Glutathione-responsive self-delivery nanoparticles assembled by curcumin dimer for enhanced intracellular drug delivery." *International Journal of Pharmaceutics* 549, no. 1–2 (2018): 230–238.

14 Bionanomaterials and Their Recent Advancements in Tissue Engineering Applications

Ponnusamy Arulpriya, Thangavel Krishnaveni, Krishnasamy Lakshmi and Krishna Kadirvelu

CONTENTS

14.1 INTRODUCTION

Human organs and tissues can often be damaged through long-term diseases (cardiovascular, diabetes, cancer, diseases of the liver or kidney, etc.) or an accidental incident (Yoshito, 2006), but no chemicals can instantly regenerate human tissue. When tissue dies (called necrosis), it cannot be brought back to life. It will affect the surrounding tissues, bones, organs and even skin if it is not removed or repaired. It cannot be completely cured by surgery or by organ transplantation. Surgical treatments may take time to mend and can cause other difficulties or challenges. In the case of organ transplantation, it is difficult to get a healthy organ and the number of donors is small.

Joining cells with natural or synthetic scaffold material in appropriate culture conditions leads to the formation of engineered tissue. Figure 14.1 clearly illustrates nanotechnology-based nanomedicine synthesis and the incorporation of drugs, nanoparticles (NPs) or bionanomaterials into the biological scaffold. The scaffold-based tissue engineering approach depends on the ability of the body to regenerate new tissues. These scaffolds can be synthesized by developing an artificial microenvironment derived from synthetic or natural materials or by removal of the cellular system from tissues. These materials degrade after implantation and will be readily replaced via extracellular matrix (ECM) proteins formed with the in-growing cells. In addition, cells can be used for treatment through injection either alone or with carriers, like hydrogels.

14.2 BIOPOLYMERS

Biopolymers are derived from natural sources such as plants and microbes. The polymeric biomolecules are larger in size (Mishra *et al.*, 2018b; Zinoviadou *et al.*, 2016). Normally "bio" implies that naturally degradable materials are obtained from living micro-organisms (Mishra *et al.*, 2018b). Biopolymers are derived from natural resources like sucrose, vegetable oils, resins, fats, amino acids, proteins, etc. (Mishra *et al.*, 2018a). Biopolymers are classified into three groups: (i) proteins (amino acids), (ii) polysaccharides (sugars) and (iii) nucleic acids (nucleotides). These are known for being biocompatible, bioplastic, biodetergent, biosurfactant, bioadhesive and bioflocculant (Tang

DOI: 10.1201/9781003231745-14

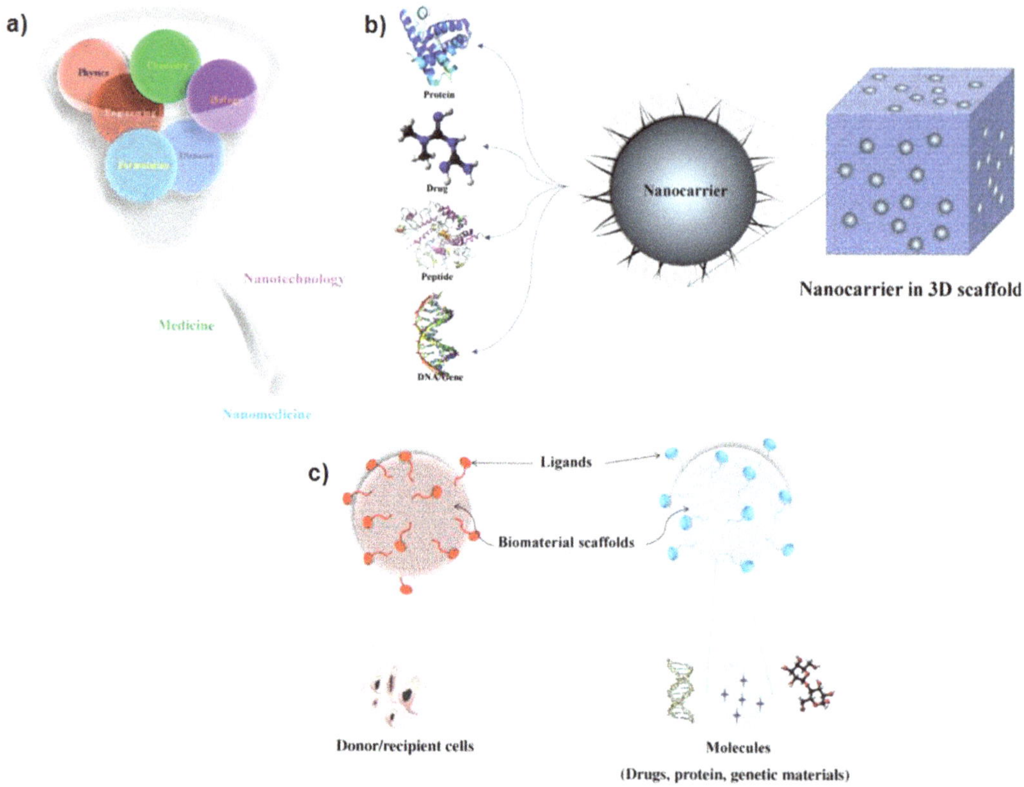

FIGURE 14.1 Schematic representation of tissue engineering fabrication. (a) Nanotechnology-based nanomedicine approach, (b) Incorporation of bionanomaterial into scaffold, (c) Bionanomaterial for tissue engineering approach. (Adapted from Sudhakar *et al.,* 2015 with permission from Elsevier 2021, license number 5053400109907.)

et al., 2012; Mekonnen *et al.*, 2013; Mishra *et al.*, 2018a). Biopolymers have different roles, such as drug release modifiers, matrix formers and gelling agents and it can be easily adjusted to fit an imperfection in a complex physiological geometry (Gopinathan and Noh 2021). Figure 14.2 illustrates the modification of biomaterial morphology through several procedures like surface graft polymerization, competitive protein binding, charges such as anionic or cationic binding, hydrophobic or hydrophilic interaction, surface free energy, roughness and degradation. These modifications help to enhance the tissue engineering regeneration process.

These polymers generally have favourable extracellular matrix properties such as collagen, alginate, chitosan, fibrin, gelatin, agarose, xanthan, heparin, cellulose and dextran etc. (Rossana *et al.*, 2018). These have been widely used for tissue engineering applications and regenerative medicine. Surface modification of biopolymers through chemical or physical processes like oxidation, quaternization and hydrolysis changes the surface property and functionality of biomaterials, which helps to enhance adsorption and cell adhesion. The incorporation of bioactive molecules like drugs or antibodies improves further surface characteristics such as immobilization, site recognition, affinities and mobility (Lee *et al.*, 1998; Lee *et al.*, 2003). It also enhances cellular interactions; integration of bioactive molecules within biomaterial scaffolds is a method to improve the result of cell-based tissue engineering.

The incorporation of metal or metallic NPs improves strength. For example, cobalt-chrome alloys, stainless steel and titanium alloys are extensively used for their high corrosion resistance in many industrial applications. Zinc oxide and titanium oxide possess catalytic properties, specifically

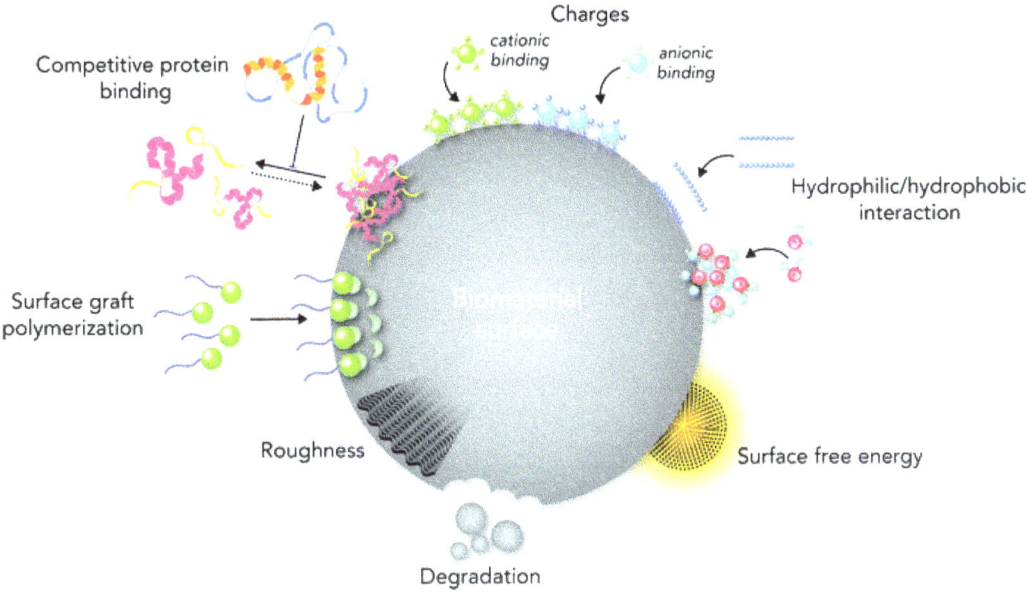

FIGURE 14.2 Schematic representation of face modification procedures of a biomaterial surface. (Adapted from Rahmati *et al.,* 2020; this article comes under RSC Creative Commons CC-BY license.)

photocatalysis, for their characteristic nature of superior band gap (Klee *et al.*, 2000). Noble metals, like gold (Au), silver (Ag) and platinum (Pt) are attractive candidates for use in dentistry due to their ductility and castability. The Au/Ag NPs are used in biomedical applications to improve antimicrobial properties and electrical conductivity. The incorporation of AuNPs in ECM revealed cellularity and inflammation for skeletal tissue engineering applications. The syntheses of biomaterial from various natural sources are given in Figure 14.3.

Chitosan, known as chitin, is polycationic. The foremost characteristics of chitosan are biodegradability and biocompatibility (Edmondson *et al.*, 2005) and it can be processed in different structural forms like hydrogels, scaffolds, films and fibres. It exhibits antimicrobial, anti-inflammatory, drug-delivery and wound-healing activities. In recent decades, chitosan has been used in tissue engineering as well as regenerative treatments as it has a natural ability to excite tissue regeneration, scaffolds, cartilage, chelation and bone regeneration and film-forming properties. Normally, chitosan alone revealed biological activity and its activity was enhanced with the incorporation of either silver or gold NPs compared to others. Researchers are exploring the potential of AuNPs and AgNPs as phototherapeutic agents in areas such as gene therapy, diagnosis and treatment of cancer, vectorization of drugs, and production of sterilized materials, artificial skin, functional contraceptive devices, bone prostheses, antibacterial surgical instruments, surface cleaners, bone coating, antimicrobial paints, food storage, automotive upholstery, etc. (Dallas *et al.*, 2011; Kumar *et al.*, 2018). AgNP incorporation with chitosan promotes antimicrobial properties, electrostatic attraction and forms stable complexes. AuNP incorporation with chitosan revealed excellent biocompatibility and improved proliferation of human fibroblasts. This incorporation of NPs altered the surface morphology size and property, which can help to improve the impact on cellular responses (Hsu *et al.*, 2011; España-Sánchez *et al.*, 2018).

Bioceramics possess high biocompatibility, strength, wear resistance and density, e.g., alumina and zirconia. The main function of alumina and zirconia are knee and hip joint replacements and for tissue engineering applications. Titanium dioxide (TiO_2) scaffold has the capability to improve the growth of vascular tissues and bone, extensive conductivity, antimicrobial activities and to promote bone regeneration. Calcium phosphates are also an ideal alternative as they can be used in the treatment of bone regeneration.

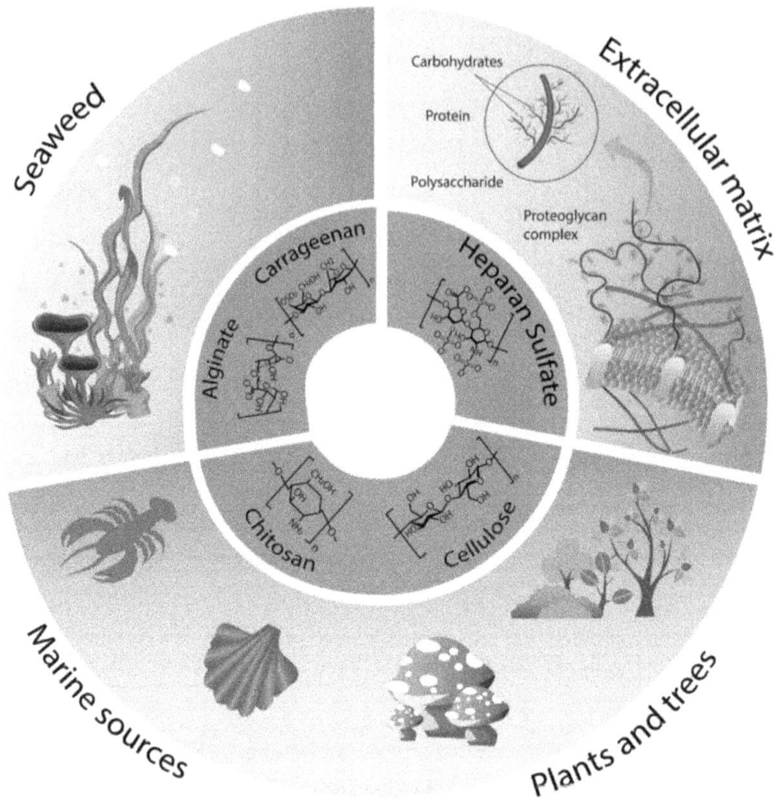

FIGURE 14.3 Natural biopolymers from different sources. (Adapted from Dinoro *et al.*, 2019 with permission from Elsevier 2021, license number 5053060910308.)

Cellulose-based nanomaterials, owing to their huge availability, structural and mechanical properties, capability to self-assemble in network structures, low cytotoxicity and biocompatibility (Abdul Khalil *et al.*, 2015; Roman, 2015; Shatkin *et al.*, 2015; Moon *et al.*, 2016), are candidates for biomedical applications, especially tissue engineering, healing (bones, skin) and medical implants (Lin *et al.*, 2014). Cellulose can be classified into three categories: (i) bacterial cellulose, (ii) cellulose nanofibrils and (iii) cellulose nanocrystals (Lin *et al.*, 2014). Bacterial-based cellulose materials are repeatedly used for vascular establishment (Schumann *et al.*, 2009). Bacterial cellulose reveals remarkable strength with the capability to be engineered at micro-, nano- and macroscales (Klemm *et al.*, 2001). It is also used for the combination of nanomaterials for biomedical applications.

Agarose is known as a carbohydrate polymer with tunable structural and mechanical properties, which promote its use in applications such as cartilage and bone tissue scaffolds. In wound healing, for example, its bond to the skin can be customized with reference to the injured area. In cell development, the gel inflexibility can be customized by the exploitation of the cell surface structure. It is also an appropriate intermediate for cell-encapsulation because of its elevated water uptake ability, spherical cell morphology and preservation of a poroelastic structure. Agarose, owing to its resemblance to ECM, self-gelation, nonimmunological features, swelling behaviour, low cost and appropriate mechanical properties, is a capable material for morphogenesis (Zarrintaj *et al.*, 2018).

Collagen is most widely used as a biopolymer for tissue engineering applications and as an ECM. Consequently, its intrinsic structural and biochemical properties improve tissue growth. The porous collagen-based scaffolds blend with a ceramic compound like hydroxyl apatite, promoting bone repair. Collagen-based biomaterials might also be exploited to produce systems for the limited delivery of beneficial agents like drugs, genes and growth factors (Sheehy *et al.*, 2018).

Lignin biopolymers can be used in a lot of fields due to their binding, dispersing, complexion and emulsion-stabilizing properties. Due to the valorization nature of lignin, lignin-based materials have wide applications in biomedicine. Lignin's morphological and structural distinctiveness make use of stem cell-based approaches in tissue regenerative medicine (Markus *et al.*, 2018).

Bionanomaterials for cardiac, urological and skin tissue-engineering applications will be discussed in Section 14.3.

14.3 CARDIOVASCULAR

Cardiovascular disorder (CVD) is one of the most dangerous diseases around the globe. CVDs such as partial or full blockage, scar formation, ischaemia, heart failure, myocardial infarction, recurrent ischaemia, cardiomyocytes, arrhythmias and coronary artery disease are not easily treated and need high-end treatments. Currently available treatments are expensive and lead to some side effects in humans, hence alternative sources are needed.

The heart is a multifaceted organ system encompassing multilayered tissues, with every layer of tissue enclosing a multiplicity of cell categories and ECM. Therefore, the process of choosing an appropriate graft site is frequently dependent on the explicit function of affected tissue and its difficulty based on the cardiac surroundings. Natural biomaterials used to treat the heart are more biocompatible and biodegradable when compared to synthetic materials. The structural and mechanical properties of biomaterials stimulate cardiomuscular tissue regeneration. Various parts in the human cardiovascular system are clearly illustrated in Figure 14.4 and can all be fabricated by tissue regeneration using biomaterial with the incorporation of biosynthesis-based nanomaterials.

Collagen, gelatin, chitosan and alginate are widely used biomaterials in cardiac applications. Alginate in a hydrogel or scaffold form promotes heart-tissue regeneration, a delivery system for

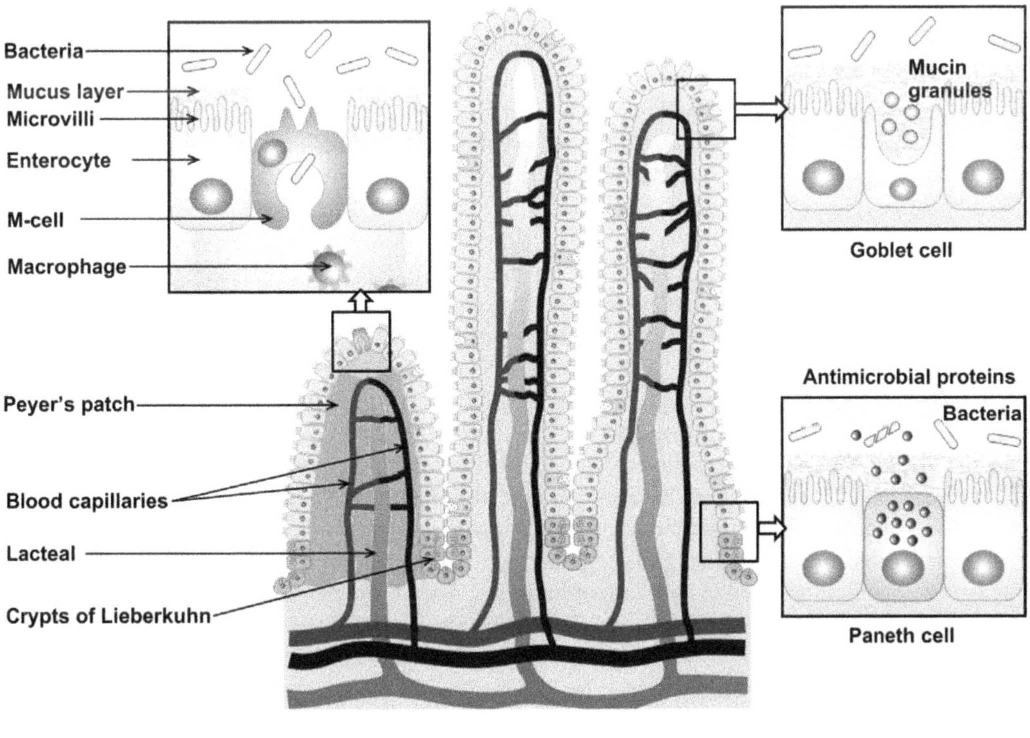

FIGURE 14.4 Schematic representation of the cardiovascular system. (Adapted from West, 2020 with permission from Elsevier 2021, license number: 5039071249171.)

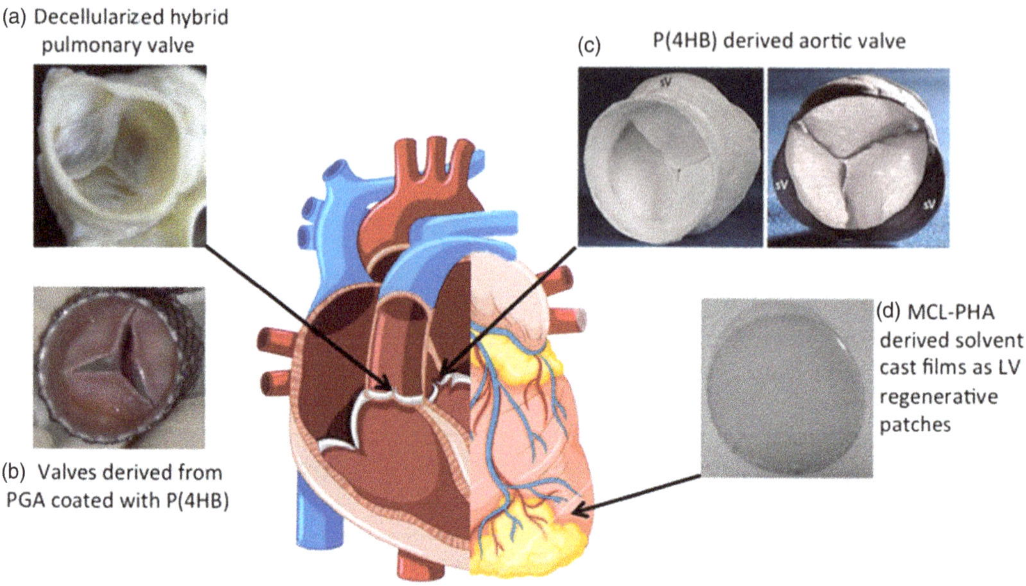

FIGURE 14.5 Schematic representations of heart valves. (Adapted from Majid *et al.*, 2020, this article comes under Frontiers Creative Commons CC-BY license.)

cardiac stem cells, provides a platform for continuous delivery of growth (Xu *et al.*, 2018; Majid *et al.*, 2020) and can produce extremely porous structures that permit the seeding of a comparatively high thickness of cells in the tissue assembly.

Polyhydroxyalkanoates (PHAs) are bioresorbable aliphatic polyesters from the fermentation of bacteria under a nutrient-restrictive situation. Due to their outstanding biocompatibility, in addition to their assorted mechanical properties, PHAs have been evaluated in diverse characteristics of cardiac tissue applications. Valve substitutes (Figures 14.5a–c) and regenerative cardiac patches (Figure 14.5d) have been examined during fabrication of inclusive scaffolds to assist the fictionalization and mechanical assets of decellularized organ xeno/homoimplants.

The advantages of AuNPs include their high electrical conductivity, superficial fabrication and amendment progression (Frens, 1973), varied nanoscale architecture (Dykman *et al.*, 2012) and reduced cytotoxicity (Kharaziha *et al.*, 2016). Several studies found that developing gold nanostructure-incorporated scaffolds – and also with alginate, chitosan, collagen and gelatin – promoted cardiac tissue engineering applications. Alginate hydrogel integrated with gold nanowires promoted electrical conductivity to sustain the functionalities of cardiac cells.

Gold nanorod (GNR)-embedded gelatin methacrylate (GelMA) fused with hydrogel-based biomaterials ultimately gave rise to superior individuality of cardiac tissue constructs and particularly enhanced tissue contractility through worse excitation threshold. The findings revealed that the nanoengineered GelMA-GNR fusion hydrogels, with admirable characteristics, offer appropriate matrices for precise applications in cardiac tissue regeneration (Navaei *et al.*, 2016). Table 14.1 shows the bionanomaterials used for cardiovascular tissue regeneration.

The main objective of cardiac tissue engineering is the expansion of functional tissue constructs that can restore the function and structure of injured/damaged myocardium. Biomaterials alone give better resolution for cardiac tissue engineering applications and the incorporation of NPs, especially AuNPs, into scaffold material can promote cardiac excitability and cellular connection. Incorporation of AuNPs improves cardiac tissue engineering (Figure 14.6) (Cohen-Karni *et al.*, 2012).

Gold nanowire incorporation with alginate scaffolds can enhance electrical contact between neighbouring cardiac cells. Chitosan-stabilized AuNPs promote viability, migration, metabolism and proliferation of stem cells. Hence, the integration of electroconductive AuNPs hooked on

TABLE 14.1
Bionanomaterials for Cardiovascular Tissue Regeneration

S. No.	Scaffolds	Remarks	References
1.	Alginate + platelet	Hydrogel scaffold that helps to improve vascularization	Hao *et al.*, 2007
2.	Gelatin, melanin + poly(L-lactide-co-e-caprolactone)	Promotes cell attachment and conductivity	Tian *et al.*, 2012
3.	AuNPs + gelatin	Scaffold that promotes cell conductivity and attachment	Shevach *et al.*, 2013
4.	Gold nanowires + alginate	Promotes conductivity and mechanical property	Dvir *et al.*, 2011
5.	Chitosan + carbon nanofibres	Enhances electrical conductivity of the cell	Shin *et al.*, 2013
6.	AuNPs + decellularized omentum	This extracellular matrix condensed immunogenicity	Shevach *et al.*, 2014
7.	AuNPs + gelatin methacrylate	Possess superior electrical and mechanical properties	Navaei *et al.*, 2017
8.	Aglinate + chitosan	Promotes improved vascularization	Stoica *et al.*, 2017
9.	Polyhydroxyalkanoates (PHAs)	Potential to keep their structural integrity	Rai *et al.*, 2011
10.	Cholecyst imitative extracellular matrix (C-ECM) + AuNPs	Scaffold supports substrate for the development and proliferation of cardiomyoblasts	Nair *et al.*, 2017

hydrogel structures promotes the properties of myocardial constructs (Li *et al.*, 2016). Gold-coated collagen nanomaterials improve biocompatibility and enhance migration of mesenchymal stem cells, proliferation and myocardial neuronal differentiation. AuNP incorporation into gelatin promotes simultaneous electrical and topographical cups, which help to improve external stimulation of cardiac muscle cells. (Navaei *et al.*, 2017).

AuNP incorporation into scaffold biomaterial can appreciably promote electrical and mechanical properties and cell addition on the exterior due to embellishment of AuNP surfaces, which is crucial for designed scaffolds in cardiac tissue engineering. AuNPs with lesser size and morphology lead to higher growth in heart and faster eradication.

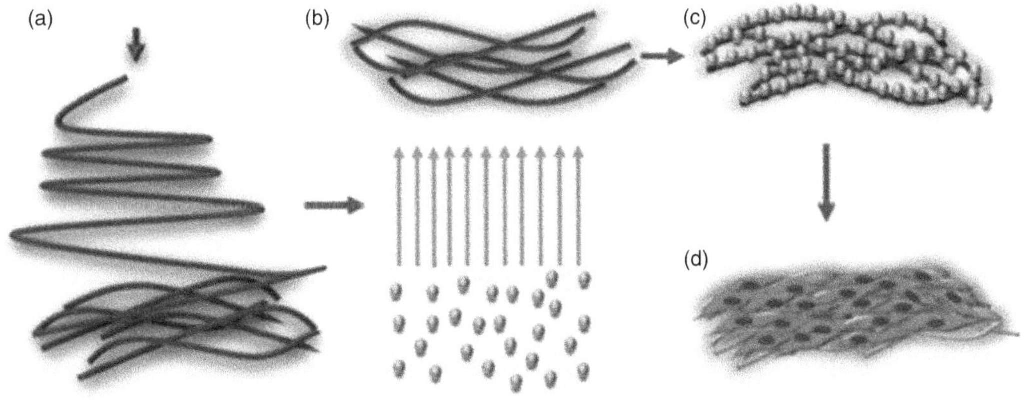

FIGURE 14.6 Fabrication of a gold nanoparticle-based hybrid scaffold for cardiac tissue regeneration. (a) Fabrication of polycaprolactone-gelatin fibre scaffolds by electrospinning, (b) Evaporation of gold nanoparticles onto the fibres to create nanocomposite scaffolds, (c & d) Cardiac cells are seeded in the nanocomposite scaffolds for engineering a functional cardiac tissue. (GN). (Adapted from Shevach *et al.*, 2013 with permission from RSC, license number 1113096-1.)

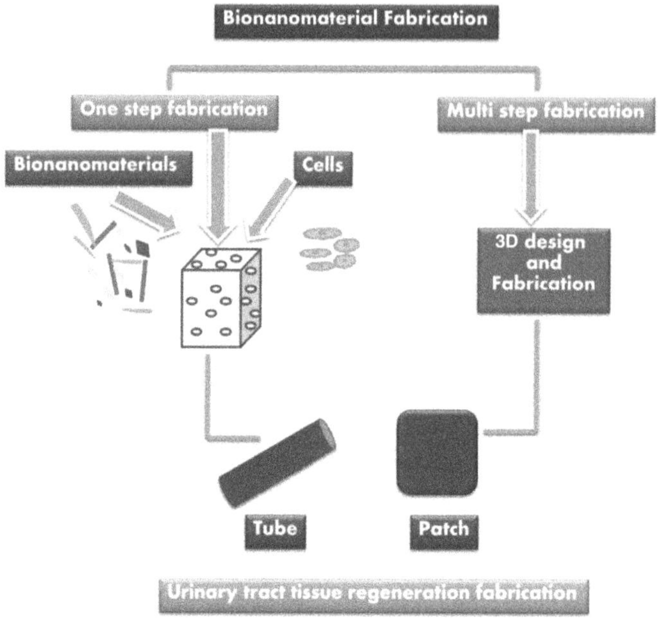

FIGURE 14.7 Approaches for biomaterial fabrication.

14.4 UROLOGICAL TISSUE REGENERATION

The human urinary system strains metabolic and end products from the blood. Once the excretory parts are damaged or injured, they are replaced by tissue-engineering applications. Urology enclosed renal tissue, nephron, kidney, bladder, ureter, urethral stricture, complicated hypospadias, neurogenic bladder dysfunction, bladder cancer, detrusor over activity, chronic inflammations, granulomas, latrogenic injuries, ureteric tumours and ureteric strictures are all parts replaced by natural biomaterials, which are biocompatible and biodegradable. These biomaterials can also enhance nerve and vascular regeneration and improve cellular differentiation; and are not lethal to the cells, have an unlimited supply and are noncarcinogenic and cost-effective. For custom-made physical and mechanical properties as well as for preparing urology tissue, these biomaterials have some specific properties such as being stretchable, soft, shape memory, watertight and resistant to encrustation and biofilm formation (Singh *et al.*, 2018).

Biopolymers like alginate, collagen, chitosan and gelatin are suitable for the regenerative process. Hence, biopolymers such as alginate, collagen, chitosan, fibrinogen, hyaluronic acid and gelatin are combined with synthetic biodegradable compounds such as poly(glycolic acid), polylactic acid, poly(ε-caprolactone) and poly(lactic-co-glycolic acid). Biopolymers alone are not effective for this tissue engineering application due to its unchangeable mechanical property. Once it combines/cross-links with biodegradable synthetic polymers, the material's mechanical property improves (Vardar, 2020). Figure 14.7 illustrates the different approaches for fabrication of biomaterials.

14.5 SKIN

All the organs of the human body are covered by skin, making it the largest organ in the body. It contains three layers, the outermost layer is thin epidermis, the middle layer is a thick dermis and the third is the deepest subcutaneous hypodermis. Damage or injury to the skin can occur and can include physiological processes or illnesses such as acute surgical interventions, trauma, chronic wounds, burns, genetic disorders or infections. If any one of the layers is affected, it can be replaced by a biopolymer, as synthetic polymers have many side effects. Biopolymers with characteristics

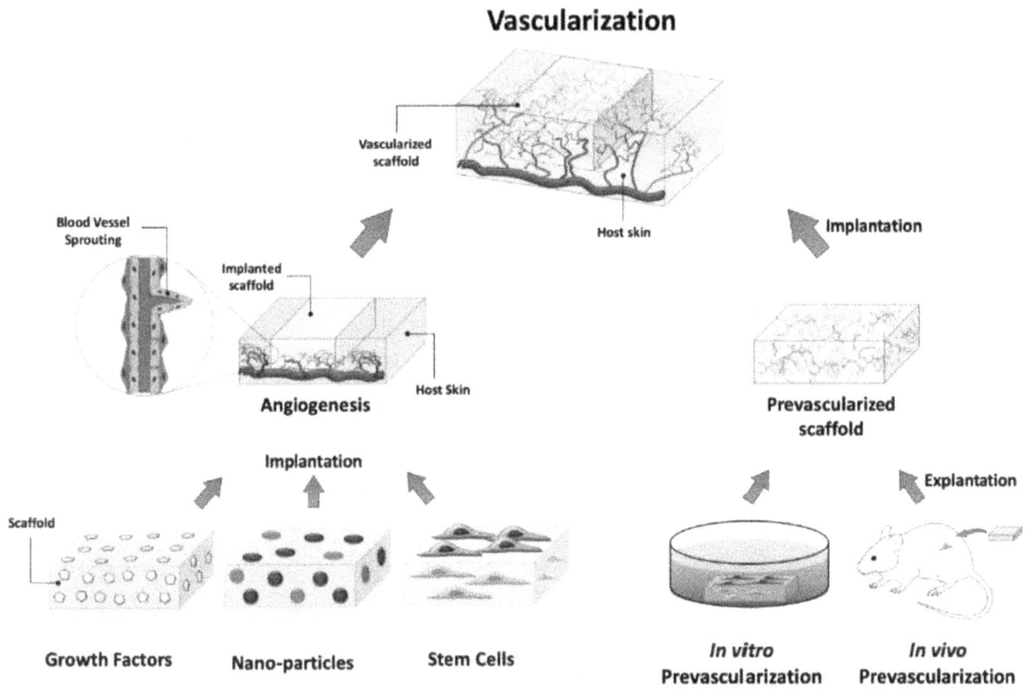

FIGURE 14.8 Schematic representation of the structure and function of skin regeneration vascularization. (Adapted from Amirsadeghi *et al.*, 2020 with permission from RSC, license number: 1113644-1.)

of biocompatibility, nontoxicity, biodegradability, drug delivery and cell adhesive properties are preferred. The steps involved in skin vascularization are illustrated in Figure 14.8.

Recently, adipose-derived stem cell (ADSC-CM)-based scaffolds were generated for skin tissue engineering applications. ADSC-CM reversed the alteration of fibroblasts in myofibroblasts and slowed extreme production of ECM biomaterials, appreciably speeding up wound closure, enhancing regeneration results and ultimately showing superior proregenerative ability. Hence ECM biomaterials are an alternative regenerative medicine with established viability and considerable biological consequences of this inventive biomaterial (Goonoo *et al.*, 2020). Bionanomaterials for skin tissue regeneration are shown in Table 14.2.

TABLE 14.2
Bionanomaterials for Skin Tissue Regeneration

S. No.	Scaffolds	Remarks	References
1.	Chitosan nanofibre	Enhanced cellular adhesion and proliferation, quicker regeneration of epidermis as well dermis	Tchemtchoua *et al.*, 2011
2.	Collagen hydrogel	Persistent cell viability, enhanced wound closing and vascularization	Wong *et al.*, 2011
3.	Collagen	Condensed wound contraction	Powell *et al.*, 2008
4.	Gelatin	Decrease in wound size, epithelialization and collagen configuration	Kanokpanont *et al.*, 2012
5.	Collagen	Improved cell viability	Lee *et al.*, 2014
6.	Chitosan	Affords better cell adherence, proliferation and viability over skin	Wu *et al.*, 2016
7.	Chitosan and collagen	Showed keratinocyte migration and wound-healing of skin	Sarkar *et al.*, 2013
8.	Chitosan	Potential material for skin scaffold preparation	Chen *et al.*, 2011

Chitosan and collagen types I, II, III and IV are the most abundant biopolymers and are widely used for skin-tissue regeneration applications (Davison-Kotler *et al.*, 2019) due to their structural and mechanical properties that help enhance all the three layers of cell growth returning to their original form. AuNP incorporation into collagen, chitosan, fibroin and hydrocolloids revealed excellent skin regeneration properties.

14.6 CONCLUSION

Much advancement in nanotechnology has brought notable changes in tissue-engineering fabrication. Amongst various biomaterials, collagen and chitosan scaffold and hydrogels are the most widely used for many tissue engineering regeneration approaches. The incorporation of drugs or nanomaterials enhances the process, especially the incorporation of gold nanomaterials into scaffold, which revealed superior regeneration in tissue formation. Based on the literature the biomaterial such as collagen or alginate along with AuNPs promotes tissue regeneration than synthetic material, not only that its also promotes wound healing very faster. This chapter portrays the importance of bionanomaterial in successful tissue engineering applications such as cardiac, urologic, and skin tissue regeneration.

REFERENCES

Abdul Khalil, H. P. S., A. H. Bhat, A. Abu Bakar, Paridah Md Tahir, I. S. M. Zaidul, and M. Jawaid. "Cellulosic nanocomposites from natural fibers for medical applications: A review." *Handbook of Polymer Nanocomposites. Processing, Performance and Application* (2015): 475–511. https://doi.org/10.1007/978-3-642-45232-1_72.

Amirsadeghi, A., A. Jafari, L. J. Eggermont, S. S. Hashemi, S. A. Bencherif, and M. Khorram. "Vascularization strategies for skin tissue engineering." *Biomaterials Science* 8, no. 15 (2020): 4073–4094.

Chen, Honglin, Jin Huang, Jiahui Yu, Shiyuan Liu, and Ping Gu. "Electrospun chitosan-graft-poly (ε-caprolactone)/poly (ε-caprolactone) cationic nanofibrous mats as potential scaffolds for skin tissue engineering." *International Journal of Biological Macromolecules* 48, no. 1 (2011): 13–19.

Cohen-Karni, Tzahi, Kyung Jae Jeong, Jonathan H. Tsui, Gally Reznor, Mirela Mustata, Meni Wanunu, Adam Graham et al. "Nanocomposite gold-silk nanofibers." *Nano Letters* 12, no. 10 (2012): 5403–5406.

Dallas, Panagiotis, Virender K. Sharma, and Radek Zboril. "Silver polymeric nanocomposites as advanced antimicrobial agents: Classification, synthetic paths, applications, and perspectives." *Advances in Colloid and Interface Science* 166, no. 1–2 (2011): 119–135.

Davison-Kotler, Evan, William S. Marshall, and Elena García-Gareta. "Sources of collagen for biomaterials in skin wound healing." *Bioengineering* 6, no. 3 (2019): 56.

Dinoro, J., M. Maher, S. Talebian, M. Jafarkhani, M. Mehrali, G. Orive, J. Foroughi, M. S. Lord, and A. Dolatshahi-Pirouz. "Sulfated polysaccharide-based scaffolds for orthopaedic tissue engineering." *Biomaterials* 214 (2019): 119214.

Dvir, Tal, Brian P. Timko, Mark D. Brigham, Shreesh R. Naik, Sandeep S. Karajanagi, Oren Levy, Hongwei Jin, Kevin K. Parker, Robert Langer, and Daniel S. Kohane. "Nanowired three-dimensional cardiac patches." *Nature Nanotechnology* 6, no. 11 (2011): 720–725.

Dykman, Lev, and Nikolai Khlebtsov. "Gold nanoparticles in biomedical applications: Recent advances and perspectives." *Chemical Society Reviews* 41, no. 6 (2012): 2256–2282.

Edmondson, Dennis, Narayan Bhattarai, Omid Veiseh, Frederick A. Matsen, and Miqin Zhang. "Electrospun chitosan-based nanofibers and their cellular compatibility." *Biomaterials* 26, no. 31 (2005): 6176–6184.

España-Sánchez, Beatriz Liliana, Martha Elena Cruz-Soto, Eduardo A. Elizalde-Peña, Samantha Sabasflores-Benítez, Adrián Roca-Aranda, Karen Esquivel-Escalante, and Gabriel Luna-Bárcenas. "Trends in tissue regeneration: Bio-nanomaterials." *Tissue Regeneration* (2018): 27: 218–28.

Frens, Gerrit. "Controlled nucleation for the regulation of the particle size in monodisperse gold suspensions." *Nature Physical Science* 241 (1973): 20–22.

Goonoo, Nowsheen and Archana Bhaw-Luximon. "Nanomaterials combination for wound healing and skin regeneration." In *Advanced 3D-Printed Systems and Nanosystems for Drug Delivery and Tissue Engineering*, pp. 159–217. Elsevier, 2020.

Gopinathan, Janarthanan, and Insup Noh. "Recent trends in metal ion based hydrogel biomaterials for tissue engineering and other biomedical applications." *Journal of Materials Science & Technology* 63, (2021): 35–53. https://doi.org/10.1016/j.jmst.2020.02.052.

Hao, Xiaojin, Eduardo A. Silva, Agneta Månsson-Broberg, Karl-Henrik Grinnemo, Anwar J. Siddiqui, Göran Dellgren, Eva Wärdell, Lars Åke Brodin, David J. Mooney, and Christer Sylvén. "Angiogenic effects of sequential release of VEGF-A165 and PDGF-BB with alginate hydrogels after myocardial infarction." *Cardiovascular Research* 75, no. 1 (2007): 178–185.

Hsu, Shan-hui, Yu-Bin Chang, Ching-Lin Tsai, Keng-Yen Fu, Shu-Hua Wang, and Hsiang-Jung Tseng. "Characterization and biocompatibility of chitosan nanocomposites." *Colloids and Surfaces B: Biointerfaces* 85, no. 2 (2011): 198–206.

Kanokpanont, Sorada, Siriporn Damrongsakkul, Juthamas Ratanavaraporn, and Pornanong Aramwit. "An innovative bi-layered wound dressing made of silk and gelatin for accelerated wound healing." *International Journal of Pharmaceutics* 436, no. 1–2 (2012): 141–153.

Kharaziha, Mahshid, Adnan Memic, Mohsen Akbari, David A. Brafman, and Mehdi Nikkhah. "Nano-enabled approaches for stem cell-based cardiac tissue engineering." *Advanced Healthcare Materials* 5, no. 13 (2016): 1533–1553.

Klee, Doris, and Hartwig Höcker. "Polymers for biomedical applications: Improvement of the interface compatibility." *Biomedical Applications Polymer Blends* 145, (2000): 1–57.

Klemm, Dieter, Dieter Schumann, Ulrike Udhardt, and Silvia Marsch. "Bacterial synthesized cellulose—artificial blood vessels for microsurgery." *Progress in Polymer Science* 26, no. 9 (2001): 1561–1603.

Kumar, Nupur, Dayananda Desagani, Girish Chandran, Narendra Nath Ghosh, Ganesh Karthikeyan, Sachin Waigaonkar, and Anasuya Ganguly. "Biocompatible agarose-chitosan coated silver nanoparticle composite for soft tissue engineering applications." *Artificial Cells, Blood Substitutes, and Biotechnology* 46, no. 3 (2018): 1–13. https://doi.org/10.1080/21691401.2017.1337021.

Lee, Jin Ho, Gilson Khang, Jin Whan Lee, and Hai Bang Lee. "Interaction of different types of cells on polymer surfaces with wettability gradient." *Journal of Colloid and Interface Science* 205, no. 2 (1998): 323–330.

Lee, Sang Jin, Gilson Khang, Young Moo Lee, and Hai Bang Lee. "The effect of surface wettability on induction and growth of neurites from the PC-12 cell on a polymer surface." *Journal of Colloid and Interface Science* 259, no. 2 (2003): 228–235.

Lee, Vivian, Gurtej Singh, John P. Trasatti, Chris Bjornsson, Xiawei Xu, Thanh Nga Tran, Seung-Schik Yoo, Guohao Dai, and Pankaj Karande. "Design and fabrication of human skin by three-dimensional bioprinting." *Tissue Engineering Part C: Methods* 20, no. 6 (2014): 473–484.

Li, Yi, Xiaoli Shi, Lei Tian, Hongyu Sun, Yujing Wu, Xia Li, Jianjun Li *et al.* "AuNP–collagen matrix with localized stiffness for cardiac-tissue engineering: Enhancing the assembly of intercalated discs by β1-integrin-mediated signaling." *Advanced Materials* 28, no. 46 (2016): 10230–10235.

Lin, Ning, and Alain Dufresne. "Nanocellulose in biomedicine: Current status and future prospect." *European Polymer Journal* 59 (2014): 302–325.

Majid, Qasim A., Annabelle T. R. Fricker, David A. Gregory, Natalia Davidenko, Olivia Hernandez Cruz, Richard J. Jabbour, Thomas J. Owen, *et al.* "Natural biomaterials for cardiac tissue engineering: A highly biocompatible solution." *Frontiers in Cardiovascular Medicine* 7 (2020): 192.

Markus, W., A. Abla, B. Michel, E. K. Basma, E. K. Stephanie, H. Dorothee, K. Birgit, K. Judith, T. Edda, and S. Margit. "Lignin-derived biomaterials for drug release and tissue engineering." *Molecules* 23 (2018): 1885.

Mekonnen, Tizazu, Paolo Mussone, Hamdy Khalil, and David Bressler. "Progress in bio-based plastics and plasticizing modifications." *Journal of Materials Chemistry A* 1, no. 43 (2013): 13379–13398.

Mishra, Raghvendra Kumar, Sung Kyu Ha, Kartikey Verma, and Santosh K. Tiwari. "Recent progress in selected bio-nanomaterials and their engineering applications: An overview." *Journal of Science: Advanced Materials and Devices* 3, no. 3 (2018a): 263–288.

Mishra, R. K., A. Sabu, and S. K. Tiwari. "Materials chemistry and the futurist eco-friendly applications of nanocellulose: Status and prospect." *Journal of Saudi Chemical Society* 22 (2018b): 949–978.

Moon, Robert J., Gregory T. Schueneman, and John Simonsen. "Overview of cellulose nanomaterials, their capabilities and applications." *Journal of the Minerals, Metals & Materials Society* 68, no. 9 (2016): 2383–2394.

Nair, Reshma S., Jimna Mohamed Ameer, Malcolm R. Alison, and Thapasimuthu V. Anilkumar. "A gold nanoparticle coated porcine cholecyst-derived bioscaffold for cardiac tissue engineering." *Colloids and Surfaces B: Biointerfaces* 157 (2017): 130–137.

Navaei, Ali, Nathan Moore, Ryan T. Sullivan, Danh Truong, Raymond Q. Migrino, and Mehdi Nikkhah. "Electrically conductive hydrogel-based micro-topographies for the development of organized cardiac tissues." *RSC Advances* 7, no. 6 (2017): 3302–3312.
Navaei, Ali, Harpinder Saini, Wayne Christenson, Ryan Tanner Sullivan, Robert Ros, and Mehdi Nikkhah. "Gold nanorod-incorporated gelatin-based conductive hydrogels for engineering cardiac tissue constructs." *Acta Biomaterialia* 41 (2016): 133–146.
Powell, Heather M., Dorothy M. Supp, and Steven T. Boyce. "Influence of electrospun collagen on wound contraction of engineered skin substitutes." *Biomaterials* 29, no. 7 (2008): 834–843.
Rahmati, M., E. A. Silva, J. E. Reseland, C. A. Heyward, and H. J. Haugen. "Biological responses to physicochemical properties of biomaterial surface." *Chemical Society Reviews* 49, no.15 (2020): 5178–5224.
Rai, Ranjana, Aldo R. Boccaccini, Jonathan C. Knowles, Nicola Mordon, Vehid Salih, and Ian C. Locke. "The homopolymer poly (3-hydroxyoctanoate) as a matrix material for soft tissue engineering." *Journal of Applied Polymer Science* 122, no. 6 (2011): 3606–3617.
Roman, Maren. "Toxicity of cellulose nanocrystals: A review." *Industrial Biotechnology* 11, no. 1 (2015): 25–33.
Rossana, Boni, Azam Ali, Amin Shavandi, and Andrew N. Clarkson. "Current and novel polymeric biomaterials for neural tissue engineering." *Journal of Biomedical Science* 25, no. 1 (2018): 1–21.
Sarkar, Soumi Dey, Brooke L. Farrugia, Tim R. Dargaville, and Santanu Dhara. "Chitosan–collagen scaffolds with nano/microfibrous architecture for skin tissue engineering." *Journal of Biomedical Materials Research Part A* 101, no. 12 (2013): 3482–3492.
Schumann, Dieter A., Jens Wippermann, Dieter O. Klemm, Friederike Kramer, Daniel Koth, Hartwig Kosmehl, Thorsten Wahlers, and Schariar Salehi-Gelani. "Artificial vascular implants from bacterial cellulose: Preliminary results of small arterial substitutes." *Cellulose* 16, no. 5 (2009): 877–885.
Shatkin, Jo Anne, and Baram Kim. "Cellulose nanomaterials: Life cycle risk assessment, and environmental health and safety roadmap." *Environmental Science: Nano* 2, no. 5 (2015): 477–499.
Sheehy, E. J., G. M. Cunniffe, and F. J. O'Brien. "Collagen-based biomaterials for tissue regeneration and repair." In *Peptides and Proteins as Biomaterials for Tissue Regeneration and Repair*, pp. 127–150. Woodhead Publishing, 2018.
Shevach, Michal, Sharon Fleischer, Assaf Shapira, and Tal Dvir. "Gold nanoparticle-decellularized matrix hybrids for cardiac tissue engineering." *Nano Letters* 14, no. 10 (2014): 5792–5796.
Shevach, Michal, Ben M. Maoz, Ron Feiner, Assaf Shapira, and Tal Dvir. "Nanoengineering gold particle composite fibers for cardiac tissue engineering." *Journal of Materials Chemistry B* 1, no. 39 (2013): 5210–5217.
Shin, S. R., S. M. Jung, M. Zalabany, K. Kim, P. Zorlutuna, S. B. Kim, M. Nikkhah et al. "Carbon-nanotube-embedded hydrogel sheets for engineering cardiac constructs and bioactuators." *ACS Nano* 7, no. 3 (2013): 2369–2380.
Singh, Anirudha, Trinity J. Bivalacqua, and Nikolai Sopko. "Urinary tissue engineering: Challenges and opportunities." *Sexual Medicine Reviews* 6, no. 1 (2018): 35–44.
Stoica, R., R. Somoghi, S. M. Doncea, and R. M. Ion. "Preparation of alginate-chitosan polyelectrolyte complexes for encapsulation of natural polyphenols." *Optoelectronics and Advanced Materials-Rapid Communications* 11, no. 1–2 (2017): 113–118.
Sudhakar, C. K., Nitish Upadhyay, Arpit Verma, Ankush Jain, R. Narayana Charyulu, and Sanjay Jain. "Nanomedicine and tissue engineering." In *Nanotechnology Applications for Tissue Engineering*, Vol. 1, pp. 1–19. William Andrew Publishing, 2015.
Tang, X. Z., P. Kumar, S. Alavi, and K. P. Sandeep. "Recent advances in biopolymers and biopolymer-based nanocomposites for food packaging materials." *Critical Reviews in Food Science and Nutrition* 52, no. 5 (2012): 426–442.
Tchemtchoua, Victor T., Ganka Atanasova, Abdel Aqil, Patrice Filée, Nancy Garbacki, Olivier Vanhooteghem, Christophe Deroanne et al. "Development of a chitosan nanofibrillar scaffold for skin repair and regeneration." *Biomacromolecules* 12, no. 9 (2011): 3194–3204.
Tian, Bozhi, Jia Liu, Tal Dvir, Lihua Jin, Jonathan H. Tsui, Quan Qing, Zhigang Suo, Robert Langer, Daniel S. Kohane, and Charles M. Lieber. "Macroporous nanowire nanoelectronic scaffolds for synthetic tissues." *Nature Materials* 11, no. 11 (2012): 986–994.
Vardar, Elif. "Tissue engineering in urology." In *Biomaterials for Organ and Tissue Regeneration*, pp. 441–455, Elsevier, 2020. https://doi.org/10.1016/B978-0-08-102906-0.00013-1
West, J. L. "Biomaterials for cardiovascular tissue engineering." In *Biomaterials Science*, pp. 1389–1397, Academic Press, 2020.

Wong, Victor W., Kristine C. Rustad, Michael G. Galvez, Evgenios Neofytou, Jason P. Glotzbach, Michael Januszyk, Melanie R. Major et al. "Engineered pullulan–collagen composite dermal hydrogels improve early cutaneous wound healing." *Tissue Engineering Part A* 17, no. 5–6 (2011): 631–644.

Wu, Yang, Gopu Sriram, Amr S. Fawzy, Jerry YH Fuh, Vinicius Rosa, Tong Cao, and Yoke San Wong. "Fabrication and evaluation of electrohydrodynamic jet 3D printed polycaprolactone/chitosan cell carriers using human embryonic stem cell-derived fibroblasts." *Journal of Biomaterials Applications* 31, no. 2 (2016): 181–192.

Xu, Zhengfan, and Mai T. Lam. "Alginate application for heart and cardiovascular diseases." In *Alginates and Their Biomedical Applications*, pp. 185–212. Springer, 2018.

Yoshito, Ikada. "Challenges in tissue engineering." *Journal of the Royal Society Interface* 3, no. 10 (2006): 589–601.

Zarrintaj, Payam, Saeed Manouchehri, Zahed Ahmadi, Mohammad Reza Saeb, Aleksandra M. Urbanska, David L. Kaplan, and Masoud Mozafari. "Agarose-based biomaterials for tissue engineering." *Carbohydrate Polymers* 187 (2018): 66–84.

Zinoviadou, K. G., Maria Gougouli, and C. G. Biliaderis. "Innovative biobased materials for packaging sustainability." In *Innovation Strategies in the Food Industry*, pp. 167–189. Academic Press, 2016.

15 Technological Innovations for the Transdermal Administration of State-of-the-Art Microneedle Arrays

A Patent Study

T. de A. Andrade, J.A.C. Nascimento Júnior, A.M. Santos, S. Shanmugam, P. dos P. Menezes, Y.M.B.G. de Carvalho, L.A. Frank and M.R. Serafini

CONTENTS

15.1 INTRODUCTION

Among the routes of drug administration, the dermal route can offer many options due to its extension throughout the human body that provides easy access for the administration of therapeutic substances (Economidou, Lamprou, and Douroumis 2018). Transdermal drug delivery systems (TDDS) present advantages, including greater patient compliance, tissue targeting, bypassing the first-pass metabolism and better drug delivery (Carter, Narasimhan, and Wang 2019). TDDS works by improving the bioavailability of the drug and excluding possible irritations and toxicity of the gastrointestinal tract. Thus, alternatives have been developed to overcome this limitation and improve the skin permeability of drugs (Gu et al. 2019). Within this context are transdermal patches and microneedles (MNs).

Transdermal patches are forms of drug delivery placed onto the skin that release a specific amount of the drug through the skin into the bloodstream. They can be found in two types: reservoir or matrix patches. These patches work by extending the pharmacological effect and eliminating changes in drug concentrations when used by the oral route (Omar, Nasr, and Rafla 2019). However, they present disadvantages such as skin irritation, obstruction of the sweat ducts (preventing vapour loss from the skin's surface), difficult application on curved surfaces, aesthetic appearance and pain on removal (Kathe and Kathpalia, 2017).

MNs are a new solution for the limitations of transdermal patches. MNs can be self-administrated and cause minimal skin trauma and increased bioavailability (Lim et al. 2018; Nguyen et al. 2018). MNs consist of micrometric projections organized in a transdermal patch that create channels to

DOI: 10.1201/9781003231745-15

facilitate the diffusion of substances into the lower layers of the skin from the viable epidermis to the deeper layers, such as the dermis. Thus, depending on the therapeutic purpose, MNs can be obtained in differents lengths and shapes and, due to the micrometric size of their needles, they are considered painless and minimally invasive devices compared to hypodermic needles (Economidou, Lamprou, and Douroumis 2018; Waghule et al. 2019).

MN systems are versatile and have numerous applications, such as insulin administration, which makes it a possible alternative to subcutaneous administration. It can be used as a device to withdraw fluid from the skin for analysis such as the analysis of blood sugar for self-regulating insulin delivery (Chowdhury 2017). Moreover, MNs can deliver active pharmaceutical ingredients (APIs), allergens, vitamins, cosmetic agents, hormones (McAllister, Praunitz, and Henry 2018) non-steroidal anti-inflammatory drugs or peptide drugs (Qingqing et al. 2019) and parathyroid hormones, (Ming and Boshen 2017) among others.

In light of these considerations, this review aims to evaluate patents that have developed arrays of MNs for drug delivery, vaccine and bioactive compounds. Thus, this review provides a comprehensive view of the methods and materials used to prepare MNs and their different geometries.

15.2 TECHNOLOGICAL PROSPECTING

In this patent review, the specialized databases World Intellectual Property Organization (WIPO) and Espacenet were used for search patents from January 2017 to December 2019. The search for patents was carried out using codes A61K and A61M from the International Patent Classification (IPC) and a combination of the descriptors "microneedles" and "drug delivery" in the title and/or in the abstract. Figure 15.1 illustrates the systematic base guidelines used for patent searching and screening of this review.

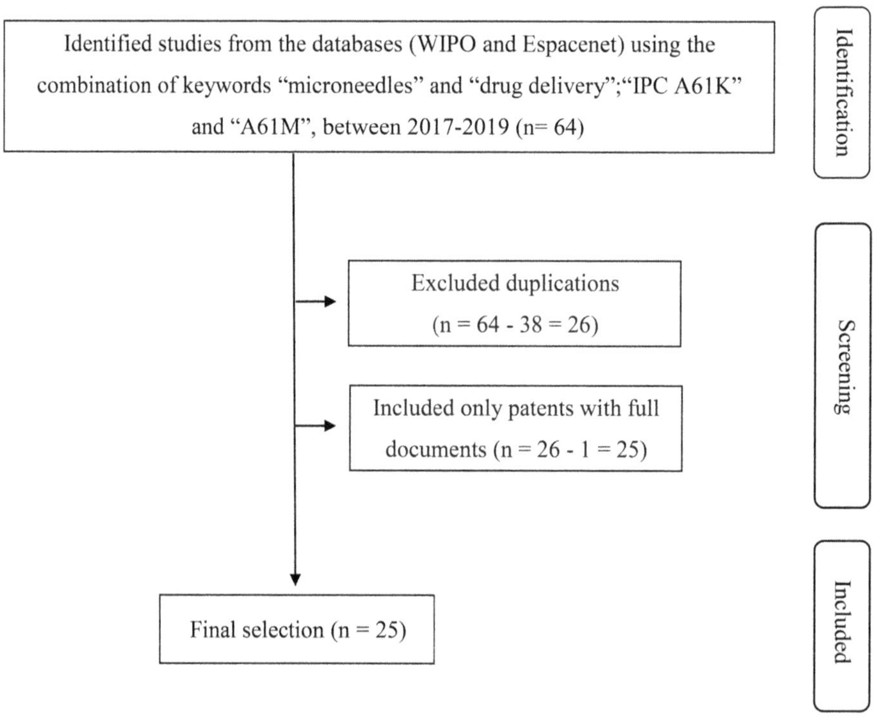

FIGURE 15.1 Flowchart of patent searching and screening.

15.3 PATENT SEARCHING AND SCREENING

The IPC is an internationally recognized patent classification system and tool that helps search for data related to patents, thus facilitating access to the technological and legal information contained therein (WIPO 2019).

In this chapter, we used the IPC A61K and A61M codes. The code A61K refers to preparations for medical, dental or toilet purposes, and A61M refers to devices for introducing media into, or onto, the body. Figure 15.2 shows that among the patents evaluated, the IPC A61M and A61K codes appeared 18 and 10 times, respectively. In addition, the patents presented other classifications than those used during the search, for example, the A61B, A61P and B29C among others. This may occur in cases where a single patent is related to multiple technical features.

The countries with the most filings were China and the United States, with nine and eight patents, respectively, as shown in Figure 15.3. Another five patents had their filings through the World Organization (WO). The processes by the WO are protected through the Patent Cooperation Treaty. The protection of the invention is not required exclusively in the country of origin of the patent application but in other countries that are part of this treaty. Multinational companies widely use this tool to expand their market with innovative and exclusive products (WIPO, 2019). Moreover, the European Patent Office (EPO) registered two patents, and Russia filed one patent.

Several entities may request the protection of their invention through a patent. The documents analysed in this review presented more processes required by the university, with 10 patents, represented in Figure 15.4. The university plays a role in research and development through the funded resources offered by the federal government to protect their discoveries in patent form.

Another applicant was the industry/company with seven patents. Among the applicants are Clearside Biomedical Inc., Intuity Medical Inc. and Sorrento Therapeutics Inc., that have

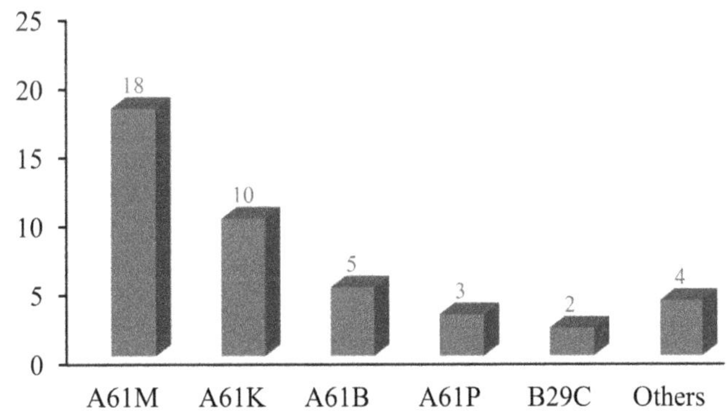

FIGURE 15.2 Final selection of patents. IPC: A61M = Devices for introducing media into or onto the body). A61K = Preparations for medical, dental, or toilet purposes. A61B = Diagnosis; surgery; identification. A61P = Specific therapeutic activity of chemical compounds or medicinal preparations. B29C = Shaping or joining of plastics; shaping of material in a plastic state not otherwise provided for; after-treatment of the shaped products (e.g. repairing). A61L = Methods or apparatus for sterilizing materials or objects in general; disinfection; sterilization or deodorisation of air; chemical aspects of bandages, dressings, absorbent pads, or surgical articles; materials for bandages, dressings, absorbent pads, or surgical articles. A61F = Filters implantable into blood vessels; prostheses; devices providing patency to, or preventing collapsing of, tubular structures of the body, e.g. stents; orthopaedic, nursing or contraceptive devices; fomentation; treatment or protection of eyes or ears; bandages, dressings or absorbent pads; first-aid kits. B33Y = Additive manufacturing, i.e. manufacturing of three-dimensional (3D) objects by additive deposition, additive agglomeration or additive layering, e.g. by 3D printing, stereolithography or selective laser sintering. G01N = Investigating or analysing materials by determining their chemical or physical properties.

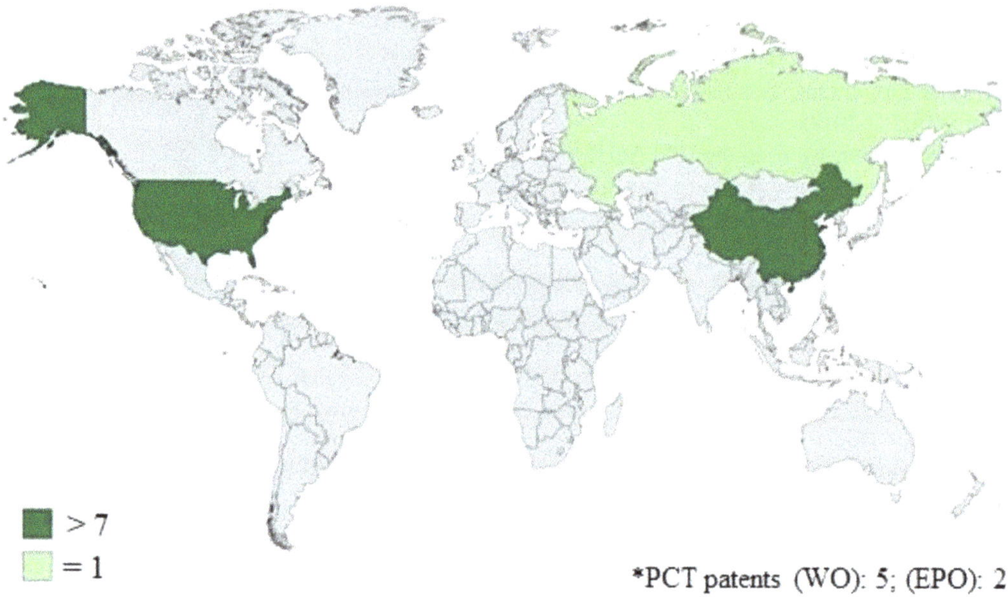

FIGURE 15.3 Final selection of patents: number of patent filings by country.

invested in research involving the development of MNs. The industries have skilled labor, sophisticated equipment and infrastructures to develop new products to be launched on the market later (Minguillo and Thelwall 2015). Independent inventors (five patents) and industry/university (two patents) deposited the other patents. Only one patent did not include the type of applicant.

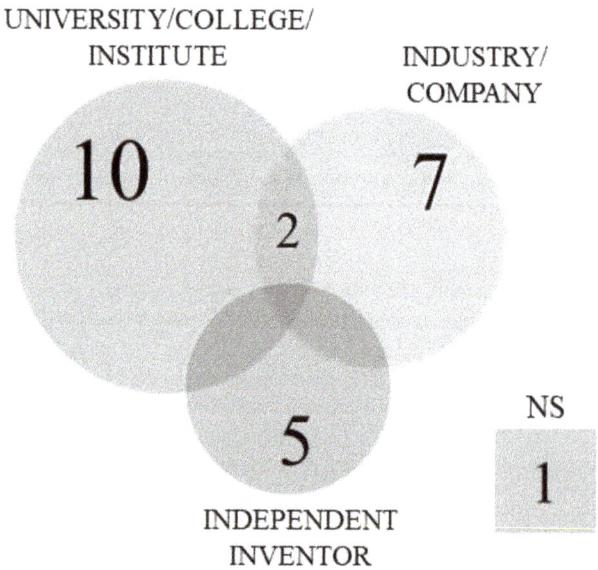

FIGURE 15.4 Final selection of patents: patent applicants. *Abbreviation*: NS = not specified.

15.4 POLYMERS AND METHODS FOR PREPARING MNs

The polymer matrix of MNs selects soluble polymers that are biocompatible, biodegradable and have appropriate mechanical properties. These properties influence the dissolution of the coating of the formulation to release the incorporated drug (Ye et al. 2018). It is recommended to insert the MN portion into the skin or mucous membrane to dissolve. In this way, the use of a water-soluble polymer is preferable as a material constituting a portion of the needle; among those most used are carboxymethyl cellulose (CMC), hyaluronic acid (HA), polyvinylpyrrolidone (PVP) and polyvinyl alcohol (PVA) (Cho, Lim, and Woodworth 2019; Lim 2019; Wang et al. 2020).

Wang et al. (2020) developed arrays for MNs to deliver therapeutic agents with PVA/PVP polymers. They conducted an *in vitro* study to determine the dissolution of the polymer. The PVA/PVP MN arrays were inserted into the agarose disc that worked as a tissue phantom. At predetermined times, the polymer MN arrays were removed and dried in the oven. Subsequently, the MNs were photographed under a microscope and quantified using Imaged. It was possible to observe that the height of the MNs decreased by 70% after 15 seconds of the experiment. After 30 seconds, the MN tips practically disappeared. This rapid dissolution of MNs may be related to the fact that PVA and PVP polymers are highly hydrophilic, thus enabling an accelerated release of the drug after the insertion of the MNs (Wang et al. 2020).

In another study developed by Zhu et al. (2009), the efficacy of CMC (500 μm) coated on metal MNs for antigen delivery was tested. After five minutes of insertion time, about 90% of the viral proteins were delivered into the stratum corneum of mice (Zhu et al. 2009).

The polymers used are those with high solubility, characterized by rapid release and dissociation of the polymer network, thus enabling an accelerated release of the drug after the insertion of the MNs. This type of polymer can be used for the incorporation of hydrophobic drugs, for example, MN arrays prepared from poly(methyl/vinyl ether/maleic anhydride) (PMVE/MA) (Donnelly and Woolfson 2017). Poly(lactic-co-glycolic acid) (PLGA) and polyglycolic acid (PGA) are biodegradable polymers that have been used to produce MN arrays for transdermal drug delivery (Ming et al. 2018; Narayan, Hall, and Zhang 2018). Table 15.1 describes other materials used.

Different technologies can obtain MN production methods, such as drawing lithography, photolithography, laser cutting, photopolymerization and moulding (Chen et al. 2018). The first stage in manufacturing MNs refers to the preparation of moulds, a method most commonly used, mainly made of ultraviolet (UV) lithography (Mogusala, Devadasu, and Venisetty 2015). However, this procedure requires a lot of time and a clean environment with advanced facilities, thus making it banned for general drug delivery. Moreover, the manufacturing of MNs becomes challenging due to the high proportion of MN structures with height between 0.5 and 3 mm and three-dimensional (3D) conic geometry (Nejad et al. 2018).

The moulding and photolithography process can make micromoulds. Chen et al. (2019) developed a master mould of MNs prepared via a micromoulding method. Pouring a polydimethylsiloxane (PDMS) solution into the custom-designed stainless steel master moulds created the micromoulds. After all the steps in the methodological process were complete, the MNs were gently peeled from the micromoulds and exposed to low-intensity UV light. The fabricated array consists of pyramidal-shaped MNs with a tip diameter of about 10 picometers (pm), a height of about 500 pm and a base width of about 250 pm (Chen et al. 2019). Another master mould of PDMS was used by Lim Geunbae et al. (2019) to produce MNs into star-shaped pyramids, cones or quadrangular pyramids (Geunbae et al. 2019).

Another method to manufacture MNs is a generic photolithographic process. The substrate needs to be covered by a photosensitive liquid polymer to be hardened by the thermal process afterward. The polymer suffers exposure to UV radiation through a matrix of circles representing the base of the needles, reproducing the design of the desired structure. UV radiation breaks or creates cross-links of the polymer, depending on the strength. The final sample is dipped in a developer solution, then the part of the polymer exposed to the radiation is removed. In the end, in the recording step, the drawing of the MNs made of the polymer is reproduced on the substrate with a wet or dry chemical application (Dardano et al. 2015).

TABLE 15.1
Patents of Microneedle Arrays Published in the Years 2017 and 2019

Inventor	Method of Preparation	Polymers and/or Metals Used	Structure/Shape	Height (μm)	Base Diameter (μm)	Diameter of Tip (μm)	Purpose of System	Active Principle
Kong et al. 2018	Injection moulding	PLGA nanoparticles/HA	Cone or pyramid	NS	NS	NS	Transdermal drug delivery of drugs and vaccines	NS
Zeng et al. 2019	Mould	PDMS	Hollow	100–700	NS	NS	Discloses an integrated transdermal drug delivery	Insulin
Huang 2017	NS	Monocrystalline silicon	Platform-shaped/ hemispherical conical/ needle-shaped	5–50	10–1500	1–100	MN array for percutaneous administration (anticracking and antiallergic effects)	NS
Prausnitz et al. 2017	Laser or similar optical energy source	Metals, glasses, ceramics, polymers, and/or nanotubes	Annular-shaped base	400 or less	3000	1000	MNs for treatment of ocular tissue	Vascular endothelial growth factor (VEGF)
Aceti et al. 2017	NS	Polymers	MNs can be either straight or curved	25–100	40–120	NS	A method of diabetes management	Various pharmaceutical agents including high-potency drugs
Yang and Dai 2018	Mould	Degradable high-molecular thermoplastic material	Conical	10 and 700	1 and 500	NS	Degradable MNs array safe to the human body and a manufacturing method and mould	Anti-inflammatory
Narayan et al. 2018	Injection moulding and drawing lithography	PGA	Hollow	50 and 900	Less than 300	NS	Transdermal drug delivery system for treating basal cell carcinoma	Itraconazole/vitamin D3
Donnelly and Woolfson 2017	Micromoulding (Micromoulds)	PMVEMA, esters thereof and PMVE/MA	Conical	400–3000	1–500	20–50	Transdermal delivery devices	Active agent having both hydrophobic and hydrophilic character
McAllister et al. 2018	NS	PDMS	NS	NS	NS	NS	MNs for the transport of therapeutic, diagnostic, cosmetic or biological molecules across tissues	Active pharmaceutical ingredients, allergens, vitamins, cosmetic agents

(Continued)

TABLE 15.1 *(Continued)*

Inventor	Method of Preparation	Polymers and/or Metals Used	Structure/Shape	Height (μm)	Base Diameter (μm)	Diameter of Tip (μm)	Purpose of System	Active Principle
Ross 2019	NS	NS	Cylindrical, rectangular or geometrically irregular shape	10~5,000 pm	NS	NS	Administration of a medicament to the lymphatic system by use of a fluid-delivery apparatus that enables the targeting of specific lymph nodes	Immune-suppressing agent
Dang and Zhang 2017	NS	2,2-dimethoxy-2-phenylacetophenone, hydroquinone (2,4,6-trimethylbenzoyl)-phosphine oxide One or more of 2-hydroxy-4'-(2-hydroxyethoxy)-2-methyl styrene	Conical or polygonal	100–2000	100–1000	100–5000	Drug administration for the treatment of osteoporosis	Parathyroid hormone
Zhu et al. 2019	Phase separation technology and application of the method	Silicon	Three-dimensional array structure	25–2000	10–10,000	NS	Drugs for transdermal administration in the treatment of diabetes, psoriasis, hair loss or superficial skin tumours, which are used in tissue fluid, blood extraction, antibiotic skin test	N,N-dimethylformamide, dimethylsulphoxide, N,N-dimethylacetamide, N-methylpyrrolidone, hexamethylphosphoramid, 1,4-bis oxyhexane, nitromethane
Baker et al. 2019	NS	Metallic materials, ceramic materials, polymer (e.g., plastic) materials	NS	NS	NS	NS	Device having an MN array for transdermal delivery of liquid formulations	Liquid formulations
Yu and Wang 2019	NS	Epoxy resin film	Inverted cones	200–3000	100–1000	600–10,000	NS	NS
Chowdhury 2017	Chemical or physical etching/standard lithographic procedures	Silicon and PDMS	Cylindrical/ hollow	1–100	NS	10	Analysis of blood sugar for self-regulating insulin delivery	Drugs, pharmacologically active agents or therapeutic agents

(Continued)

TABLE 15.1 *(Continued)*
Patents of Microneedle Arrays Published in the Years 2017 and 2019

Inventor	Method of Preparation	Polymers and/or Metals Used	Structure/Shape	Height (μm)	Base Diameter (μm)	Diameter of Tip (μm)	Purpose of System	Active Principle
Wang et al. 2019	NS	Carbopol, monomeric polymers or copolymers of vinylpyrrolidone and its derivatives	Conical, conical cylindrical or polygonal pyramid	100–1000	NS	NS	MNs for treating rheumatoid arthritis	Nonsteroidal anti-inflammatory drugs, peptide drugs
Kim et al. 2019	NS	HA	Microstructures with blunt edges and heterogeneous shapes	NS	NS	NS	To overcome the disadvantages of an injection, present research on invention being conducted in several areas of drug delivery, blood collection, biosensors and skin care	Drugs, skin care ingredients, or a combination of both
Ruckh 2019	NS	Polyvinyl alcohol, dextran, hyaluronan, chitosan or another water-soluble polymer	NS	NS	NS	NS	Therapeutic drug delivery and/or diagnostic information about the tissue	Active pharmaceutical ingredient
Falo and Erdos 2018	NS	PLGA	Square, rectangular, trapezoidal or polygonal	NS	NS	NS	MNs with bioactive components to exhibit specific release kinetics when the MNs array is inserted into the skin of a patient	Analgesic, anaesthetic, antiasthmatic, antibiotics, antidepressant, antidiabetic, anti-inflammatory
Wang et al. 2020	NS	PVA and PVP	Pyramidal master structures	600 mm	200 mm	500 mm	Therapeutic treatment by delivering protein locally to a target site	Binding proteins
Lim 2019	Mould	HA	Pyramid	600	200	NS	Increased drug-loading capacity and delivery accuracy	Sodium diclofenac
Baker et al. 2017	NS	Silicon	Conical, pyramidal or cylindrical in transitioning to a conical or pyramidal/hollow	290	NS	NS	MNs for delivering drug formulations through the skin	NS

(Continued)

TABLE 15.1 *(Continued)*

Inventor	Method of Preparation	Polymers and/or Metals Used	Structure/Shape	Height (μm)	Base Diameter (μm)	Diameter of Tip (μm)	Purpose of System	Active Principle
Chen et al. 2019	Micromoulding (PDMS) Master-moulds (stainless steel)	Outer layer (MeHA) Inner core (HA)	Pyramidal/ hollow	500 pm	250 pm	10 pm	Polymeric patch with an array of needle-shaped and double-layered (DL-MN) microreservoirs for ocular drug delivery	Neutralizing antibody (IgGs) and anti-inflammatory compound (diclofenac)
Cho et al. 2019	NS	Carboxymethylcellulose or HA	Concave/ hollow	250–2500	NS	NS	Sustained and controlled intratonsillar drug delivery using a needle array	Steroid (mometasone, an antibiotic, or a therapeutic compound)
Lim et al. 2019	Master mould	PDMS	Star-shaped pyramid/ cone and quadrangular pyramids	NS	NS	NS	Transdermal drug delivery patch and a method for manufacturing the drug with a MN having a large surface-area-to-volume while increasing the absorption efficiency of the drug while maintaining a high aspect ratio	NS

Abbreviations: Hyaluronic acid (HA); Methacrylated hyaluronic acid (MeHA); Microneedle (MN); Not specified (NS); Poly(lactic-co-glycolic acid) (PLGA); Poly(methyl/vinyl ether/maleic anhydride) (PMVE/MA); Poly(methyl-vinyl-ether/maleic acid) (PMVEMA); Polyglycolic acid (PGA); Polydimethylsiloxane (PDMS); Polyvinyl alcohol (PVA); Polyvinylpyrrolidone (PVP).

In the photolithographic manufacturing process, the intensity of UV light influenced the geometry of MNs. According to studies by Kathuria et al. (2015), the intensity of 6.44 W/cm^2 gives MNs an ideal mechanical resistance for skin applications. In addition, with each increase in UV intensity, the length of the MNs grows. The advantage of this manufacturing method without using a mould is that it is fast. However, it is not possible to predict timing when it comes to manufacturing on a large scale (Kathuria et al. 2015).

15.5 MICRONEEDLE DESIGN

MNs can be fabricated in different shapes, structures and geometries such as lengths, base and tip diameter. Regarding the dimensions, MNs can present heights ranging from 25 to 2000 μm and base and tip diameters ranging 50–250 μm and 1–25 μm, respectively. Under these conditions, MNs presented adequate dimensions to reach deeper layers of the skin without reaching the nerves present in the dermis (Alkilani, McCrudden, and Donnelly 2015; Economidou, Lamprou, and Douroumis 2018).

As for shape and structure, MNs are pyramidal or conical, solid, coated, hollow, soluble and hydrogel-forming. Solid MNs are combined with a conventional formulation containing the drug and a gel, solution or transdermal patch. The application happens in two steps. First, the MNs are applied to the skin and create micron-sized pores. Then, the conventional formulation is applied and the drug penetrates the skin layers through the passive diffusion process. Solid MNs also may be coated with a drug-containing solution or dispersion (coated MNs). In this case, after the insertion of the MNs into the skin, the coating material begins to dissolve, and the drug is delivered instantly. However, the amount of drug to be coated is a limitation due to the small surface area of the MNs. Thus, another type of MN used to deliver larger amounts of drug is the hollow MN (Larrañeta et al. 2016; Economidou, Lamprou, and Douroumis 2018).

Hollow MNs have a similar structure to conventional hypodermic needles with empty spaces inside the needle filled with a fluid formulation containing the drug. Unlike conventional needles, they are micron-sized. This type of MN is used primarily for high molecular-weight compounds and the delivery of the bioactive compound may occur by pressure or electrically driven flow or diffusion. Other types include soluble and hydrogel-forming MNs. Soluble MNs consist of soluble matrices composed of biodegradable polymers containing the drug. After insertion of the MNs, the tips of the needles meet the skin's interstitial fluid, begin to dissolve and release the drug. Thus, drug release is controlled according to the dissolution of the polymer used in the composition of MNs. Hydrogel-forming MNs are composed of a polymer matrix associated with a transdermal patch containing the drug. When the polymer encounters the skin's interstitial fluid, it swells. Subsequently, the drug contained in the patch is released through the swollen microprojections (Larrañeta et al. 2016; Ita 2018; Lim et al. 2018; Waghule et al. 2019).

15.6 GENERAL CONCLUSIONS

MNs arrays are promising technologies for drug delivery, vaccines and bioactive compounds. To translate the research on MNs from the bench to the patient, the most important criteria are the safety and the efficacy of MNs demonstrated in clinical trials. According to Nguyen and Park (2018), participants in various clinical trials using MNs have reported little or no pain, even when using different MN lengths. Promising studies have shown that the patients received the MNs and researchers evaluated their self-administration (Nguyen and Park 2018).

The field of therapeutic alternatives for drug administration has been growing every year. Undoubtedly, more devices containing MN technology are already arriving in the market, bringing new forms of clinical management and diseases treatment to help patient quality of life.

REFERENCES

Aceti, John Gregory; McBride, Sterling Eduard; Moroney III, Richard Morgan; Gregory, Christopher Carter; and Zanzucchi, Peter John, inventors; Intuity Medical Inc, assignee. "Autonomous, ambulatory analyte monitor or drug delivery device." US2017319121 (A1). 2017.

Alkilani, Ahlam Zaid; McCrudden, Maelíosa T. C.; and Donnelly, Ryan F. "Transdermal drug delivery: Innovative pharmaceutical developments based on disruption of the barrier properties of the stratum corneum." *Pharmaceutics* 7, no. 4 (2015): 438–470.

Baker, Andrew T.; Gadsby, Elizabeth Deibler; Ross, Russell Frederick; and Hagan, Luke, inventors; Kimberly-Clark Worldwide, Inc.; Sorrento Therapeutics, Inc, assignee. "Receptacle portion of transdermal drug delivery apparatus." EP3137150 (A1). 2017.

Baker, Andrew T.; Ross, Russell Frederick; Gadsby, Elizabeth Deibler; and Hagan, Luke, inventors; Sorrento Therapeutics Inc, assignee. "Microneedle array assembly, drug delivery device and method for administering liquid across a broad area at low pressure." US20190143090. 2019.

Carter, Prerana; Narasimhan, Balaji; and Wang, Qun. "Biocompatible nanoparticles and vesicular systems in transdermal drug delivery for various skin diseases." *International Journal of Pharmaceutics* 555 (2019): 49–62.

Chen, Peng; Wang, Xiao Meng; Xu, Chenjie; and Than, Aung, inventors; Nanyang Technological University, assignee. "Self-implantable micro-drug-reservoirs for localized and controlled ocular drug delivery." WO 2019143293. 2019.

Chen, Xiang; Wang, Li; Yu, Haojie; Li, Chengjiang; Feng, Jingyi; Haq, Fazal; Khan, Amin; and Khan, Rizwan Ullah. "Preparation, properties and challenges of the microneedles-based insulin delivery system." *Journal of Controlled Release* 288 (2018): 173–188.

Cho, Do-Yeon; Lim, Dong-Jin; and Woodworth, Bradford A. "Sustained intratonsillar drug delivery with needle array." EP3573697 (A1). 2019.

Chowdhury, Dewan Fazlul Hoque, inventor; Chowdhury, Dewan Fazlul Hoque, assignee. "Microneedle transdermal delivery device." United States US2017312488 (A1). 2017.

Dardano, Principia; Caliò, Alessandro; Di Palma, Vincenza; Bevilacqua, Maria Fortuna; Di Matteo, Andrea; and De Stefano, Luca. "A photolithographic approach to polymeric microneedles array fabrication." *Materials* 8, no. 12 (2015): 8661–8673.

Donnelly, Ryan and Woolfson, David, inventors; The Queen's University of Belfast, assignee. "Drug delivery device." WO2017191457 (A1). 2017.

Economidou, Sophia N.; Lamprou, Dimitrios A. and Douroumis, Dennis. "3D printing applications for transdermal drug delivery." *International Journal of Pharmaceutics* 544, no. 2 (2018): 415–424.

Falo, Louis D. Jr. and Erdos, Geza, inventors; University of Pittsburgh - Of the Commonwealth System of Higher Education, assignee. "Multi-component bio-active drug delivery and controlled release to the skin by microneedle array devices." US2018304062 (A1). 2018.

Geunbae, Lim; Lee, Jungho; Choi, Byoungsun; and Woo, Hyeonsu, inventors; Postech Academy-Industry Foundation, assignee. "Transdermal drug delivery patch and method for manufacturing same." WO 2019221319. 2019.

Gu, Yongwei; Tang, Xiaomeng; Yang, Meng; Yang, Dishun; and Liu, Jiyong. "Transdermal drug delivery of triptolide-loaded nanostructured lipid carriers: Preparation, pharmacokinetic, and evaluation for rheumatoid arthritis." *International Journal of Pharmaceutics* 554, (2019): 235–244.

Huang, Zhiqiang, inventor. Huang Zhiqiang, assignee. "Anti-allergy monocrystalline silicon microneedle." CN106581849 (A). 2017.

Ita, Kevin. "Ceramic microneedles and hollow microneedles for transdermal drug delivery: Two decades of research." *Journal of Drug Delivery Science and Technology* 44 (2018): 314–322.

Kathe, Kashmira, and Kathpalia, Harsha. "Film forming systems for topical and transdermal drug delivery." *Asian Journal of Pharmaceutical Sciences* 12, no. 6 (2017): 487–497.

Kathuria, Himanshu; Kochhar, Jaspreet Singh; Fong, Michelle Hui Min; Hashimoto, Michinao; Iliescu, Ciprian; Yu, Hanry; and Kang, Lifeng. "Polymeric microneedle array fabrication by photolithography." *Journal of Visualized Experiments* 105 (2015): e52914.

Kim, Jae Soo; Kwon, Soon Chan; and Park, San Jin, inventors. "Microstructure comprising a cross-linked hydrogel of hyaluronic acid, and a method for production thereof." RU2692261C2. 2019.

Kong, Ming; Hong, Qin; Lu, Wenjing; and Du, Yutang, inventors; Du Yutang, assignee. "A multi-layer soluble microneedle array and a preparing method thereof." CN109010257 (A). 2018.

Larrañeta, Eneko; Lutton, Rebecca E. M.; Woolfson, A. David; and Donnelly, Ryan F. "Microneedle arrays as transdermal and intradermal drug delivery systems: Materials science, manufacture and commercial development." *Materials Science and Engineering: R: Reports* 104 (2016): 1–32.

Lim, Chee Yen, inventor; Lim, Chee Yen, assignee. "Pyramidal microneedles with enhanced drug loading capacity and method for manufacturing." CN110475581 (A). 2019.

Lim, Dong-Jin; Vines, Jeremy B.; Park, Hansoo; and Lee, Soo-Hong. "Microneedles: A versatile strategy for transdermal delivery of biological molecules." *International Journal of Biological Macromolecules* 110 (2018): 30–38.

McAllister, Devin V.; Prausnitz, Mark R.; and Henry, Sebastien, inventors; Georgia Tech Res Corporation, assignee. "Drug delivery devices having separable microneedles." US20180133447. 2018.

Ming, Dang and Boshen, Zhang, inventors; Dang Ming, assignee. "High-polymer microneedle and preparation and application methods thereof." CN107320841 (A). 2017.

Minguillo, David, and Thelwall, Mike. "Which are the best innovation support infrastructures for universities? Evidence from R&D output and commercial activities." *Scientometrics* 102, no. 1 (2015): 1057–1081.

Mogusala, Nikita Reddy; Devadasu, Venkat Ratnam and Venisetty, Raj Kumar. "Fabrication of microneedle molds and polymer based biodegradable microneedle patches: A novel method." *American Journal of Drug Delivery and Therapeutics* 2, no. 2 (2015): 060–071.

Narayan, Roger; Hall, Russell and Zhang, Yunyan, inventors; North Carolina State University, assignee. "Devices and methods for transdermal treatment of basal cell carcinoma." US2018369136 (A1). 2018.

Nejad, Hojatollah Rezaei; Sadeqi, Aydin; Kiaee, Gita; and Sonkusale, Sameer. "Low-cost and cleanroom-free fabrication of microneedles." *Microsystems & Nanoengineering* 4, no. 1 (2018): 1–7.

Nguyen, Hiep X.; Bozorg, Behnam Dasht; Kim, Yujin; Wieber, Alena; Birk, Gudrun; Lubda, Dieter; and Banga, Ajay K. "Poly (vinyl alcohol) microneedles: Fabrication, characterization, and application for transdermal drug delivery of doxorubicin." *European Journal of Pharmaceutics and Biopharmaceutics* 129 (2018): 88–103.

Nguyen, Thuy Trang, and Park, Jung Hwan. "Human studies with microneedles for evaluation of their efficacy and safety." *Expert Opinion on Drug Delivery* 15, no. 3 (2018): 235–245.

Omar, Samia M.; Nasr, Mohamed; and Rafla, Diana A. "Transdermal patches loaded with L-cysteine HCL as a strategy for protection from mobile phone emitting electromagnetic radiation hazards." *Saudi Pharmaceutical Journal* 27, no. 1 (2019): 112–125.

Prausnitz, Mark R.; Edelhauser, Henry F.; Zarnitsyn,Vladimir G.; Patel, Samirkumar R.; and Grossniklaus, Hans, inventors; Clearside Biomedical, Inc; Georgia Tech Research Corporation; Emory University, assignee. "Apparatus and methods for drug delivery using microneedles." US2017273827 (A1). 2017.

Ross, Russell Frederick,inventor; Sorrento Therapeutics, Inc, assignee. "Drug delivery methods targeting the lymphatic system." WO2019232265 (A1). 2019.

Ruckh, Timothy, inventor; Verily Life Sciences LLC, assignee. "Micro-structures with magnetic removal capability and optionally clear optical path." US10292734 (B1). 2019.

Waghule, Tejashree; Singhvi, Gautam; Dubey, Sunil Kumar; Pandey, Murali Monohar; Gupta, Gaurav; Singh, Mahaveer; and Dua, Kamal. "Microneedles: A smart approach and increasing potential for transdermal drug delivery system." *Biomedicine & Pharmacotherapy* 109 (2019): 1249–1258.

Wang, Qingqing; Liu, Hao; Chen, Minglong; Wei, Yingmei; Zhao, Weiman; and Dong, Zhiyong, inventors; Bengbu Medical College, assignee. "Microneedle transdermal drug delivery paste for treating rheumatoid arthritis and preparation method of microneedle transdermal drug delivery paste." CN 109528695 (A). 2019.

Wang, Yong; Coyne, James; Kauffman, David; and Brandon, Davis, inventors; The Penn State Research Foundation, assignee. Polymer Microneedle Mediated Drug Delivery. WO2020046863 (A1). 2020.

World Intellectual Property Organization (WIPO). "World Intellectual Property Indicators 2019." Available in: <https://www.wipo.int/edocs/pubdocs/en/wipo_pub_941_2019.pdf>. 2019.

Yang, Cheng and Dai, Wanyu, inventors; Tsinghua University Shenzhen Graduate School, assignee. "Degradable microneedle array safe to human body and manufacturing method and mold thereof." CN108939283 (A). 2018.

Ye, Yanqi; Yu, Jicheng; Wen, Di; Kahkoska, Anna R.; and Gu, Zhen. "Polymeric microneedles for transdermal protein delivery." *Advanced Drug Delivery Reviews* 127 (2018): 106–118.

Yu, Haining and Wang, Mengke, inventors; Dalian University of Technology, assignee. "Microneedle template and preparation method and application thereof." CN109278219 (A). 2019.

Zeng, Liuchan; Xian, Yiwen; Zhang, Xu; Chen, Yongli; Cui, Shufen; and Wang, Jinlin, inventors; Shenzhen Polytechnic, assignee. "An integrated transdermal drug delivery device based on drug storage and microneedles." CN209075839 (U). 2019.

Zhu, Jintao; Lju, Pei; and Zhang, Lianbin, inventors; Huazhong University of Science and Technology, assignee. "Method for preparing porous polymer microneedle based on phase separation technology and application of method." CN110115707 (A). 2019.

Zhu, Qiyun; Zarnitsyn, Vladimir G.; Ye, Ling; Wen, Zhiyuan; Gao, Yulong; Pan, Lei; Skountzou, Ioanna et al. "Immunization by vaccine-coated microneedle arrays protects against lethal influenza virus challenge." *Proceedings of the National Academy of Sciences* 106, no. 19 (2009): 7968–7973.

16 Medicinal Plants and Natural Products from Caatinga Biome with Anti-Inflammatory Activity

The Chemistry behind Biological Activity

Jackson Roberto Guedes da Silva Almeida, Ana Paula de Oliveira, Juliane Cabral Silva, Mariana Gama e Silva, Raimundo Gonçalves de Oliveira Junior and Vítor Prates Lorenzo

CONTENTS

16.1 INTRODUCTION

Natural products are produced by various life forms as specialized metabolites that control cellular processes and drive biology. They are also some of the most prolific sources of therapeutics. Academic and other scientists continue to explore the natural world for molecules with interesting chemistries, for they are the drivers of biology and are important to biotechnology, biofuels, agriculture and medicine (Davies and Ryan, 2012).

The interest in medicinal products derived from higher plants (also known as herbal remedies or herbal medicines or phytomedicines) has increased significantly worldwide. This interest is seen especially in developed countries, mainly in some European countries and in the United States (Dutra et al., 2016). Notably, plant-derived compounds are currently employed in modern therapy in addition to playing an important role for the synthesis of some more complex molecules. It is estimated that about 30% of the available therapeutic medications are derived from natural sources, notably from plants and micro-organisms. In some therapeutic areas, such as oncology, the amount of plant-derived medicines achieves 60% (Newman and Cragg, 2020).

DOI: 10.1201/9781003231745-16

The Brazilian population has a long tradition in the use of medicinal plants for the treatment of different acute and chronic diseases, such as pain and inflammation. In the last years, the number of preclinical (nonclinical) studies carried out by Brazilian researchers related to anti-inflammatory effects and immunomodulatory medicinal plants is astonishing (Dutra et al., 2016). Among these studies, we can highlight those carried out with medicinal plants from the Caatinga biome, a semi-arid region of Brazil.

Beginning the study of traditional medicinal plants from the Caatinga biome, we have some questions: Why study these species? What kinds of biological activities do these species possess? What kinds of molecules are present in these species?

Located entirely within northeast semiarid Brazil, the Caatinga region harbours the largest seasonally dry tropical forests and woodlands nucleus, representing approximately 31% of the total area of that biome in the neotropics (Queiroz et al., 2017). Many medicinal plant species from Caatinga are widely known and used in folk medicine to treat several diseases. The main uses of these plants are as sedative, gastroprotective, antioxidant, cytotoxic, antimicrobial, antinociceptive and anti-inflammatory.

Studies carried out by our research group at the Federal University of Vale do São Francisco (UNIVASF) have shown that extracts, essential oils and isolated compounds obtained from several species of medicinal plants of the Caatinga have anti-inflammatory activity. We have selected species from the families *Annonaceae*, *Bromeliaceae*, *Euphorbiaceae*, *Fabaceae*, *Lamiaceae*, *Moraceae*, *Passifloraceae* and *Selaginellaceae*. The phytochemistry of these families is characterized by the presence of terpenoids, cinnamic acid derivatives, lignans, alkaloids and flavonoids. We can highlight the species *Hymenaea martiana* (Pacheco et al., 2021), *Croton rhamnifolioides* (Martins et al., 2020), *Passiflora cincinnata* (Lavor et al., 2018), and *Annona vepretorum* (Silva et al., 2015) as a source of bioactive compounds.

The purpose of this chapter is to present results of research carried out at UNIVASF with the species *Hymenaea martiana* and *Passiflora cincinnata*, typical species of the Caatinga biome popularly used to treat pain and inflammation, and to show how the flavonoids identified in these species are involved in the mechanism of action of the anti-inflammatory activity presented by the extracts of the plants.

Natural products play a significant role in human health in relation to the prevention and treatment of inflammatory conditions. A fairly large number of plants known to contain flavonoids are used in folk medicine, in some cases as anti-inflammatory agents. A variety of *in vitro* and *in vivo* experiments have shown that selected flavonoids possess anti-inflammatory activity. The mechanism by which flavonoids exert their anti-inflammatory effects involves the inhibition of eicosanoid biosynthesis, neutrophil degranulation and inhibition of lipoxygenase (LOX) and cyclooxygenase (COX) activities (Yuan et al., 2006).

16.2 FLAVONOIDS

Flavonoids are phenolic constituents commonly found in a variety of fruits, vegetables and medicinal plants. They add colour, flavour and aroma to plants and play an important role in protection mechanisms against pathogens, ultraviolet radiation and herbivores. Flavonoids comprise an important class of secondary metabolites with numerous possibilities of chemical structures (Al-Jadidi et al., 2016).

Most flavonoids are derived from benzo-γ-pyrone and contain a C_6-C_3-C_6 basic skeleton structure. According to their substitution pattern of A, B and C rings, unsaturation degree and arrangement of the basic skeleton, flavonoids can be classified into a number of subtypes, including chalcones (isoliquiritigenin), flavans, catechins (epicatechin, catechin, and gallocatechin), leucocyanidins (flavan-3,4-diol), flavanones (naringenin, eriodictyol and hesperidin), flavones (luteolin, chrysin and apigenin), flavanonols (dihidrokaempferol), flavonols (kaempferol), anthocyanidins (pelargonidin, cyanidin and malvidin), aurones and isoflavones (genistein, daidzein and glycitein). Some flavonoids occur in plants in free form, while others occur as conjugated *O*- or *C*-glycosides.

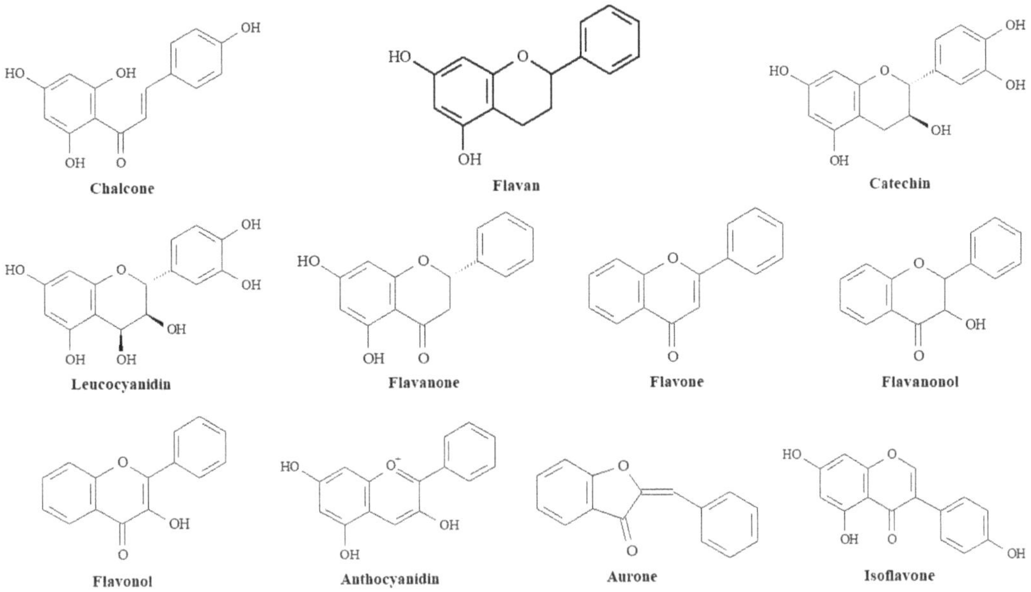

FIGURE 16.1 Chemical structures of different classes of flavonoids.

For *O*-glycosides, the sugar residues are usually linked to 3, 7 or 4′- hydroxyl groups, while for *C*-glycosides, the sugar residues are linked directly to C-6 or C-8. The sugar residues include glucose, galactose, rhamnose, apiose, etc. (Qiao et al., 2011). The most common classes of flavonoids are shown in Figure 16.1.

Flavonoids are found in numerous plants, fruits and vegetables and are known as the most common phytochemicals that possess a multiple range of pharmacological effects. These secondary metabolites have been described as potent antioxidants, free radical scavengers and metal chelators that also present anticholinesterase, antiageing, neuroprotective and anti-inflammatory properties (Teles et al., 2018).

16.3 EXTRACTION AND IDENTIFICATION OF FLAVONOIDS

Several extraction procedures have been used to extract flavonoids from plant materials. Methods can range from classic procedures such as maceration, percolation and soxhlet to more modern procedures such as ultrasound-assisted extraction, supercritical fluid extraction, microwave-assisted extraction and accelerated solvent extraction.

The main extraction method used in our laboratory is still maceration with ethanol (EtOH) 95%. Maceration is a simple extraction method that, although valuable, presents the disadvantage of long extraction time and low efficiency. This procedure consists of soaking the raw material (coarse or powdered) in a selected solvent at room temperature for at least three days, with frequent agitation. After the extraction, the solvent is removed from the mixture, frequently by vacuum evaporation, to concentrate the product. The vital point of this method is the solvent choice, which delineates the classes of compounds recovered from the samples, and also enables the maceration use for the extraction of thermolabile components (Mazzutti et al., 2021).

After obtaining the crude ethanolic extract, the extract is solubilized in a mixture of methanol:water (MeOH:H_2O, 3:7) and subjected to a liquid-liquid partition with the solvents hexane, chloroform and ethyl acetate in ascending order of polarity to obtain the respective fractions, as shown in Figure 16.2.

After this procedure, the crude ethanolic extract and fractions are subjected to chromatographic methods such as high-performance liquid chromatography coupled to diode array detector (HPLC-DAD) and liquid chromatography coupled to mass spectrometry (LC-MS) with electrospray

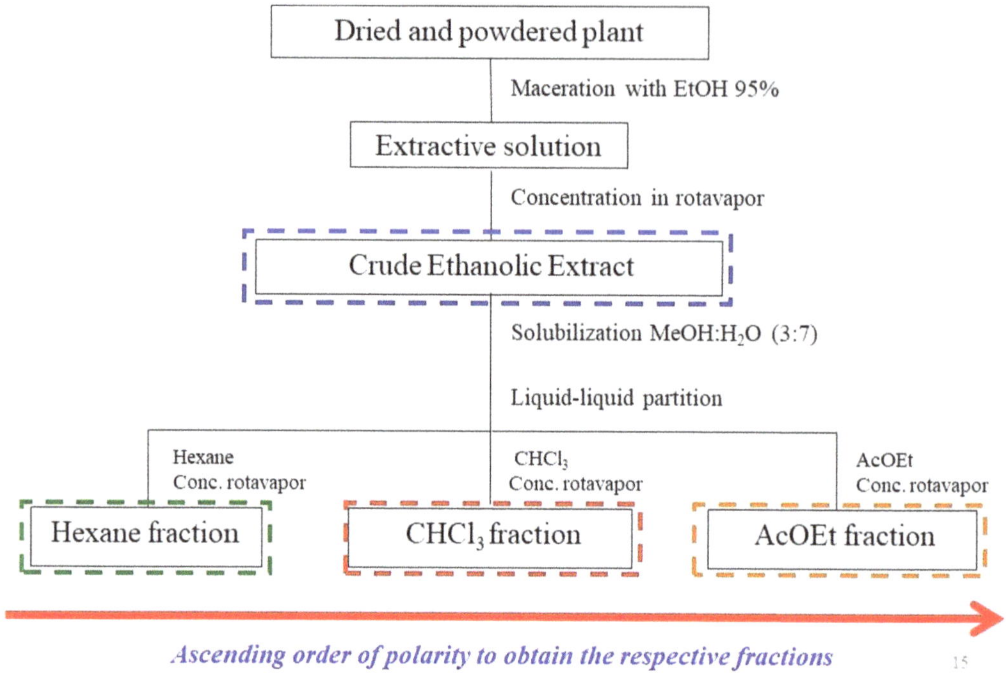

FIGURE 16.2 Method of extraction of chemical constituents from the plant material.

ionization (ESI) for analysis of the chemical composition. The identification of the compounds is based on the retention time, ultraviolet and mass spectra of the compounds in comparison with standards and data in the literature.

16.4 ANTI-INFLAMMATORY ACTIVITY OF FLAVONOIDS

Inflammation is a complex (patho)physiological process, which is regulated by multiple signalling pathways, requires the interaction of different cell types and modulates a broad spectrum of cellular responses including immune cell maturation and function as well as tissue homeostasis (Medzhitov, 2008). Pharmacological strategies to suppress inflammation focus on agonists of the glucocorticoid receptor (glucocorticoids), interference with eicosanoid biosynthesis (nonsteroidal anti-inflammatory drugs [NSAIDs]) and the blockade of proinflammatory cytokine signalling (biologics targeting tumour necrosis factor [TNF-α] and interleukin [IL-1] signalling) (Rainsford, 2007).

Inhibition of cyclooxygenases (COX-1 and COX-2 by NSAIDs [e.g., indomethacin]) is one of the main targets for anti-inflammatory therapy. However, due to the adverse effects of COX-1 inhibition by these drugs, such as renal and gastrointestinal toxicity, selective COX-2 inhibitors have been developed. Among NSAIDs, piroxicam, meloxicam, naproxen and nimesulide are first-generation selective COX-2 inhibitors, while celecoxib, rofecoxib, etoricoxib, valdecoxib, parecoxib and lumiracoxib are highly selective second-generation alternatives (Chagas-Paula et al., 2015).

Several natural products have demonstrated anti-inflammatory activity, acting as multitarget drugs. Among them, we can highlight curcumin, (-)-epigallocatechin-3-gallate (EGCG), *trans*-resveratrol, salicylic acid and quercetin, acting on NF-κB, COX-1, 5-lipoxygenase and COX-2 (Koeberle and Werz, 2014). Other compounds include parthenolide, cucurbitacins, 1,8-cineole, pseudopterosins, lyprinol, bromelain, flavonoids, saponins, marine sponge natural products and *Boswellia serrata* gum resin (Yuan et al., 2006).

Several flavonoids such as hesperidin, apigenin, luteolin, and quercetin are reported to possess analgesic and anti-inflammatory effects. It has been reported that flavonoids are able to inhibit expression of isoforms of inducible nitric oxide synthase, cyclooxygenase and lipoxygenase, which

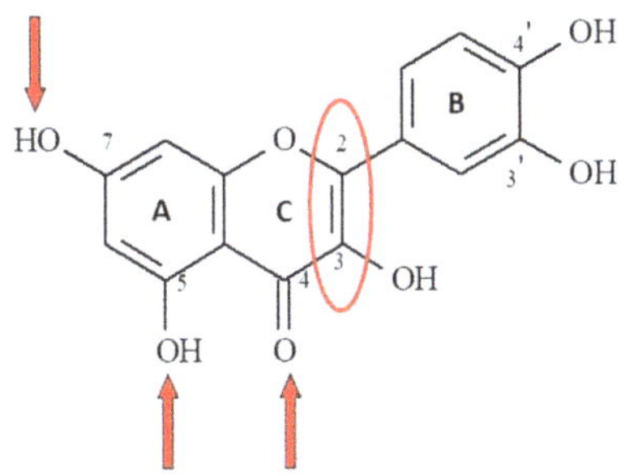

FIGURE 16.3 Structural requirements observed for anti-inflammatory activity of flavonoids. (Adapted from Coutinho et al., 2009.)

are responsible for the production of a great amount of nitric oxide, prostanoids, leukotrienes and other mediators of the inflammatory process such as cytokines, chemokines or adhesion molecules (Kumar and Pandey, 2013). The flavones 7-hydroxy-flavone, 5-hydroxyflavone, chrysin, baicalein, apigenin and luteolin showed COX-inhibitory activity (Verma and Pratap, 2010). Selective flavonoids such as quercetin inhibited both COX and LOX activities (Yuan et al., 2006).

Due to the wide range of biological activities of flavones, their structure-activity relationships have generated interest among medicinal chemists. The currently accepted structural requirements for anti-inflammatory activity of flavonoids are shown in Figure 16.3.

Thus, among the important structural factors for anti-inflammatory activity of flavonoids, the unsaturation in ring C stands out (positions 2–3), the number and position of hydroxyl groups (for example: catechol pattern in ring B flavonoid, that is, ring 3′,4′-dihydroxylated), carbonyl at C-4 (ring B) and the nonglycosylation of the molecule. However, subclasses of flavonoids that do not have one of these patterns in their structure, such as the aglycone kaempferol, also stand out for their activity on enzymes in the inflammation cascade (Coutinho et al., 2009).

16.5 ANTI-INFLAMMATORY ACTIVITY OF TWO MEDICINAL PLANTS FROM THE CAATINGA BIOME

Hymenaea martiana Hayne is a native tree known in the northeast region of Brazil as *jatobá*. Previous studies have showed antinociceptive and anti-inflammatory activities related with the hydroalcoholic extract obtained from the bark. Carneiro et al. (1993) demonstrated the presence of glycoside compounds such as astilbin, eucryphin and engeletin in the bark of *H. martiana*.

In another study, we carried out an investigation about the effects of the crude ethanolic extract (Hm-EtOH) and the ethyl acetate fraction (Hm-AcOEt) of *H. martiana* in models of nociception and inflammation in mice. According to the results of this study, both Hm-EtOH and Hm-AcOEt have antinociceptive and anti-inflammatory activities, which could be related with the presence of flavonoid in the extracts (Pacheco et al., 2021). The chemical composition of the species also was investigated. HPLC-DAD-ESI/MS^n analysis of the ethanolic extract and ethyl acetate fraction revealed the presence of 11 peaks in the chromatograms, and all these peaks were identified: taxifolin, eucryphin, astilbin and 3 diastereoisomers, engeletin and 2 diastereoisomers, quercitrin and 2,6,3′,4′-tetrahydroxy-2-benzylcoumaran-3-one. The ethyl acetate fraction had a higher astilbin concentration (151.87 mg/mL) than the ethanolic extract (40.13 mg/mL) (Pacheco et al., 2019).

Passiflora cincinnata Mast. is a native tree known in the northeast region of Brazil as *maracujá do mato* and *maracujá da Caatinga*. Our results have showed antinociceptive and anti-inflammatory activities related with ethanolic extract obtained from the aerial parts (Lavor et al., 2018). The extracts obtained from different parts of this species were analysed by modern chromatographic methods such as HPLC-DAD, HPLC-DAD-MS and HPLC-DAD-MS/MS analysis, leading to the identification of 14 secondary metabolites in *P. cincinnata* extracts, mainly derivatives of vitexin, isovitexin, orientin and isoorientin (Leal et al., 2020).

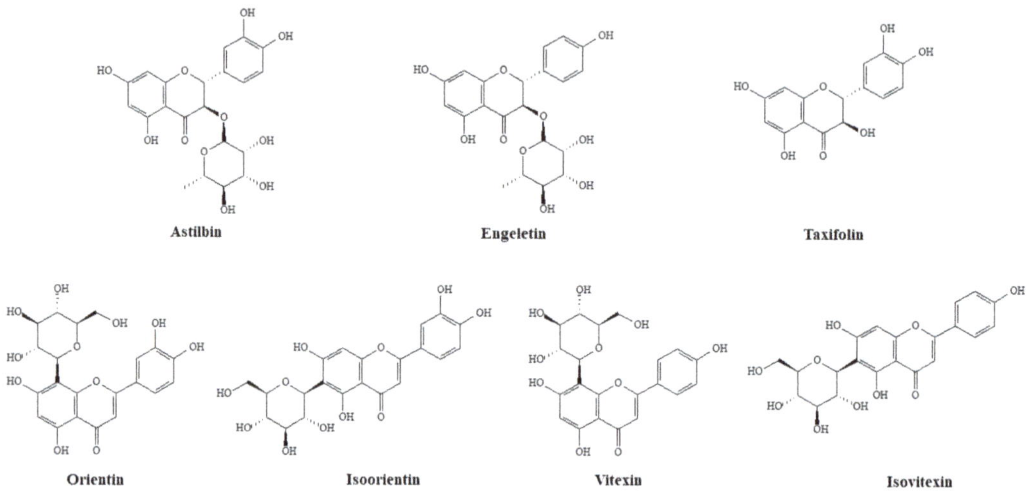

FIGURE 16.4 Chemical structures of chemical markers of *Hymenaea martiana* and *Passiflora cincinnata*.

Regarding the pharmacological activity of two species, taken together, the results showed the antinociceptive and anti-inflammatory activity of the plant extracts. The exact mechanism involved in the antinociceptive and anti-inflammatory activities is not completely understood but, at least in part, there is the participation of opioid receptors (antinociceptive effect) and inhibition of cyclooxygenase-2 enzyme (anti-inflammatory effect).

To confirm the hypothesis that the identified flavonoids can bind to the enzyme cyclooxygenase, a molecular docking study was carried out with the flavonoids considered to be chemical markers for both species (astilbin, taxifolin and engeletin to *H. martiana* and orientin, isoorientin, vitexin and vitexin to *P. cincinnata*). Figure 16.4 shows the chemical structure of these flavonoids.

Considering the popular use of the plants for treatment of pain and inflammation, as well as the presence of flavonoids in plant extracts, we decided to investigate the action of these flavonoids on the human cyclooxygenase-2 enzyme. Animal studies showed that the extracts have anti-inflammatory activity and that the likely mechanism of action is the inhibition of COX-2. Our key question was: Do identified flavonoids have an effect on human COX-2?

For this, the X-ray crystallographic structure of human COX-2 enzyme complexed with rofecoxib was obtained from the RCSB Protein Data Bank (PDB ID: 5KIR) (Orlando and Malkowski, 2016). Figures 16.5 and 16.6 show the crystallographic structure of human COX-2 enzyme. In Figure 16.7,

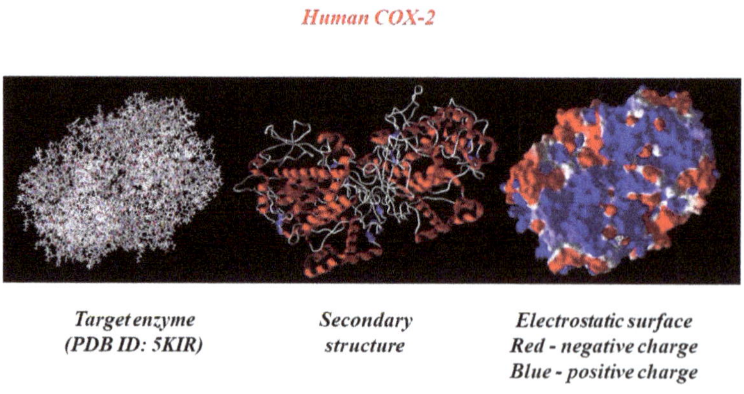

FIGURE 16.5 Crystallographic structure of human cyclooxygenase-2. (Obtained from PDB.)

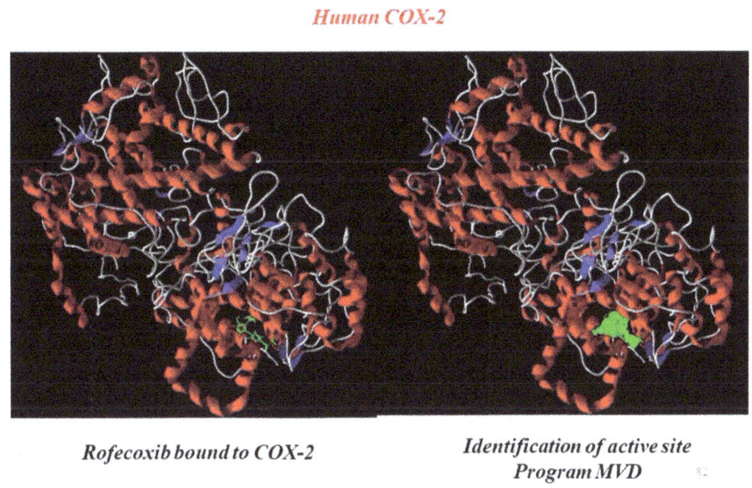

FIGURE 16.6 Crystallographic structure of rofecoxib bound to human cyclooxygenase-2. (Obtained from PDB.)

we have the validation of the methodology by redocking, overlapping the crystallographic structure (in green) to the structure obtained by the docking (in purple). It is possible to observe a good overlap of the structures of rofecoxib.

Values of the interaction energy of flavonoids with the target enzyme has been calculated (see Table 16.1). All flavonoids evaluated had lower interaction energy than the original ligand rofecoxib, demonstrating the pharmacological potential of these compounds on the COX-2 enzyme. Astilbin and orientin were more active than other flavonoids. Figure 16.8 shows the binding of rofecoxib and flavonoids in the active site of the enzyme.

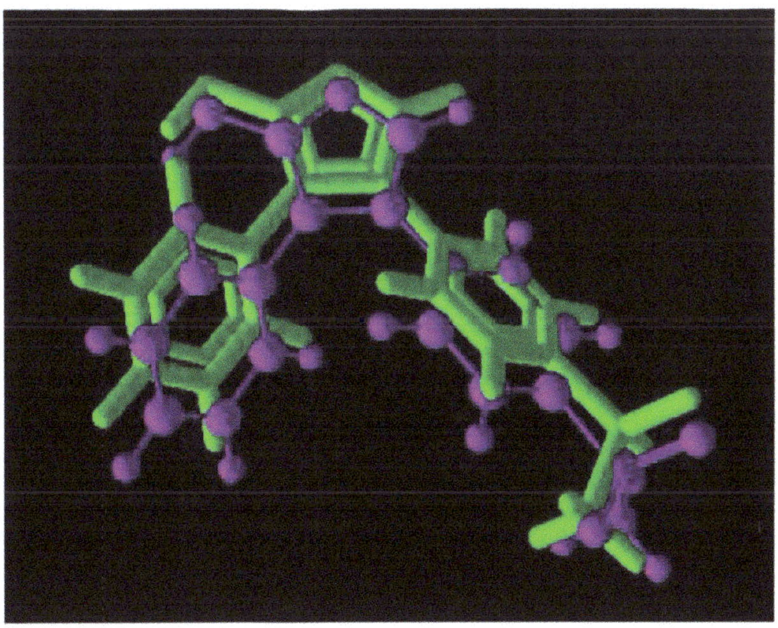

FIGURE 16.7 Validation by redocking.

TABLE 16.1
Values of the Interaction Energy of Flavonoids with the Target Enzyme Cyclooxigenase-2 (COX-2) from Humans

Binding Molecule	Energy
Astilbin	–170.642
Orientin	–166.532
Engeletin	–158.866
Vitexin	–154.691
Isoorientin	–154.292
Taxifolin	–149.727
Isovitexin	–147.069
Rofecoxib	–146.046

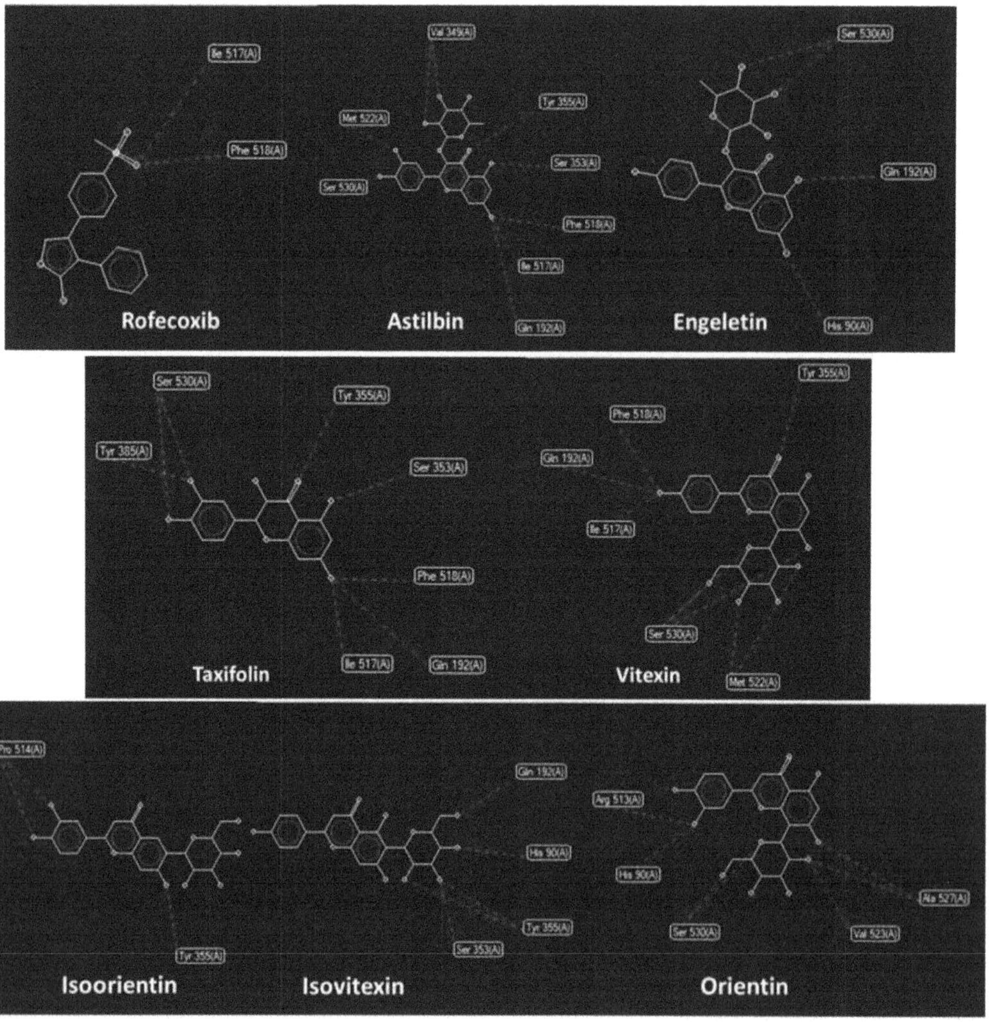

FIGURE 16.8 Binding of rofecoxib and flavonoids in the active site of the enzyme cyclooxygenase-2.

16.6 CONCLUDING REMARKS

In summary, species from the Caatinga biome present an interesting profile of chemical constituents and biological activities, mainly antinociceptive and anti-inflammatory activity. The chemical constituents present in the extracts can act by several mechanisms of action (multitarget compounds), but with strong selectivity for the enzyme COX-2. The flavonoids may be responsible for the biological activity presented by the extracts. Pharmacological and chemical studies are continuing in order to characterize the chemical constituents as well as the mechanism responsible for these effects.

ACKNOWLEDGEMENTS

The authors thanks to Fundação de Amparo à Ciência e Tecnologia do Estado de Pernambuco – FACEPE (Process APQ-0309-4.03/15) and Conselho Nacional de Desenvolvimento Científico e Tecnológico – CNPq (Process 431566/2018-6 and 307782/2017-4) for financial support.

REFERENCES

Al-Jadidi, Hoda Salim Khamis, and Mohammad Amzad Hossain. "Determination of the total phenols, flavonoids and antimicrobial activity of the crude extracts from locally grown neem stems." *Asian Pacific Journal of Tropical Disease* 6, no. 5 (2016): 376–379.

Carneiro, Eliane, João B. Calixto, Franco Delle Monache, and Rosendo A. Yunes. "Isolation, chemical identification and pharmacological evaluation of eucryphin, astilbin and engelitin obtained from the bark of *Hymenaea martiana*." *International Journal of Pharmacognosy* 31, no. 1 (1993): 38–46.

Chagas-Paula, Daniela A., Tong Zhang, Fernando B. Da Costa, and Ru Angelie Edrada-Ebel. "A metabolomic approach to target compounds from the Asteraceae family for dual COX and LOX inhibition." *Metabolites* 5, no. 3 (2015): 404–430.

Coutinho, Marcela AS, Michele F. Muzitano, and Sônia S. Costa. "Flavonoides: Potenciais agentes terapêuticos para o processo inflamatório." *Revista Virtual de Química* 1, no. 3 (2009): 241–256.

Davies, Julian, and Katherine S. Ryan. "Introducing the parvome: Bioactive compounds in the microbial world." *ACS Chemical Biology* 7, no. 2 (2012): 252–259.

Dutra, Rafael C., Maria M. Campos, Adair RS Santos, and João B. Calixto. "Medicinal plants in Brazil: Pharmacological studies, drug discovery, challenges and perspectives." *Pharmacological Research* 112, (2016): 4–29.

Koeberle, Andreas, and Oliver Werz. "Multi-target approach for natural products in inflammation." *Drug Discovery Today* 19, no. 12 (2014): 1871–1882.

Kumar, Shashank, and Abhay K. Pandey. "Chemistry and biological activities of flavonoids: An overview." *Scientific World Journal* 2013 (2013): 1–16.

Lavor, E. M., A. E. B. P. Leal, A. W. C. Fernandes, F. P. R. A. Ribeiro, J. M. Barbosa, M. G. Silva, R. B. A. Teles, L. F. S. Oliveira, J. C. Silva, and L. A. Rolim. "Ethanolic extract of the aerial parts of *Passiflora cincinnata* Mast. (Passifloraceae) reduces nociceptive and inflammatory events in mice." *Phytomedicine* 47 (2018): 58–68.

Leal, Ana Ediléia Barbosa Pereira, Ana Paula de Oliveira, Raira Feitosa dos Santos, Juliana Mikaelly Dias Soares, Erica Martins de Lavor, Michelle Cruz Pontes, Julianeli Tolentino de Lima et al. "Determination of phenolic compounds, in vitro antioxidant activity and characterization of secondary metabolites in different parts of *Passiflora cincinnata* by HPLC-DAD-MS/MS analysis." *Natural Product Research* 34, no. 7 (2020): 995–1001.

Martins, Anita Oliveira Brito Pereira Bezerra, Almir Gonçalves Wanderley, Isabel Sousa Alcântara, Lindaiane Bezerra Rodrigues, Francisco Rafael Alves Santana Cesário, Maria Rayane Correia de Oliveira, Fyama Ferreira Castro et al. "Anti-inflammatory and physicochemical characterization of the *Croton rhamni folioides* essential oil inclusion complex in β-cyclodextrin." *Biology* 9, no. 6 (2020): 114.

Mazzutti, Simone, Rozangela Curi Pedrosa, and Sandra Regina Salvador Ferreira. "Green processes in foodomics. Supercritical fluid extraction of bioactives." *Comprehensive Foodomics* 2021 (2021): 725–743.

Medzhitov, Ruslan. "Origin and physiological roles of inflammation." *Nature* 454, no. 7203 (2008): 428–435.

Newman, David J., and Gordon M. Cragg. "Natural products as sources of new drugs over the nearly four decades from 01/1981 to 09/2019." *Journal of Natural Products* 83, no. 3 (2020): 770–803.

Orlando, Benjamin J., and Michael G. Malkowski. "Crystal structure of rofecoxib bound to human cyclooxygenase-2." *Acta Crystallographica Section F: Structural Biology Communications* 72, no. 10 (2016): 772–776.
Pacheco, A. G. M., E. J. Pacheco, L. A. R. O. Macedo, J. C. Silva, S. R. G. Lima-Saraiva, V. P. Barros, R. G. Oliveira-Junior et al. "Antinociceptive and anti-inflammatory activities of *Hymenaea martiana* Hayne (Fabaceae) in mice." *Brazilian Journal of Biology* 82 (2022): e240359.
Pacheco, Alessandra Gomes Marques, Alexsandro Branco, Celso Amorim Câmara, Tânia Maria Sarmento Silva, Telma Maria Guedes Silva, Ana Paula de Oliveira, Alan Diego da Conceição Santos et al. "Identification of flavonoids in *Hymenaea martiana* Hayne (Fabaceae) by HPLC-DAD-MSn analysis." Natural Product Research 35, no. 14 (2021): 2414–2419.
Qiao, Xue, Wen-zhi Yang, De-an Guo, and Min Ye. "Extraction, separation, detection, and structural analysis of flavonoids." *Current Organic Chemistry* 15, no. 15 (2011): 2541–2566.
Queiroz, Luciano Paganucci, Domingos Cardoso, Moabe Ferreira Fernandes, and Marcelo Freire Moro. "Diversity and evolution of flowering plants of the Caatinga domain." In *Caatinga*, pp. 23–63. Springer, Cham, 2017.
Rainsford, K. D. "Anti-inflammatory drugs in the 21st century." *Subcellular Biochemistry* 42 (2007): 3–27.
Silva, Juliane C., Camila de S Araújo, Sarah Raquel G. de Lima-Saraiva, Raimundo G. de Oliveira-Junior, Tâmara C. Diniz, Carlos Wagner de S. Wanderley, Raimundo C. Palheta-Júnior et al. "Antinociceptive and anti-inflammatory activities of the ethanolic extract of *Annona vepretorum* Mart. (Annonaceae) in rodents." *BMC Complementary and Alternative Medicine* 15, no. 1 (2015): 1–10.
Teles, R. B. A., T. C. Diniz, T. C. C. Pinto, R. G. Oliveira-Júnior, M. G. Silva, E. M. Lavor, A. W. C. Fernandes, A. P. Oliveira, F. P. R. A. Ribeiro, A. A. M. Silva, T. C. F. Cavalcante, L. J. Quintans-Júnior, J. R. G. S. Almeida. "Flavonoids as therapeutic agents in Alzheimer's and Parkinson's diseases: A systematic review of preclinical evidences." *Oxidative Medicine and Cellular Longevity* 2018 (2018): 7043213.
Verma, Alok Kumar, and Ram Pratap. "The biological potential of flavones." *Natural Product Reports* 27, no. 11 (2010): 1571–1593.
Yuan, Gaofeng, Mark L. Wahlqvist, Guoqing He, Min Yang, and Duo Li. "Natural products and anti-inflammatory activity." *Asia Pacific Journal of Clinical Nutrition* 15, no. 2 (2006).

17 Plant-Based Nanoparticle Synthesis for Human Health

Current Status and Future Prospects

Gurusaravanan Packiaraj, Vinoth Sathasivam and Arun Muthukrishnan

CONTENTS

17.1 INTRODUCTION

Nanotechnology is an emerging field with wide applications in medicine and its related areas. Nanotechnology offers special benefits to humans because of its unique size, shape, physical, chemical and biological properties. This technology transfer helps to produce small particles that range from 1 to 100 nm, hence their classification as nanoparticles (NPs) (Sun et al. 2014). Today, the synthesized small particles are used in various fields including agriculture, food production, molecular diagnostics, drug delivery, biomedical and catalytic applications and therapeutics such as anticancer, antimicrobial, antioxidant and antidiabetic therapies (Mashwani et al. 2015). Physical, chemical and biological methods are employed for the synthesis of metal nanoparticles. The effect of these synthesized metal nanoparticles in various fields mainly depends on their shape and size. Based on their morphologies, these particles may be considered as potential and valuable products with different applications especially in the fields of healthcare, environment, electronics, agriculture, etc. Among the different methods of synthesis, the biological method of NP synthesis has gained much attention from researchers.

Green synthesized NPs mainly focus on diagnosis and treatment of disease due to the easy process of separation and alteration of molecules without any pressure, energy, temperature or toxic chemicals (Rajan et al. 2015). In the last few years, researchers and scientists have paid more attention to biosynthesized or green synthesized nanomaterials by different biological agents, i.e., extracts of plant, algae, fungi, micro-organisms, oligo and polysaccharides, which are shown in Figure 17.1.

In the last decade, biological methods of metallic nanomaterial synthesis using silver (Ag), zinc (Zn), gold (Au), copper (Cu), selenium (Se), etc. have been explored in different fields of application.

DOI: 10.1201/9781003231745-17

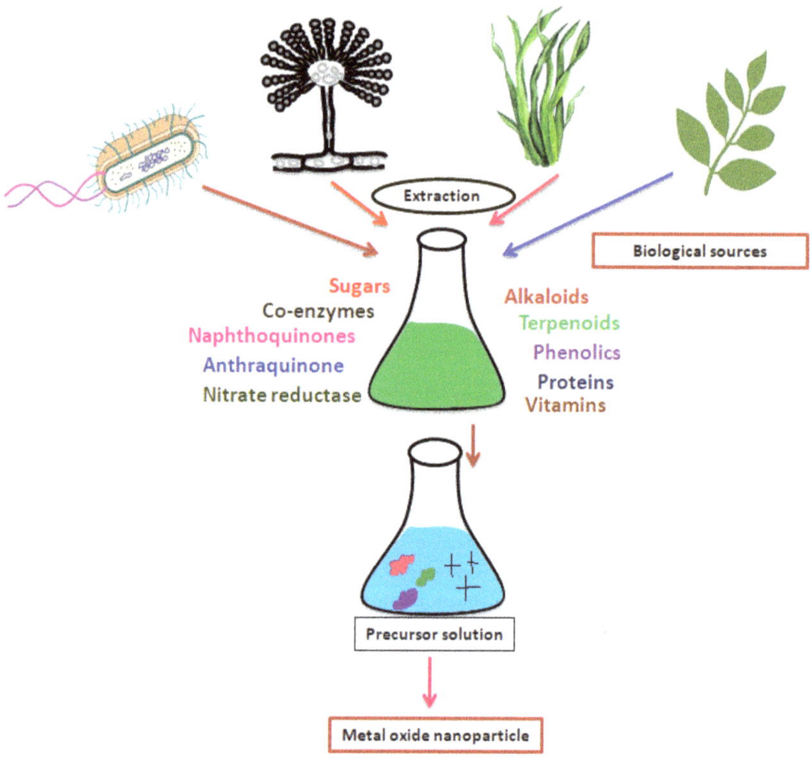

FIGURE 17.1 Synthesis of nanoparticles from biological sources.

Recently, plant-based synthesis of metal NPs has proven to be a highly significant method due to its unique properties such as slower kinetics, better manipulation and easy reduction of metal ions, etc. (Ventola 2017; Boomi et al. 2019; Khan et al. 2020). In the biological method of NP synthesis, phytocompounds and macromolecules are used instead of chemical compounds and organic solvents (Ebrahiminezhad et al. 2018). The efficiency and significant effects of various plant-based nanomaterials mainly depend on physicochemical characteristics (see Figure 17.2). The reaction conditions (temperature, time and concentration of plant extracts) and chemical specification of the plant extracts (concentration of metals, pH and phytocompounds) might influence the production of nanomaterials (Harshiny et al. 2015; Jassal et al. 2016). Green synthesis of metal NPs using plant extracts is less hazardous, has lower manufacturing costs and is versatile and ecofriendly. Currently, green synthesized metallic NPs by plant extract play a crucial role in the pharmaceutical industry for curing various diseases, especially in cancer therapy. Metal NPs conjugated with phytocompounds can efficiently deliver drugs to target sites (Gnanavel et al. 2017). Therefore, the production of plant-based nanomaterials or nanomedicines has the potential to have a big impact on human healthcare (see Figure 17.3). These types of green synthesized NPs have a high potential for therapeutic effects against infectious or chronic diseases. In this chapter, we discuss the current information and future prospects of plant-based metallic NPs and their mechanistic actions are elaborated in therapeutic responses.

17.2 SYNTHESIS AND CHARACTERIZATION OF PLANT-BASED NANOPARTICLES

Today, the biosynthesis of NPs has gained more attention and several researchers have investigated the actual mechanism of metal ions in NP formation with plant-derived molecules. Plant extracts contain various secondary metabolites, which are involved in the reduction of metal ions by redox

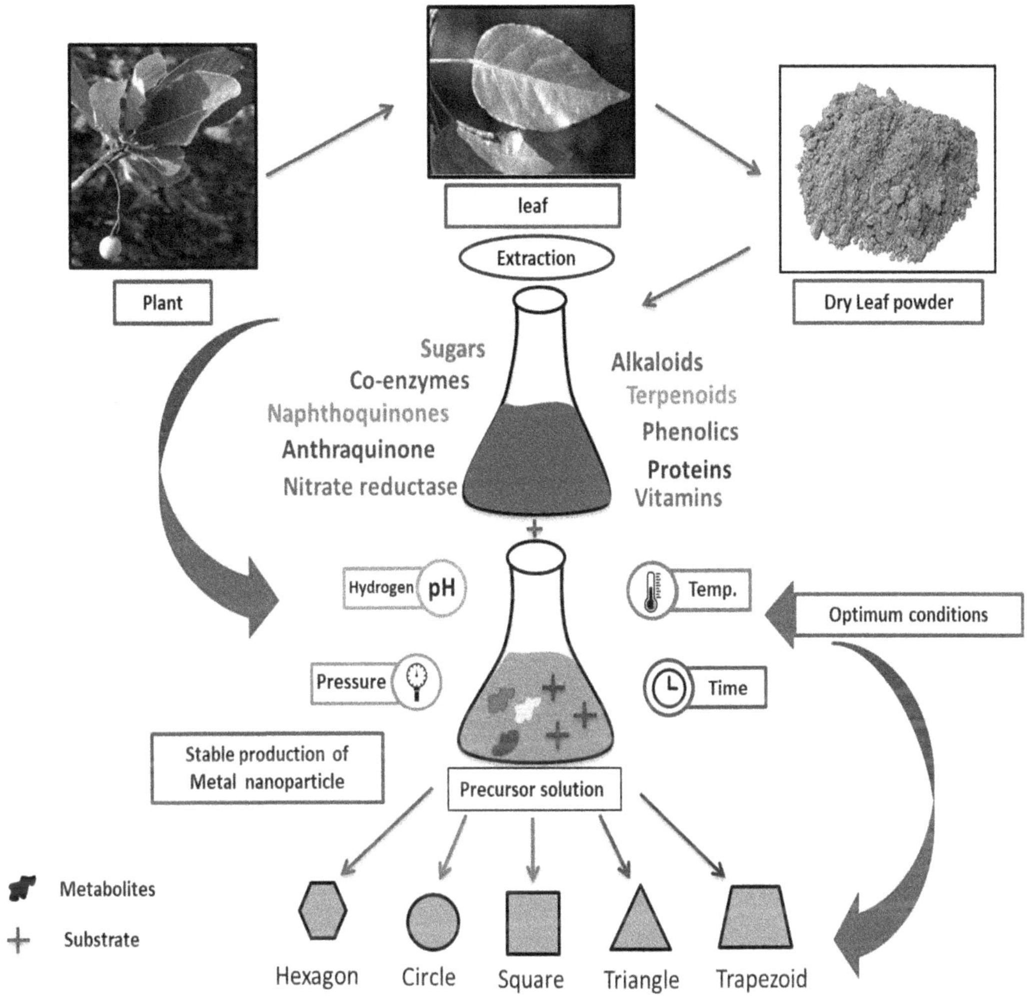

FIGURE 17.2 Plant-based nanoparticles synthesis and optimization of physicochemical factors.

chemical reactions. During the redox matrix, optimized conditions can manipulate the capping and stabilization of NPs. The functional group of plant secondary metabolites envelops the core metal ions and facilitates the stability of NPs for longer periods. The redox chemical reaction and optimized conditions help the synthesis of metal NP formation in plants with compatible natures and therapeutic significance (Javed et al. 2021).

Plant-based NPs are cost effective and ecofriendly compared with the chemical and physical synthetic approaches. The potential efficacies of green synthesized nanomaterials in therapeutic actions depends on their nanostructure, especially size and shape (see Figure 17.4). The significant effects of physicochemical reactions such as temperature, pH, concentration of metal ions and plant extracts are manipulated by the size, shape and potential biological activities of the NPs. Different researchers reported that plant-based NPs and their unique size and shape make them ideal to treat various diseases (Javed et al. 2020). The various shape and size of the NPs were characterized by ultraviolet visible spectrum (UV-Vis), X-ray diffraction (XRD), scanning electron microscopy (SEM), transmission electron microscopy (TEM), Fourier-transform infrared (FTIR) and atomic force microscopy (AFM). These characterization analysis methods showed the presence of large-sized to smaller and aggregated NPs (Bar et al. 2009). The SEM and TEM analyses also confirmed the presence of different shapes, namely spherical, oval or triangular, cubic/hexagonal, quasi-spherical, and irregular formation of NPs (Singh et al. 2016b; Velmurugan et al. 2015).

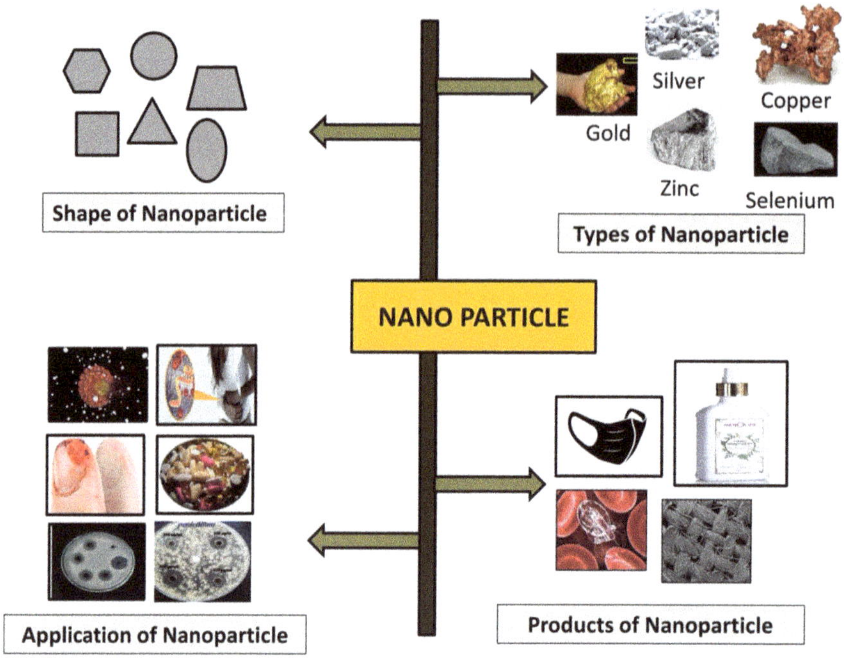

FIGURE 17.3 Biological synthesis of metal nanoparticles and its biomedical applications.

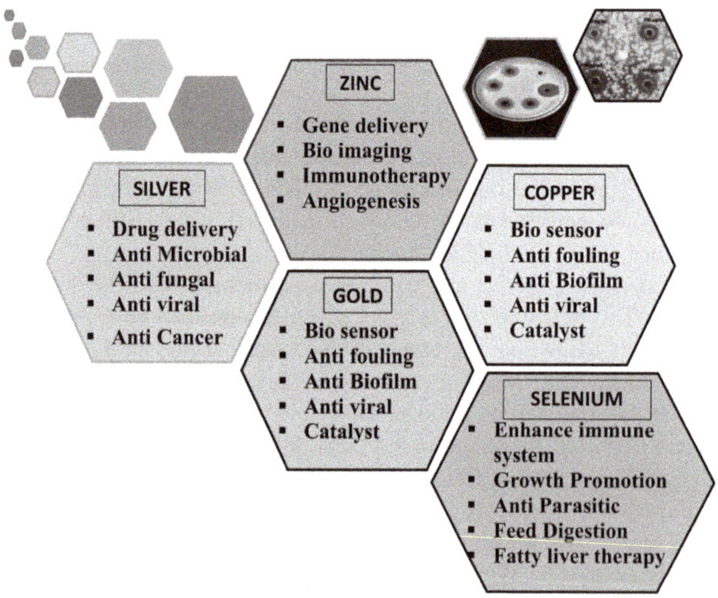

FIGURE 17.4 Various structures of nanoparticles and their uses.

17.3 GREEN SYNTHESIS OF PLANT-BASED SILVER NANOPARTICLES AND THEIR APPLICATIONS IN MEDICINE

Scientists have been successfully able to synthesize silver NPs (AgNPs) by using extracts of plant parts like roots, stems, leaves, flowers and fruits. The plant-based synthesized AgNPs are in high demand in the field of medicine and life sciences because of their capping and stabilizing properties (Singh et al. 2016a). Plant secondary metabolites and other molecules such as proteins, enzymes, amino acids, vitamins and polysaccharides have played an important role in the metal salt reduction reaction. Furthermore, plant-based synthesized metal NPs contain anticancer, antidiabetic, antimicrobial and antioxidant potential (Ghotekar et al. 2020). Many researchers are continuously exploring novel strategies to produce minute particles with NP dimensions ranging from 1 to 100 nm. Recently, biogenic plant-based AgNPs have attracted considerable attention because of their controlled physicochemical properties. These particles constitute effective antioxidant, anticancer, antidiabetic, antimalarial and antimicrobial agents (Sukirtha et al. 2012; Daisy and Saipriya 2012; Singh et al. 2021). AgNPs have antimicrobial potential and can extinguish microbes through membrane damage and the oxidation of cellular materials (Huang et al. 2014).

Carissa carandas L. leaf extracts contain active phytocompounds used to synthesize AgNPs. *C. carandas* leaf extract is traditionally used as an antibacterial agent and it is rich in flavonoids, vitamin C and iron (Arif et al. 2016). Based on its traditional usage, scientists used *C. carandas* leaf extract as a reducing and capping agent for synthesized AgNPs, which showed high efficiency in antioxidant and antimicrobial activity. The maximum antioxidant potential of AgNPs using plant leaf extracts shows 90% 2,2-Diphenyl-1-picrylhydrazylradical (DPPH) scavenging activity radical scavenging activity. In addition, the effects of synthesized AgNPs using *C. carandas* leaf extract showed maximum growth of inhibition against *Shigella flexineri* (24 mm) compared with the plant extract (Singh et al. 2021). Different researchers reported that green synthesis of AgNPs using plant extracts such as *Boerhaavia diffusa* (Kumar et al. 2014b), *Azadirachta indica* (Ahmed et al. 2016) and *Ocimum basilicum (*Khan et al. 2017*)* acts as a good antimicrobial agent against various human and animal pathogens. The synthesized AgNPs from plant secondary metabolites have an essential role in drug delivery due to their remarkable physicochemical properties. The AgNps synthesized from the methonilic extract of *Vitex negundo* (Prabhu et al. 2013) and *Acalypha indica* (Krishnaraj et al. 2014) have been administrated against human cancer cells, and results revealed that the AgNPs showed fewer adverse effects.

17.4 GREEN SYNTHESIS OF PLANT-BASED ZINC NANOPARTICLES AND THEIR PHARMACEUTICAL APPLICATIONS

Nanotechnology dealing with the synthesis of a variety of metal NPs is gaining lot of significance due to its suitability in the field of medicine. There are several fine metal NPs synthesized using plant extracts that have distinctive properties such as stability, capping, catalytic, antimicrobial and anticancer traits (Table 17.1). Among the various noble metal NPs, zinc oxide (ZnO) nanoparticles (ZnONPs) have received a special focus for the past few years due to their notable properties, like chemical stability, being rich in -OH groups and the fact that ZnO allows slow dissolution in both acidic and strong basic conditions if the surface is in direct contact with the solution (Zhou et al. 2006). The biosynthesis of ZnONPs and their properties focuses more attention onto their biomedical applications, as they play an important role in medicine as anti-inflammatory, anticancer and antimicrobial agents (Liu et al. 2008; Bisht and Rayamajhi 2016). Recent reports confirmed that green synthesis of ZnONPs using plant extracts possesses significant anticancer activity. The NP synthesis using aqueous leaf extracts of *C. auriculata* with zinc acetate can enhance the chemical stability of reducing and capping agents by a continuous stirring process at room temperature. The synthesized ZnONPs are hexagonal nanostructured with an average diameter of 20 nm (Prasad et al. 2020).

TABLE 17.1
Plant-Based Nanoparticle Synthesis and Their Biological Applications

Plant Name	Compound	Nanoparticle	Size	Shape	Application	Reference
Memecylon umbellatum Burm F.	4-N-methyl-benzoic acid	Silver	7–22 nm	Spherical	Antimicrobial, antioxidant and antitumour activity	[AlSalhi et al. 2019]
Peganum harmala L.	Alkaloid	Platinum and Palladium	Pt NPs-20.3 ± 1.9nm; PD NPs-22.5 ± 5.7	Spherical with smooth margins	Antitumour activity	[Fahmy et al. 2021]
Ocimum sanctum	Quercetin	Silver	11.35 nm	Spherical	Antibacterial activity	[Jain and Mehata 2017]
Syzygium a lternifolium (Wt.) Walp	Ascorbic acid	Silver	4–48 nm	Spherical	Antimicrobial activity	[Yugandhar et al. 2015]
Mimusops elengi L.	Gallic acid, pinocembrin, quercetin and chlorogenic acid	Gold	9–14 nm	Spherical	Catalytic activity	[Majumdar et al. 2016]
Terminalia arjuna (Roxb. Ex DC.) Wight & Arn	Polyphenols (flavonoids), terpenoids, ketones, aldehydes	Copper	~23 nm	Spherical	Antimicrobial and antioxidant property	[Yallappa et al. 2013]
Syzygiumjambos (L.) Alston	Saccharides and phenolics	Silver Gold	8.51 ± 1.63 nm 10.34 ± 1.01 nm	Spherical Ellipsoidal	Antimalarial activity	[Dutta et al. 2017]
Saraca asoca (Roxb.)	Hydroxylamine and carboxyl groups	Silver	2–10 nm	Spherical	Antibacterial activity	[Banerjee and Nath 2015]
Allium cepa L.	Vitamin C	Gold	~100 nm	Spherical and cubic	Probing cellular process through nanoparticle-mediated imaging	[Parida et al. 2011]
Justicia gendarussa Burm F.	Polyphenol and flavonoid	Gold	27 nm	Spherical, triangle, truncated triangle and hexagonal	Biomedical applications	[Fazaludeen et al. 2012]
Andrographis paniculata	Terpenoid (TAP) fractions	Zinc oxide (ZnO)	20.23 nm	Hexagonal	Drug delivery processes	[Kavitha et al. 2017]
Peganum harmala L.	Alkaloids	Silver	22.5–66.2 nm	Spherical	Growth inhibitor	[Almadiy et al. 2018]

Scientists reported that the sizes of the ZnONPs influenced tumour reduction potential. Different sized ZnO particles (20, 60, 80 and 100 nm) were tested in leukaemia K562 and adriamycin-resistant K562/A02 cells. A synergistic cytotoxic effect was observed and K562/A02 cells were found to be more sensitive to ZnONPs (Guo et al. 2008). The synthesized ZnONPs enhanced the tumouricidal potential against cancer cells and inhibited select multidrug resistance pathogens (Prasad et al. 2020). Furthermore, ZnONPs were synthesized from *Cassia fistula* plant extract by using a fuel-employed solution combustion method. The significant metabolites effectively acted as reducing agents and led to synthesis of ZnONPs. The synthesized ZnONPs were hexagonal nanostructures and the formation NP size was found to be ~5–15 nm. These ZnONPs revealed excellent antioxidant activity and enhanced the antimicrobial effects against *Escherichia coli* and multidrug-resistant *Klebsiella pneumoniae.*

The synthesis of ZnONPs using plant extracts is of particular interest in pharmaceutical industries due to its cost effectiveness, high particle yield, high stability and solubility in water. Aqueous leaf extract of *Hibiscus subdariffa* has been used for the synthesis of ZnONPs under optimum concentration of plant extracts and temperature conditions. As a result, spherical shaped NPs with a size range of 12–16 nm were synthesized and were effective antibacterial agents on *Escherichia coli* and *Staphylococcus aureus*. The small size of the NPs stabilized by plant bioactive compounds showed effective antidiabetic activity. The ZnONPs can induce the function of Th1 and Th2 cells and other genes of the pancreas (Bala et al. 2015). Thus, the ZnONPs synthesized using plant extracts can be used in pharmaceutical industries for their potential antibacterial effects and antidiabetic activity.

17.5 GREEN SYNTHESIS OF PLANT-BASED SELENIUM NANOPARTICLES AND THEIR THERAPEUTIC APPLICATIONS

Selenium nanoparticles (SeNPs) are very attractive due to their versatile biocompatibility, low toxicity and use as a drug delivery agent. SeNPs have been proven a good carrier platform to deliver drugs in the targeting ligand's active site (Sperling and Parak 2010). SeNPs have attractive anticancer and immunomodulatory properties and they are used in medical research for treatment, diagnosis, monitoring biological systems, etc. (Moghimi et al. 2005). The green synthesized SeNPs using plant extract possess active phytoconstituents, which help to enhance the efficiency of biological properties. Its effects on biopharmaceutical and pharmacokinetic uses are largely unknown but it solves many pharmaceutical problems associated with drugs in different diseases (Khurana et al. 2019). All synthesized metal NPs like silver, zinc, gold and copper possess a significant role due to their unique properties. Among different metal NPs, SeNPs serve to reduce the toxicity, improve biocompatibility, deliver the drugs in target sites and regulate the enzymatic activities in biological systems. Green synthesized SeNPs have an excellent anticancer potential and since selenium is a trace element, it is incorporated into selenoproteins as selenocysteine, which regulates enzymatic activities (Khurana et al. 2019). SeNPS have been explored in various human disease conditions including cancer, diabetes, inflammatory disorders, liver fibrosis and arthritis caused by oxidative stress and inflammation (Huang et al. 2013; Kumar et al. 2014a). The synthesized SeNPs inhibit the multiplication of acute myeloid leukaemia cells due to their excellent scavenging-free radical activity. Scavenging of free radicals by SeNPs *in vitro* is dependent on particle size. The small size of plant-based SeNPs (~5–15 nm) has reported anticancer potential, involves effective removal of free radicals and prevented oxidation damage in DNA (Huang et al. 2003).

Recently, green synthesized SeNPs using the leaf extract of *Withania somnifera* have tested against human alveolar basal epithelial cancer cells and demonstrated their potential anticancer activity. The fabricated phytocompounds dramatically trigger chromosomal instability and arrest the human alveolar basal epithelial cancer cells' mitotic division (Alagesan and Venugopal 2019). The phytofabrication of SeNPs is found to possess significant antioxidant and antiproliferative effects. In addition, SeNPs have attractive antibacterial activity on *Staphylococcus aureus*,

Klebsiella pneumoniae and *Bacillus subtilis* (Alagesan and Venugopal 2019). The SeNPs synthesized using *Carica papaya* latex controlled the growth of breast cancer cells by apoptotic changes. The plant-based SeNPs caused cytotoxicity in cancer cells by the formation of nuclear condensation and fragmentation (Rajasekar and Kuppusamy 2021). The phytofabrication of SeNPs using fruit extract of *Emblica officinalis* shows a wide range of antimicrobial and anticancer applications and is highly efficient against foodborne pathogens including fungi. The biocompatibility of facile phytofabricated nanoparticles (PF-SeNPs) was tested on N2a cell lines (*Mus musculus* neuroblastoma) and results revealed that PF-SeNPs inhibited the cell viability much more than sodium selenite (control). Authors also proved that PF-SeNPs were less toxic and safer compared to sodium selenite and that they could act as antimicrobial and anticancer agents in the biomedical and pharmaceutical industries (Gunti et al. 2019).

17.6 CONCLUSION

NPs play a major role in the field of agriculture, medicine, precise drug delivery systems and diagnosing disease. The biocompatibility of NPs from biological sources, especially medicinal plants and their phytochemicals, offers more options for potential anticancer, antimicrobial and antilarvicidal activity. Green synthesis of metallic NPs has more advantages in biomedical applications because of its cost-effectiveness, speed and eco-friendly qualities. Based on several research findings, the green synthesis route for NP formation has realistic evidence in human health. Thus, NP synthesis through biological sources like plant, seaweed and micro-organisms plays a significant role in the pharmaceutical industry for use as nanomedicine material in the near future.

REFERENCES

Ahmed, Shakeel, Saif Ullah, Mudasir Ahmad, Babu Lal Swami, and Saiqa Ikram. "Green synthesis of silver nanoparticles using *Azadirachta indica* aqueous leaf extract." *Journal of Radiation Research and Applied Sciences* 9, no. 1 (2016): 1–7.

Alagesan, Venkatesan, and Sujatha Venugopal. "Green synthesis of selenium nanoparticle using leaves extract of *Withania somnifera* and its biological applications and photocatalytic activities." *Bionanoscience* 9, no. 1 (2019): 105–116.

Almadiy, Abdulrhman A., Gomah E. Nenaah, and Dalia M. Shawer. "Facile synthesis of silver nanoparticles using *Harmala* alkaloids and their insecticidal and growth inhibitory activities against the khapra beetle." *Journal of Pest Science* 91, no. 2 (2018): 727–737.

AlSalhi, Mohamad S., Kannan Elangovan, Amirtham Jacob A. Ranjitsingh, P. Murali, and Sandhanasamy Devanesan. "Synthesis of silver nanoparticles using plant derived 4-N-methyl benzoic acid and evaluation of antimicrobial, antioxidant and antitumor activity." *Saudi Journal of Biological Sciences* 26, no. 5 (2019): 970–978.

Arif, Muhammad, Mehnaz Kamal, Talha Jawaid, Mohammad Khalid, Kuldeep Singh Saini, Arun Kumar, and Mohammad Ahmad. "*Carissa carandas* Linn. (Karonda): An exotic minor plant fruit with immense value in nutraceutical and pharmaceutical industries." *Asian Journal of Biomedical and Pharmaceutical Sciences* 6, no. 58 (2016): 14–19.

Bala, Niranjan, S. Saha, M. Chakraborty, M. Maiti, S. Das, R. Basu, and P. Nandy. "Green synthesis of zinc oxide nanoparticles using *Hibiscus subdariffa* leaf extract: Effect of temperature on synthesis, antibacterial activity and anti-diabetic activity." *RSC Advances* 5, no. 7 (2015): 4993–5003.

Banerjee, P., and D. Nath. "A phytochemical approach to synthesize silver nanoparticles for non-toxic biomedical application and study on their antibacterial efficacy." *Nanoscience Technology* 2, no. 1 (2015): 1–14.

Bar, Harekrishna, Dipak Kr Bhui, Gobinda P. Sahoo, Priyanka Sarkar, Santanu Pyne, and Ajay Misra. "Green synthesis of silver nanoparticles using seed extract of *Jatropha curcas*." *Colloids and Surfaces A: Physicochemical and Engineering Aspects* 348, no. 1–3 (2009): 212–216.

Bisht, Gunjan, and Sagar Rayamajhi. "ZnO nanoparticles: A promising anticancer agent." *Nanobiomedicine* 3, no. Godište 2016 (2016): 3–9.

Boomi, P., R. M. Ganesan, G. Poorani, H. Gurumallesh Prabu, S. Ravikumar, and J. Jeyakanthan. "Biological synergy of greener gold nanoparticles by using *Coleus aromaticus* leaf extract." *Materials Science and Engineering: C* 99 (2019): 202–210.

Daisy, P., and K. Saipriya. "Biochemical analysis of *Cassia fistula* aqueous extract and phytochemically synthesized gold nanoparticles as hypoglycemic treatment for diabetes mellitus." *International Journal of Nanomedicine* 7 (2012): 1189.

Dutta, Partha P., Manobjyoti Bordoloi, Kabita Gogoi, Sonali Roy, Bardwi Narzary, Dibya R. Bhattacharyya, Pradyumna K. Mohapatra, and Bhaskar Mazumder. "Antimalarial silver and gold nanoparticles: Green synthesis, characterization and in vitro study." *Biomedicine & Pharmacotherapy* 91 (2017): 567–580.

Ebrahiminezhad, Alireza, Alireza Zare-Hoseinabadi, Ajit K. Sarmah, Saeed Taghizadeh, Younes Ghasemi, and Aydin Berenjian. "Plant-mediated synthesis and applications of iron nanoparticles." *Molecular Biotechnology* 60, no. 2 (2018): 154–168.

Fahmy, Sherif Ashraf, Iten M. Fawzy, Basma M. Saleh, Marwa Y. Issa, Udo Bakowsky, and Hassan Mohamed El-Said Azzazy. "Green synthesis of platinum and palladium nanoparticles using *Peganum harmala* L. seed alkaloids: biological and computational studies." *Nanomaterials* 11, no. 4 (2021): 965.

Fazaludeen, Mohammad Feroze, Chinna Manickam, Ibraheem M. A. Ashankyty, Mohammed Qumani Ahmed, and Quaser Zafar Beg. "Synthesis and characterizations of gold nanoparticles by *Justicia gendarussa* Burm F leaf extract." *Journal of Microbiology and Biotechnology Research* 2, no. 1 (2012): 23–34.

Ghotekar, Suresh, Trupti Pagar, Shreyas Pansambal, and Rajeshwari Oza. "A review on green synthesis of sulfur nanoparticles via plant extract, characterization and its applications." *Advanced Journal of Chemistry-Section B* 2, no. 3, (2020): 128–143.

Gnanavel, V., V. Palanichamy, and Selvaraj Mohana Roopan. "Biosynthesis and characterization of copper oxide nanoparticles and its anticancer activity on human colon cancer cell lines (HCT-116)." *Journal of Photochemistry and Photobiology B: Biology* 171 (2017): 133–138.

Gunti, Lokanadhan, Regina Sharmila Dass, and Naveen Kumar Kalagatur. "Phytofabrication of selenium nanoparticles from *Emblica officinalis* fruit extract and exploring its biopotential applications: Antioxidant, antimicrobial, and biocompatibility." *Frontiers in Microbiology* 10 (2019): 931.

Guo, Dadong, Chunhui Wu, Hui Jiang, Qingning Li, Xuemei Wang, and Baoan Chen. "Synergistic cytotoxic effect of different sized ZnO nanoparticles and daunorubicin against leukemia cancer cells under UV irradiation." *Journal of Photochemistry and Photobiology B: Biology* 93, no. 3 (2008): 119–126.

Harshiny, Muthukumar, Chandrasekaran Nivedhini Iswarya, and Manickam Matheswaran. "Biogenic synthesis of iron nanoparticles using *Amaranthus dubius* leaf extract as a reducing agent." *Powder Technology* 286 (2015): 744–749.

Huang, Yanyu, Lizhen He, Wen Liu, Cundong Fan, Wenjie Zheng, Yum-Shing Wong, and Tianfeng Chen. "Selective cellular uptake and induction of apoptosis of cancer-targeted selenium nanoparticles." *Biomaterials* 34, no. 29 (2013): 7106–7116.

Huang, Yunpeng, Yue-E. Miao, Shanshan Ji, Weng Weei Tjiu, and Tianxi Liu. "Electrospun carbon nanofibers decorated with Ag–Pt bimetallic nanoparticles for selective detection of dopamine." *ACS Applied Materials & Interfaces* 6, no. 15 (2014): 12449–12456.

Huang, Bo, Jinsong Zhang, Jingwu Hou, and Chang Chen. "Free radical scavenging efficiency of Nano-Se in vitro." *Free Radical Biology and Medicine* 35, no. 7 (2003): 805–813.

Jain, Siddhant, and Mohan Singh Mehata. "Medicinal plant leaf extract and pure flavonoid mediated green synthesis of silver nanoparticles and their enhanced antibacterial property." *Scientific Reports* 7, no. 1 (2017): 1–13.

Jassal, Vidhisha, Uma Shanker, and Sweta Gahlot. "Green synthesis of some iron oxide nanoparticles and their interaction with 2-amino, 3-amino and 4-aminopyridines." *Materials Today: Proceedings* 3, no. 6 (2016): 1874–1882.

Javed, Bilal, Muhammad Ikram, Fatima Farooq, Tahira Sultana, Zia-ur-Rehman Mashwani, and Naveed Iqbal Raja. "Biogenesis of silver nanoparticles to treat cancer, diabetes, and microbial infections: A mechanistic overview." *Applied Microbiology and Biotechnology* 105, no. 6 (2021): 2261–2275.

Javed, Bilal, Naveed Iqbal Raja, and Akhtar Nadhman. "Understanding the potential of bio-fabricated non-oxidative silver nanoparticles to eradicate Leishmania and plant bacterial pathogens." *Applied Nanoscience* 10, no. 6 (2020): 2057–2067.

Kavitha, S., M. Dhamodaran, Rajendra Prasad, and M. Ganesan. "Synthesis and characterisation of zinc oxide nanoparticles using terpenoid fractions of *Andrographis paniculata* leaves." *International Nano Letters* 7, no. 2 (2017): 141–147.

Khan, Tariq, Bilal Haider Abbasi, Muhammad Siddique Afridi, Faouzia Tanveer, Ikram Ullah, Samina Bashir, and Christophe Hano. "Melatonin-enhanced biosynthesis of antimicrobial AgNPs by improving the phytochemical reducing potential of a callus culture of *Ocimum basilicum* L. var. *thyrsiflora*." *RSC Advances* 7, no. 61 (2017): 38699–38713.

Khan, Nida, Deepak Kumar, and Pramendra Kumar. "Silver nanoparticles embedded guar gum/gelatin nanocomposite: green synthesis, characterization and antibacterial activity." *Colloid and Interface Science Communications* 35 (2020): 100242.

Khurana, Amit, Sravani Tekula, Mohd Aslam Saifi, Pooladanda Venkatesh, and Chandraiah Godugu. "Therapeutic applications of selenium nanoparticles." *Biomedicine & Pharmacotherapy* 111 (2019): 802–812.

Krishnaraj, C., P. Muthukumaran, R. Ramachandran, M. D. Balakumaran, and P. T. Kalaichelvan. "*Acalypha indica* Linn: Biogenic synthesis of silver and gold nanoparticles and their cytotoxic effects against MDA-MB-231, human breast cancer cells." *Biotechnology Reports* 4 (2014): 42–49.

Kumar, Goru Santosh, Apoorva Kulkarni, Amit Khurana, Jasmine Kaur, and Kulbhushan Tikoo. "Selenium nanoparticles involve HSP-70 and SIRT1 in preventing the progression of type 1 diabetic nephropathy." *Chemico-Biological Interactions* 223 (2014a): 125–133.

Kumar, P. P. N. Vijay, S. V. N. Pammi, Pratap Kollu, K. V. V. Satyanarayana, and U. Shameem. "Green synthesis and characterization of silver nanoparticles using *Boerhaavia diffusa* plant extract and their anti bacterial activity." *Industrial Crops and Products* 52 (2014b): 562–566.

Liu, Dongfang, Wei Wu, Yongfu Qiu, Shihe Yang, Si Xiao, Qu-Quan Wang, Lu Ding, and Jiannong Wang. "Surface functionalization of ZnO nanotetrapods with photoactive and electroactive organic monolayers." *Langmuir* 24, no. 9 (2008): 5052–5059.

Majumdar, Rakhi, Braja Gopal Bag, and Pooja Ghosh. "*Mimusops elengi* bark extract mediated green synthesis of gold nanoparticles and study of its catalytic activity." *Applied Nanoscience* 6, no. 4 (2016): 521–528.

Mashwani, Zia-ur-Rehman, Tariq Khan, Mubarak Ali Khan, and Akhtar Nadhman. "Synthesis in plants and plant extracts of silver nanoparticles with potent antimicrobial properties: Current status and future prospects." *Applied Microbiology and Biotechnology* 99, no. 23 (2015): 9923–9934.

Moghimi, S. Moein, A. Christy Hunter, and J. Clifford Murray. "Nanomedicine: Current status and future prospects." *FASEB Journal* 19, no. 3 (2005): 311–330.

Parida, Umesh Kumar, Birendra Kumar Bindhani, and Padmalochan Nayak. "Green synthesis and characterization of gold nanoparticles using onion (*Allium cepa*) extract." *World Journal Nano Science and Engineering* 1, no. 4 (2011): 93.

Prabhu, D., C. Arulvasu, G. Babu, R. Manikandan, and P. Srinivasan. "Biologically synthesized green silver nanoparticles from leaf extract of *Vitex negundo* L. induce growth-inhibitory effect on human colon cancer cell line HCT15." *Process Biochemistry* 48, no. 2 (2013): 317–324.

Prasad, Kollur Shiva, Shashanka K. Prasad, Mohammad Azam Ansari, Mohammad A. Alzohairy, Mohammad N. Alomary, Sami Al Yahya, Chandrashekar Srinivasa, Mahadevamurthy Murali, Veena Malligere Ankegowda, and Chandan Shivamallu. "Tumoricidal and bactericidal properties of ZnONPs synthesized using *Cassia auriculata* leaf extract." *Biomolecules* 10, no. 7 (2020): 982.

Rajan, Ramachandran, Krishnaraj Chandran, Stacey L. Harper, Soon-Il Yun, and P. Thangavel Kalaichelvan. "Plant extract synthesized silver nanoparticles: An ongoing source of novel biocompatible materials." *Industrial Crops and Products* 70 (2015): 356–373.

Rajasekar, Srinivasan, and Santhi Kuppusamy. "Eco-friendly formulation of selenium nanoparticles and its functional characterization against breast cancer and normal cells." *Journal of Cluster Science* 32, no. 4 (2021): 907–915.

Singh, Reetika, Christophe Hano, Gopal Nath, and Bechan Sharma. "Green biosynthesis of silver nanoparticles using leaf extract of *Carissa carandas* L. and their antioxidant and antimicrobial activity against human pathogenic bacteria." *Biomolecules* 11, no. 2 (2021): 299.

Singh, Priyanka, Yeon Ju Kim, Chao Wang, Ramya Mathiyalagan, Mohamed El-Agamy Farh, and Deok Chun Yang. "Biogenic silver and gold nanoparticles synthesized using red ginseng root extract, and their applications." *Artificial Cells, Nanomedicine, and Biotechnology* 44, no. 3 (2016a): 811–816.

Singh, Priyanka, Yeon Ju Kim, and Deok Chun Yang. "A strategic approach for rapid synthesis of gold and silver nanoparticles by *Panax ginseng* leaves." *Artificial Cells, Nanomedicine, and Biotechnology* 44, no. 8 (2016b): 1949–1957.

Sperling, Reisa A., and Wolfgang J. Parak. "Surface modification, functionalization and bioconjugation of colloidal inorganic nanoparticles." *Philosophical Transactions of the Royal Society A: Mathematical, Physical and Engineering Sciences* 368, no. 1915 (2010): 1333–1383.

Sukirtha, Raman, Kandula Manasa Priyanka, Jacob Joe Antony, Soundararajan Kamalakkannan, Ramar Thangam, Palani Gunasekaran, Muthukalingan Krishnan, and Shanmugam Achiraman. "Cytotoxic effect of green synthesized silver nanoparticles using *Melia azedarach* against in vitro HeLa cell lines and lymphoma mice model." *Process Biochemistry* 47, no. 2 (2012): 273–279.

Sun, Qian, Xiang Cai, Jiangwei Li, Min Zheng, Zuliang Chen, and Chang-Ping Yu. "Green synthesis of silver nanoparticles using tea leaf extract and evaluation of their stability and antibacterial activity." *Colloids and Surfaces A: Physicochemical and Engineering Aspects* 444 (2014): 226–231.

Velmurugan, Palanivel, Jung-Hee Park, Sang-Myeong Lee, Jum-Suk Jang, Kui-Jae Lee, Sang-Sub Han, Sang-Hyun Lee, Min Cho, and Byung-Taek Oh. "Synthesis and characterization of nanosilver with antibacterial properties using *Pinus densiflora* young cone extract." *Journal of Photochemistry and Photobiology B: Biology* 147 (2015): 63–68.

Ventola, C. Lee. "Progress in nanomedicine: Approved and investigational nanodrugs." *Pharmacy and Therapeutics* 42, no. 12 (2017): 742.

Yallappa, S., J. Manjanna, M. A. Sindhe, N. D. Satyanarayan, S. N. Pramod, and K. Nagaraja. "Microwave assisted rapid synthesis and biological evaluation of stable copper nanoparticles using *T. arjuna* bark extract." *Spectrochimica Acta Part A: Molecular and Biomolecular Spectroscopy* 110 (2013): 108–115.

Yugandhar, Pulicherla, Reddla Haribabu, and Nataru Savithramma. "Synthesis, characterization and antimicrobial properties of green-synthesised silver nanoparticles from stem bark extract of *Syzygium alternifolium* (Wt.) Walp." *3 Biotech* 5, no. 6 (2015): 1031–1039.

Zhou, Jun, Ning Sheng Xu, and Zhong L. Wang. "Dissolving behavior and stability of ZnO wires in biofluids: A study on biodegradability and biocompatibility of ZnO nanostructures." *Advanced Materials* 18, no. 18 (2006): 2432–2435

18 Argan Oil
Phytochemistry, Biological Activities and Potential Development of Nanocapsules for Drug Delivery

Hicham Mechqoq and Noureddine El Aouad

CONTENTS

18.1 INTRODUCTION

Aromatic and medicinal plants have always been a rich source of natural products. They were used for centuries due to their biological properties. Some plant species are endemic and have very important socio-economical values dating back centuries. Endemic plants constitute an important source of biomolecules for drug discovery, nutraceutical, textile and flavour research fields (Rostagno and Prado 2013).

DOI: 10.1201/9781003231745-18

In Morocco, especially in the southwest region, we can find argan forests. This biosphere covers a vast intramontane plain of more than 2.5 million hectares, bordered by the High Atlas and Anti-Atlas mountains and open to the Atlantic in the west. The area is known for more than 1200 plant and animal species, 140 of which are endemic.

In this unique zone, argan trees have been cultivated for centuries and combine agricultural biodiversity, resilient ecosystems and valuable cultural heritage. This area received many recognitions from various United Nations organizations. The United Nations Educational Scientific and Cultural Organization (UNESCO) designated the area as the Arganeraie Biosphere Reserve. In 2014, the argan tree was inscribed on the UNESCO Representative List of the Intangible Cultural Heritage of Humanity. In December 2018, the Food and Agriculture Organization of the United Nations (FAO) recognized the argan-based agro-sylvo-pastoral system within the area of Ait Souab-Ait Mansour in Morocco as a Globally Important Agricultural Heritage System. In 2021, the United Nations General Assembly proclaimed 10 May International Day of Argania.

Argan (*Argania spinose*) is not only important in terms of conservation, but also for research and socio-economic development. The argan tree is widely exploited by the berber population for their daily uses and for its nutritional, cosmetic and medicinal proprieties. Recently, argan oil was used to develop more elaborate medicines as skin and hair serums, food supplements and nanocapsules (Zoubida et al. 2001, Rosset et al. 2012).

Nanocapsules are a form of medicine where the drug is stored inside a nanometre-sized shell; this preparation method improves the oral administration and absorption of the drug.

The main objective of this chapter is to report traditional uses of the argan tree as well as the chemical composition, biological effects, and high-value products developed from argan oil.

18.2 PRESENTATION OF THE ARGAN TREE

18.2.1 Botanic Aspects and Geographic Region

Argania spinosa (L.) Skeels or argan formerly named *Sideroxylon spinosum*, of Sapotaceae family, (Morton and Voss 1987) is an endemic plant of Morocco, mostly found in the southwestern areas. It is a xero-thermophilic species perfectly adapted to the arid climate. Argan trees present an erect or weeping port with a short trunk with a height from 1 to 2 metres that can reach up to 10 metres (Mechqoq et al. 2021a) (see Figure 18.1).

Argan trees are present in the arid and semiarid areas of Morocco covering about 800,000 hectares with a density of 10–50 trees/ha. It is located in the south region, along Atlantic coastline, from the mouth of Tensift River in the north to the mouth of the Drâa River in the south (Msanda et al. 2005).

18.2.2 Ethnobotany of the Argan Tree

The argan tree has been subjected to plenty of ethnobotanical investigations. The majority of those studies were interested in argan oil due to its large panel of uses and biological proprieties. Figure 18.2 summarizes the traditional uses of the argan tree.

Argan oil is externally used for skin hydration and to treat dermatitis like juvenile acne, burns and mange (Charrouf and Guillaume 1999). It is also orally used alone or in association with other plants such as mint to treat rheumatisms, cholesterolemia, lung infections, gastrointestinal disorders and diabetes (El Hafian et al. 2014, Khallouki et al. 2017).

Argan oil has been a culinary staple in Morocco for centuries, used for its nutty flavour and wide array of potential health benefits (Yaghmur et al. 2001). The leaves and bark are mainly used for medicinal purposes, as decoction and infusion for gastritis, ulcer and dysentery (Ouhaddou et al. 2014). The flowers are the least exploited part due to their very small size, and their collection reduces the number of fruits produced, however, they proved to have potential in facial skin care (Moukal 2004). The roots are prepared as decoction in association with juniper, oregano, fenugreek and caraway to treat diabetes (Abouri et al. 2012).

FIGURE 18.1 The argan tree. (a) Fresh fruit, (b) fruit cores and (c) kernels.

Finally, the wood is widely used by natives for carpentry, tanning leathers (Abouri et al. 2012) and heating (Khallouki et al. 2017).

18.2.3 Argan Oil Extraction

Argan oil can be extracted using different methods. The traditional hand mill and the mechanical cold press techniques are most used (Charrouf and Guillaume 1999). Prior to extraction, argan kernels can be subjected to the roasting process in order to obtain culinary oil with a darker colour and

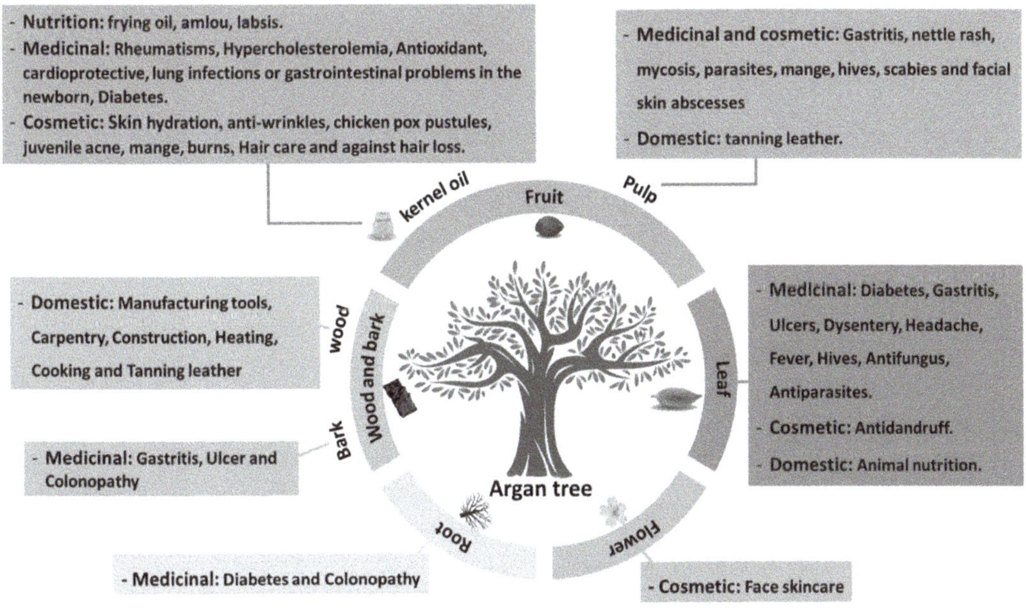

FIGURE 18.2 Traditional uses of different parts of an argan tree.

stronger nutty taste; suppression of the roasting process leads to an oil being labeled as cosmetic grade (El Monfalouti 2013).

The kernels are grind-milled and the obtained paste is then hand pressed to extract the oil (El Monfalouti 2013). This technique presents many disadvantages since it is time consuming and the extraction yield is generally above 30%. Additionally, the storage time is limited to a few weeks due to the use of water during the extraction process (Charrouf and Guillaume 2008).

The mechanical cold press technique presents a very good alternative to traditional methods. The kernels are pressed with a hydraulic press, which makes the extraction process faster and leads to a higher quality oil with a yield of 55% (Charrouf 1984).

Argan oil can also be extracted by soxhlet (Monfalouti et al. 2010) where kernels are ground and extracted with a nonpolar hydrocarbon solvent such as cyclohexane, heptane or petroleum ether, and the extraction process is carried out using a stainless steel or glass apparatus (Adlouni 2010). One or two extraction cycles are enough to obtain an oil with a yield value up to 57% (Guillaume and Charrouf 2011). However, this type of oil has some disadvantages if solvent was used as it not easily removed. Argan oil extracted following this method is exclusively intended for cosmetology (Adlouni 2010).

Recently, ecofreindly extraction methods have been used for argan oil extraction. Taribak et al. (2013) reported the use of supercritical fluid extraction to obtiain argan oil and according to the authors, the highest obtained extraction yield was 48% when the mechanical cold-pressing yield was around 45%, which is lower than previously cited sources.

More recently, enzymatic extraction was performed on argan kernels and seemed to be very promising, providing an oil with very good quality parameters in comparison with cold pressed oil and an extraction yield of 65% (Mechqoq et al. 2021b).

18.3 CHEMICAL COMPOSITION

Many studies have been made on the chemical composition of argan oil, composed of a glyceridic fraction (99%) and an unsaponifiable fraction (1%). The glyceridic fraction consists of triacylglycerols, diacylglycerols, monoacylglycerols and free fatty acids (Khallouki et al. 2003). Unsaponified matter contains mainly polyphenols, tocopherols and sterols. Argan pulp, press-cake, leaves and wood are composed of polyphenols, sterols and saponins.

18.3.1 Triacylglycerols

Triacyclglycerols (TAGs) are typical constituents of argan oil, constituting 95% of the oil (Maurin et al. 1992). The triacylglycerols of argan oil have been identified as OOL (16–20%), OLL (14–17%), and POO (7–16%) (with molecular species oleic (O), linoleic (L) and palmitic acyl residue (P)) (Charrouf and Guillaume 2008).

18.3.2 Fatty Acids

Fatty acids (FA) are also exclusive to argan oil, mostly unsaturated (80%), and represented by oleic acid (46–48%) and linoleic acid (32–33%). The main saturated fatty acids of argan oil are palmitic and stearic acids with 12–14% and 5–7%, respectively. Other traces of fatty acids such as myristic, palmitoleic, arachidic and linolenic acids have been reported (Taribak et al. 2013, Taneva et al. 2020).

18.3.3 Sterols and Triterpene Alcohols

Argan oil differs from other vegetable oils by the presence of two sterols: schottenol and spinasterol with 46–48% and 35–39%, respectively (Farines et al. 1982, Khallouki et al. 2003, Hilali et al.

2005, Charrouf and Guillaume 2008). Also, triterpene alcohol has been identified as butyrospermol (18.1%), tirucallol (27.9%) and β-amyrin (27.3%) (Farines et al. 1984).

18.3.4 Tocopherols

The unsaponified matter contains significant isomers of tocopherols. The γ-tocopherol had the highest value with 460–590 mg/kg, followed by δ-tocopherol with 50–190 mg/kg and α-tocopherol with 40–75 mg/kg (Gharby et al. 2011).

18.3.5 Polyphenols

Polyphenols of argan oils were reported as ferulic acid with 30–31.5%, syringic acid with 25–28.5% and vanillic and caffeic acids with 11–13% and 8–9%, respectively (Kamal et al. 2016). The leaf's main polyphenols are myricitrin with 16.8 mg/g of dry weight and quercetin with 5.3 mg/g of dry weight (Tahrouch et al. 2000). Argan press-cakes present phenolic compounds similar to argan oil such as epicatechin, syringic and vanillic acids. The highest value was observed in epicatechin (54.72%), followed by vanillic acid (16.3%) and protocatechuic acid (15.2%) (Rojas et al. 2005).

18.3.6 Saponins

Saponins of the argan tree formerly named arganines present exclusively in kernels, pulp and wood (saponin A, B, C, D, E, F, G, H, J, K and misaponin A) (Charrouf et al. 1992). The press-cake and oil contain mainly arganine A (39%) and arginine B (8%). The pulp features only arganine K (0.02%) (Alaoui et al. 2001). Wood contains traces of arganines G, H and J (Oulad-Ali et al. 1996).

18.3.7 Terpenes

Leaves and pulp also contain volatile compounds. The chemical evaluation of leaf essential oils obtained by hydrodistillation showed the presence cubenol (35.7%), bornyl acetate (9.3%) and camphor (8.6%) (El Amrani et al., 2015). The essential oils of fruit pulp were identified as camphor with 34–36% and 1,8-cineole with 7–16% (Harhar et al. 2011).

In conclusion, all argan oils samples are composed mainly by oleic, linoleic, palmitic and stearic acids. Studies of the chemical composition of argan oil showed that geographic origin has significantly affected fatty acid composition. This variability can be explained in our point of view by extraction and climatic conditions, genetic factors and ripening grade of fruits.

In addition to the chemical investigation conducted on the Moroccan argan oil, many others described the chemical profiles of Algerian and Tunisian oils. They all proved to have similar lipid compositions, phytosterols, tocopherols and phenolic contents with slight variations (Kouidri 2008).

Many studies compared argan oil with different vegetal oils. The most recent one published by Taneva et al. (2021) compared the chemical profiles of both argan and cactus (*Opuntia ficus-indica* L.) seed oils. Determination of fatty acids profiles showed that argan oil has higher amounts of oleic, palmitic and stearic acids with 46%, 14% and 6%, respectively, and cactus oil contained more linoleic acid with a percentage of 60%. According to the same study, argan oil contained more α-tocopherols when the cactus oil had more γ-tocopherols. Additionally, Khallouki et al. (2017) investigated the chemical difference between argan and olive (*Olea europaea* L. cv. Moroccan picholine) oils. In spite of having different chemical compositions, with olive oil richer in oleic acid, both oils presented very interesting profiles with high amounts of unsaturated fatty acids and good chemical stability due to polyphenols and other antioxidants found in these oils.

All investigations demonstrated that despite this rarity, argan oil is a good substitute for other vegetable oils as its oxidation stability makes its storage and use relatively easy.

18.4 BIOLOGICAL ACTIVITIES

18.4.1 Antimicrobial

Argan oil showed moderate antibacterial activity due to the low percentage of terpens and polyphenols, which are known for their strong antibacterial activity. However, the combination of argan oil with hydrogen peroxide resulted in a synergetic effect inhibiting *Pseudomonas aeruginosa* with the same strength as antibiotics such as amikacin, tobramycin, ceftazidime, aztreonam, norfloxacin and gentamycin (Naher et al. 2014).

The organic extracts of argan leaves have antibacterial activity against methicillin-sensitive and methicillin-resistant *Staphylococcus aureus* strains (Bonvicini et al. 2017). The leaf's essential oils El Kabouss et al. (2002) showed an antibacterial activity against Gram-positive (*Staphylococcus aureus*, *Bacillus* sp. and *Micrococci* sp.) and Gram-negative strains (*Escherichia coli*, *Klebsiella*, *Enterobacter* and *Pseudomonas*). The antifungal activity has been reported of essential oils against *Aspergillus flavus*, *Aspergillus niger*, *Candida* sp. and *Saccharomyces cerevisiae*.

18.4.2 Antioxidant Activity

Antioxidants protect cells from oxidative stress mainly caused by active free radicals (Halliwell 1996). Argan oil has a 2,2-diphenyl-1-picrylhydrazyl (DPPH)-scavenging capacity similar to virgin olive oil with a value ranging between 0.41–0.34 mmol Trolox/kg oil. 2, 2′-Azino-Bis-3-Ethylbenzothiazoline-6-Sulphonic Acid (ABTS) and Ferric Reducing Antioxidant Power (FRAP) assays gave significant results as well (Marfil et al. 2011). Leaf polyphenols exhibited an IC_{50} value 125.59 ± 1.87 μg, largely superior to positive control Trolox, which had an IC_{50} of 50.55 ± 0.7 μg. Those findings suggest that the leaf polyphenols had lower activity in comparison with positive control (Dakiche et al. 2016). Moreover, press-cake and pulp extracts showed low-to-moderate antioxidant activity (El Monfalouti et al. 2012).

18.4.3 Hypolipemiant and Antidiabetic Activities

The argan oil tocopherol monounsaturated fatty acids (MUFA) and polyunsaturated fatty acids (PUFA) play a very big role in the prevention of cardiovascular diseases such as hyperlipidemia (Drissi et al. 2004), hypercholesterolemia (El Kharrassi et al. 2014) and high blood pressure-related diseases (Charrouf and Guillaume 2008).

18.4.4 Anti-Inflammatory Activity

Carrageenan-induced paw oedema is one of the most popular tests used to evaluate anti-inflammatory activity. This test is highly sensitive for nonsteroidal anti-inflammatory drugs and is used to detect orally active anti-inflammatory agents. The arganins of an argan press-cake showed strong activity — they reduced rat-induced inflammation from doses of 10 mg/kg. This activity was similar to indomethacin, a commonly used nonsteroidal anti-inflammatory drug (Bouroud 1995).

18.4.5 Skin Hydration and Hair Protection

Daily application and consumption of argan oil by post-menopausal women can enhance skin hydration by maintaining water retention capacity, which improves skin elasticity (Boucetta et al. 2014).

Argan oil proved to inhibit two key enzymes of melanin biosynthesis (tyrosinase and dopachrometautomerase). The daily application of argan oil decreased the enzyme expression by 60% and 80%, respectively (Villareal et al. 2013). Faria et al. (2013) demonstrated the hair-regenerating effect of argan oil as a hair dye treatment by showing that the oil decreased damage caused by the colouring process.

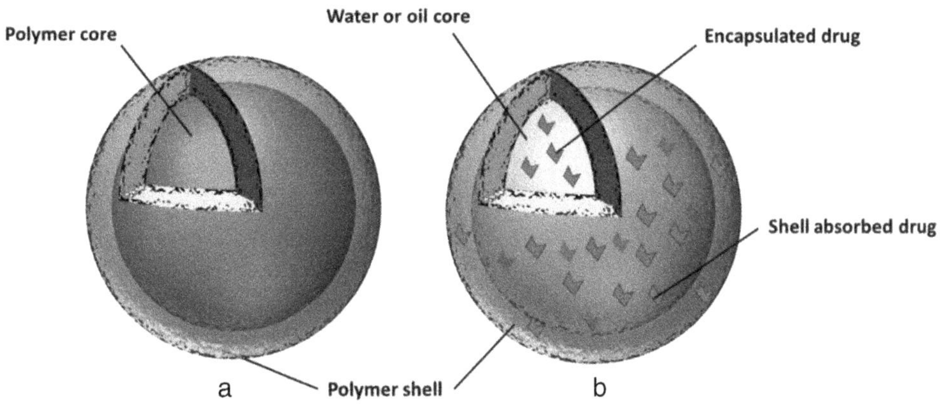

FIGURE 18.3 Different nanocapsule structures. (a) Polymer core and (b) oil or aqueous core with drug. (Modified from Erdoğar et al. 2018.)

18.5 ARGAN OIL AND NANOPHYTOMEDECINE

18.5.1 Presentation of Nanocapsules

According to the International Union of Pure and Applied Chemistry (IUPAC), a nanoparticle is a drug delivery system composed of a core space entrapping a substance surrounded by a solid shell. The internal core contains the active substance in liquid or solid forms (Devissaguet et al. 1991) and the core can be hydrophobic or lipophilic, depending of the drug type and nanoencapsulation technique. Nanocapsules can also carry the active substance on shell surface membranes. The shell membrane is made of a polymeric biodegradable membrane due to its internal use in biological systems. Typical nanocapsule size varies from 10 to 1000 nm (Nagavarma et al. 2012).

In general, polymeric nanocapsules are composed of a drug, oil, solvent, surfactant and polymeric membrane (see Figure 18.3). Generally, the choice of components depends on the nanoencapsulation technique used. The oils present many advantages such as high drug solubility, absence of toxicity, and easy absorption by the organism (Blouza et al. 2006). The most-used oils are Miglyol® and capric/caprylic triglyceride types due to their wide range of solubility for active substances. Other oils such as olive oil, argan oil and soybean oil, are also used and show promise (Mora-Huertas et al., 2010). For a polymer solvent, acetone and ethanol are widely used. Surfactants are amphiphilic molecules, which means they have hydrophilic and hydrophobic sides that form a layer around the core to give it rigidity (Morsy 2014).

18.5.2 Nanocapsules Developed from Argan Oil

Argan oil has been used to develop many nutritional and cosmetic products. It has also been used to develop more elaborate medicinal solutions such as nanocapsules. Argan oil is mainly used as a drug carrier as it is found to solubilize high quantities of drugs in comparison with conventional long-chain triglycerides. Its biological proprieties provide a secondary action complementary to the lead drug. To date, three argan oil-based nanocapsules have been reported for use with tacrolimus, naproxen and indomethacin as drugs. Tacrolimus is used primarily for prevention of allograft rejection after organ transplantation. It is also used in dermatology for eczema treatment and as an inflammation suppressor (Nassar et al. 2009, Rosset et al. 2012). Naproxen is a non-steroidal drug usually used as a painkiller and anti-inflammation drug; it is also used to relieve osteoarthritis as well as for other symptoms (Rosset et al., 2012). Whereas, indomethacin drug is used for rheumatoid arthritis and collagen disease (Tsvetkova et al. 2012).

18.6 CONCLUSION

This chapter describes the chemical composition and biological activities of *Argania spinosa* oil as well as the other tree parts. The argan oil is known for its dietary proprieties, which justify its culinary use by the Moroccan berber populations for a long time. The other tree parts such as leaves, wood and roots have been exploited in the form of decoction, infusion, maceration and powders to treat diabetes, rheumatism, dermatitis and gastritis. Many investigations revealed the presence of many pharmaceuticals properties such as the antibacterial, antifungal, antioxidant, anti-diabetic, anti-inflammatory and anticancer activities of oil and different extracts of *A. spinosa*. Those biological activities offer many perspectives for the use of argan oils and organic extracts in pharmaceutical preparations.

Recently, the nanotechnology boom gave rise to a trend of using natural products for drugs preparation. This is mainly due to the absence of toxicity of natural compounds, and the biological properties that can be added to the main active agent. The argan oil is known for its large panel of biological proprieties ranging between skin hydration, cholesterol stabilization and prevention of cardiovascular proprieties. Many studies reported the use of argan oil for the preparation of nanocapsules for therapeutic or cosmetic applications in dermatology. Those studies proved that the argan oil is a good drug carrier, permits a better drug encapsulation and guarantee the stability of the preparation. Those studies are very promising for new trends in the field of dermatology. However, it would be very interesting to use the other argan tree bioactive natural compounds for the elaboration of nanocapsules.

REFERENCES

Abouri, Mustapha, Abdelhamid El Mousadik, Fouad Msanda, Hassan Boubaker, Baha Saadi, and Khalil Cherifi. "An ethnobotanical survey of medicinal plants used in the Tata Province, Morocco." *International Journal of Medicinal Plants Research* 1, no. 7 (2012): 99–123.

Adlouni, A. "L'huile d'argan, de la nutrition à la santé." *Phytothérapie* 8, no. 2 (2010): 89–97.

Alaoui, A., Z. Charrouf, G. Dubreucq, E. Maes, J. C. Michalski, and M. Soufiaoui. "Saponins from the pulp of the fruit of *Argania spinosa* (L.) Skeels (Sapotaceae)." In *International Symposium of the Phytochemical Society: Lead Compounds from Higher Plants*. Lausana, Suiza, pp. 137–141. 2001.

Blouza, I. Limayem, C. Charcosset, S. Sfar, and H. Fessi. "Preparation and characterization of spironolactone-loaded nanocapsules for paediatric use." *International Journal of Pharmaceutics* 325, no. 1–2 (2006): 124–131.

Bonvicini, F., F. Antognoni, M. Mandrone, M. Protti, L. Mercolini, M. Lianza, G. A. Gentilomi, and F. Poli. "Phytochemical analysis and antibacterial activity towards methicillin-resistant *Staphylococcus aureus* of leaf extracts from *Argania spinosa* (L.) Skeels." *Plant Biosystems* 151, no. 4 (2017): 649–656.

Boucetta, Kenza Qiraouani, Zoubida Charrouf, Abdelfattah Derouiche, Younes Rahali, and Yahya Bensouda. "Skin hydration in postmenopausal women: Argan oil benefit with oral and/or topical use." *Przeglad menopauzalny = Menopause review* 13, no. 5 (2014): 280.

Bouroud, H. "Etude de l'activité anti-inflammatoire des saponines d'*Argania spinosa*." PhD diss., Thèse de la faculté de Médecine et de pharmacie de Rabat, 1995.

Charrouf, M'hamed. "Contribution à l'étude chimique de l'huile d'*Argania spinose* (L.) (Sapotaceae)." PhD diss., 1984.

Charrouf, Zoubida, and Dominique Guillaume. "Argan oil: Occurrence, composition and impact on human health." *European Journal of Lipid Science and Technology* 110, no. 7 (2008): 632–636.

Charrouf, Zoubida, and Dominique Guillaume. "Ethnoeconomical, ethnomedical, and phytochemical study of *Argania spinosa* (L.) Skeels." *Journal of Ethnopharmacology* 67, no. 1 (1999): 7–14.

Charrouf, Z., J. M. Wieruszeski, S. Fkih-Tetouani, Y. Leroy, M. Charrouf, and B. Fournet. "Triterpenoid saponins from *Argania spinosa*." *Phytochemistry* 31, no. 6 (1992): 2079–2086.

Dakiche, Hadjira, Mustapha Khali, Ahmed Khaled Abu-el-Haija, Ammar Al-Maaytah, and Qosay Ali Al-Balas. "Biological activities and phenolic contents of *Argania spinosa* L. (Sapotaceae) leaf extract." *Tropical Journal of Pharmaceutical Research* 15, no. 12 (2016): 2563–2570.

Devissaguet, Jean-Philippe, Hatem Fessi, and Francis Puisieux. Process for the preparation of dispersible colloidal systems of a substance in the form of nanocapsules. Google Patents, 1991.

Drissi, Anas, Josefa Girona, Mounia Cherki, Gemma Godàs, Abdelfettah Derouiche, Mariame El Messal, Rachid Saile et al. "Evidence of hypolipemiant and antioxidant properties of argan oil derived from the argan tree (*Argania spinosa*)." *Clinical Nutrition* 23, no. 5 (2004): 1159–1166.

El Amrani, Abdelaziz, José Antonio Cayuela Sánchez, and Jamal Eddine. "Chemical composition of moroccan *Argania spinosa* leaf essential oils isolated by supercritical CO_2, microwave and hydrodistillation." *Journal of Essential Oil Bearing Plants* 18, no. 5 (2015): 1138–1147.

El Hafian, Mohamed, N. Benlandini, Houda Elyacoubi, Lahcen Zidane, and Atmane Rochdi. "Étude floristique et ethnobotanique des plantes médicinales utilisées au niveau de la préfecture d'Agadir-Ida-Outanane (Maroc)." *Journal of Applied Biosciences* 81 (2014): 7198–7213.

El Kabouss, Adil, Zoubida Charrouf, Mohamed Faid, François-Xavier Garneau, and Guy Collin. "Chemical composition and antimicrobial activity of the leaf essential oil of *Argania spinosa* L. Skeels." *Journal of Essential Oil Research* 14, no. 2 (2002): 147–149.

El Kharrassi, Youssef, Mohammad Samadi, Tatiana Lopez, Thomas Nury, Riad El Kebbaj, Pierre Andreoletti, Hammam I. El Hajj et al. "Biological activities of Schottenol and Spinasterol, two natural phytosterols present in argan oil and in cactus pear seed oil, on murine miroglial BV2 cells." *Biochemical and Biophysical Research Communications* 446, no. 3 (2014): 798–804.

El Monfalouti, Hanae. "Contribution à la détermination des propriétés photo-protectrices et anti-oxydantes des dérives de l'arganier: études chimiques et physiologiques." PhD diss., Reims, 2013.

El Monfalouti, Hanae, Zoubida Charrouf, Simona Belviso, Daniela Ghirardello, Bernardo Scursatone, Dominique Guillaume, Clément Denhez, and Giuseppe Zeppa. "Analysis and antioxidant capacity of the phenolic compounds from argan fruit (*Argania spinosa* (L.) Skeels)." *European Journal of Lipid Science and Technology* 114, no. 4 (2012): 446–452.

Erdoğar, Nazli, Safiye Akkın, and Erem Bilensoy. "Nanocapsules for drug delivery: An updated review of the last decade." *Recent Patents on Drug Delivery & Formulation* 12, no. 4 (2018): 252–266.

Faria Pamella Mello, Luciana Neves Camargo, Regina Siqueira Haddad Carvalho, Luis Antonio Paludetti, Maria Valéria Robles Velasco, and Robson Miranda da Gama. "Hair protective effect of Argan oil (*Argania spinosa* Kernel oil) and Cupuassu butter (*Theobroma grandiflorum* seed butter) post treatment with hair dye." *Journal of Cosmetics, Dermatological Sciences and Applications* 3, no.03 (2013): 40.

Farines, M., M. Charrouf, and J. Soulier. "Study of the chemical composition of argan seed oil." *Comptes rendus Congrès National Sociétés Savantes, Section Sciences* 106 (1982): 327–333.

Farines, M., J. Soulier, M. Charrouf, and R. Soulier. "Study on the seed oil from *Argania spinosa* (L.) Sapotaceae. 1. The glyceride fraction." *Revue Francaise des Corps Gras (France)* (1984).

Gharby, Saïd, Hicham Harhar, Dominique Guillaume, Aziza Haddad, Bertrand Matthäus, and Zoubida Charrouf. "Oxidative stability of edible argan oil: A two-year study." *LWT-food Science and Technology* 44, no. 1 (2011): 1–8.

Guillaume, Dom, and Zoubida Charrouf. "Argan oil: Monograph." *Alternative Medicine Review: A Journal of Clinical Therapeutic* 16, no. 3 (2011): 275–279.

Halliwell, Barry. "Antioxidants in human health and disease." *Annual Review of Nutrition* 16, no. 1 (1996): 33–50.

Harhar, Hicham, Said Gharby, Badr Eddine Kartah, Hanae El Monfalouti, Nadia Maata, Dominique Guillaume, and Zoubida Charrouf. "Les huiles essentielles des feuilles et de la pulpe du fruit de l'arganier: composition chimique et activité microbiologique." 2011.

Hilali, Miloudi, Zoubida Charrouf, Abd El Aziz Soulhi, Larbi Hachimi, and Dominique Guillaume. "Influence of origin and extraction method on argan oil physico-chemical characteristics and composition." *Journal of Agricultural and Food Chemistry* 53, no. 6 (2005): 2081–2087.

Kamal, R., M. Kharbach, J. D. Imig, M. Eljmeli, Z. Doukkali, H. Naceiri Mrabti, H. Elmsellem, A. Bouklouze, Y. Cherrah, and K. Alaoui. "Antioxidant activities, total polyphenolic compounds and Hplc/Dad/Ms phenolic profile of argan oil derived from two different methods of extractions." *Journal of Chemical and Pharmaceutical Research* 8 (2016): 173–179.

Khallouki, Farid, Mohamed Eddouks, Akdad Mourad, Andrea Breuer, and Robert Wyn Owen. "Ethnobotanic, ethnopharmacologic aspects and new phytochemical insights into *Moroccan argan* fruits." *International Journal of Molecular Sciences* 18, no. 11 (2017): 2277.

Khallouki, F., C. Younos, R. Soulimani, T. Oster, Z. Charrouf, Bartsch Spiegelhalder, H. Bartsch, and R. W. Owen. "Consumption of argan oil (Morocco) with its unique profile of fatty acids, tocopherols, squalene, sterols and phenolic compounds should confer valuable cancer chemopreventive effects." *European Journal of Cancer Prevention* 12, no. 1 (2003): 67–75.

Kouidri, Mohamed. "Extraction et caractérisation physico-chimique de l'huile d'argan provenant d'arbres cultivés dans deux régions de l'algérie (Tindouf et Mostaganem)." PhD diss., Université de Chlef-Hassiba Benbouali, 2008.

Marfil, Rocío, Rafael Giménez, Olga Martínez, Paula R. Bouzas, José A. Rufián-Henares, Marta Mesías, and Carmen Cabrera-Vique. "Determination of polyphenols, tocopherols, and antioxidant capacity in virgin argan oil (*Argania spinosa*, Skeels)." *European Journal of Lipid Science and Technology* 113, no. 7 (2011): 886–893.

Maurin, R., K. Fellat-Zarrouck, and M. Ksir. "Positional analysis and determination of triacylglycerol structure of *Argania spinosa* seed oil." *Journal of the American Oil Chemists' Society* 69, no. 2 (1992): 141–145.

Mechqoq, Hicham, Mohamed El Yaagoubi, Abdallah El Hamdaoui, Svetlana Momchilova, J. R. G. da Silva Almeida, Fouad Msanda, and Noureddine El Aouad."Ethnobotany, phytochemistry and biological properties of argan tree (*Argania spinosa* (L.) Skeels)(Sapotaceae)—A review." *Journal of Ethnopharmacology* 281 (2021a): 114528.

Mechqoq, Hicham, Mohamed El Yaagoubi, Svetlana Momchilova, Fouad Msanda, and Noureddine El Aouad. "Comparative study on yields and quality parameters of argan oils extracted by conventional and green extraction techniques." *Grain & Oil Science and Technology* 4, no. 3 (2021b): 125–130.

Monfalouti, Hanae El, Dom Guillaume, Clément Denhez, and Zoubida Charrouf. "Therapeutic potential of argan oil: A review." *Journal of Pharmacy and Pharmacology* 62, no. 12 (2010): 1669–1675.

Mora-Huertas, C. E., H. Fessi and A. Elaissari. "Polymer-based nanocapsules for drug delivery." *International Journal of Pharmaceutics* 385, no. 1–2 (2010): 113–142.

Morsy, Salwa MI. "Role of surfactants in nanotechnology and their applications." *International Journal of Current Microbiology and Applied Science* 3, no. 5 (2014): 237–260.

Morton, Julia F., and Gilbert L. Voss. "The argan tree (*Argania sideroxylon,* Sapotaceae), a desert source of edible oil." *Economic Botany* 41, no. 2 (1987): 221–233.

Moukal, A. "L'arganier, *Argania spinosa* L. (Skeels), usage thérapeutique, cosmétique et alimentaire." *Phytothérapie* 2, no. 5 (2004): 135–141.

Msanda, Fouad, Ahmed El Aboudi, and Jean-Paul Peltier. "Biodiversité et biogéographie de l'arganeraie marocaine." *Cahiers Agricultures* 14, no. 4 (2005): 357–364.

Nagavarma, B. V. N., Hemant K. S. Yadav, A. V. L. S. Ayaz, L. S. Vasudha, and H. G. Shivakumar. "Different techniques for preparation of polymeric nanoparticles—A review." *Asian Journal of Pharmaceutical and Clinical Research* 5, no. 3 (2012): 16–23.

Naher, Habeeb Sahib, H. O. Al-Dahmoshi, N. S. Al-Khafaji, A. K. Al-Saffar, and H. K. Al-Saffar. "Anti-pseudomonal effect of argan oil on *Pseudomonas aeruginosa* recovered from burn patients in Hilla City, Iraq." *Journal of Infectious Diseases. Photon* 113 (2014): 270–276.

Nassar, Taher, Alona Rom, Abraham Nyska, and Simon Benita. "Novel double coated nanocapsules for intestinal delivery and enhanced oral bioavailability of tacrolimus, a P-gp substrate drug." *Journal of Controlled Release* 133, no. 1 (2009): 77–84.

Ouhaddou, H., Hassan Boubaker, Fouad Msanda, and Abdelhamid El Mousadik. "An ethnobotanical study of medicinal plants of the Agadir-Ida-Ou Tanane Province (southwest Morocco)." *Journal of Applied Biosciences* 84 (2014): 7707–7722.

Oulad-Ali, Ali, Valérie Kirchner, Annelise Lobstein, Bernard Weniger, Robert Anton, Dominique Guillaume, and Zoubida Charrouf. "Structure elucidation of three triterpene glycosides from the trunk of *Argania spinosa*." *Journal of Natural Products* 59, no. 2 (1996): 193–195.

Rojas, Luis B., Stéphane Quideau, Patrick Pardon, and Zoubida Charrouf. "Colorimetric evaluation of phenolic content and GC-MS characterization of phenolic composition of alimentary and cosmetic argan oil and press cake." *Journal of Agricultural and Food Chemistry* 53, no. 23 (2005): 9122–9127.

Rosset, Valentina, Naveed Ahmed, I. Zaanoun, Barbara Stella, Hatem Fessi, and Abdelhamid Elaissari. "Elaboration of argan oil nanocapsules containing naproxen for cosmetic and transdermal local application." *Journal of Colloid Science and Biotechnology* 1, no. 2 (2012): 218–224.

Rostagno, Mauricio A., and Juliana M. Prado, eds. *Natural Product Extraction: Principles and Applications.* No. 21. Royal Society of Chemistry, 2013.

Tahrouch, Saïda, Claude Andary, Sylvie Rapior, Laurence Mondolot, Annick Gargadennec, and Alain Fruchier. "Polyphenol investigation of *Argania spinosa* (Sapotaceae) endemic tree from Morocco." *Acta Botanica Gallica* 147, no. 3 (2000): 225–232.

Taneva, Sabina, Hicham Mechqoq, Iskra Totseva, Yana Nikolova, Mariana Kamenova-Nacheva, Mohamed El Yaagoubi, Fouad Msanda, Noureddine El Aouad, Vladimir Dimitrov, and Svetlana Momchilova. "Lipid composition and oxidative stability of argan and cactus *Opuntia ficus indica* seed oils from Morocco—assessment of two extraction methods." *Journal of Chemical Technology and Metallurgy* 56, no. 3 (2021): 548–560.

Taribak, Chouaa, Lourdes Casas, Casimiro Mantell, Zoubaida Elfadli, Rédouane E. Metni, and Enrique J. Martínez de la Ossa. "Quality of cosmetic argan oil extracted by supercritical fluid extraction from *Argania spinosa* L." *Journal of Chemistry* 2013 (2013): 1–9.

Tsvetkova, B., I. Pencheva, A. Zlatkov, and P. Peikov. "High performance liquid chromatographic assay of indomethacin and its related substances in tablet dosage forms." *International Journal of Pharmacy and Pharmaceutical Sciences* 4, Supplement no. 3 (2012): 549–552.

Villareal, Myra O., Sayuri Kume, Thouria Bourhim, Fatima Zahra Bakhtaoui, Kenichi Kashiwagi, Junkyu Han, Chemseddoha Gadhi, and Hiroko Isoda. "Activation of MITF by argan oil leads to the inhibition of the tyrosinase and dopachrome tautomerase expressions in B16 murine melanoma cells." *Evidence-Based Complementary and Alternative Medicine* 2013 (2013):340107.

Yaghmur, A., A. Aserin, Y. Mizrahi, A. Nerd, and N. Garti. "Evaluation of argan oil for deep-fat frying." *LWT-Food Science and Technology* 34, no. 3 (2001): 124–130.

Zoubida, Charrouf, Henry Florence, Moser Philippe, and Pauly Gilles. *Cosmetic and/or Dermopharmaceutical Preparations Containing Native Proteins from the Plant Argania spinosa*. Cognis France S.A., 2001.

19 Photoactive Nanoparticles of Plant Origin in Cancer Treatment

Rahul Chandran, Blassan P. George and Heidi Abrahamse

CONTENTS

19.1 INTRODUCTION

Nanotechnology and nanomedicine offer a platform to deliver drugs by increasing their bioavailability and increased accumulation in cancer tissues to promote efficient cancer therapy. Hence, multidisciplinary scientific approaches are in progress to improve the therapy of cancer. Nanoparticles (NPs) can be exploited for different applications, ranging from diagnosis to therapy (Martinelli et al., 2019). Plants with diverse origins and genetic variations are one of the best sources of biologically active compounds that can reduce metal ions to NPs with less effort (Malik et al., 2014). Silver and gold nanoparticle synthesis from plants has always been a focus of interest with many other metals being introduced for their cost effectiveness and high efficiency. There are reports of plant extracts being used for nanoparticle synthesis for innovative medicinal applications (Park et al., 2011).

With new and efficient drugs being introduced against cancer, challenges to improving bioavailability, stability, half-life in humans and poor specificity still persist (Cho et al., 2008). Multidrug resistance mediated by P-glycoprotein, a member of adenosine-5′-triphosphate (ATP) binding cassette (ABC) transporter superfamily, plays a crucial role in drug efflux resulting in reduced accumulation at target sites. However, nanocarriers offer a promising therapy through target-specific drug delivery via various mechanisms. Hence, phytocompounds for synthesis of efficient NPs can enhance the therapeutic index of drugs and reduce side effects are of great interest (Thanki et al., 2013).

Photodynamic therapy (PDT) is an interesting field that offers the application of light in the presence of photosensitizers (PSs) to kill cancer cells through the generation of reactive oxygen species (ROS). PSs are synthetic or natural compounds that, upon exposure to light of a specific wavelength, excite electrons and initiate a chain of reactions that generate free radicals. These free radicals induce apoptosis, necrosis or autophagy through various signalling pathways. Hypericin from *Hypericum* sp. is one such plant-derived natural PS that is widely used in cancer PDT. There could be hundreds of such compounds in plants that could act as PSs and play significant roles in

DOI: 10.1201/9781003231745-19

nanoparticle synthesis. A thorough search for relevant articles was done through PubMed, Google Scholar, ScienceDirect, Wiley Online Library and SpringerLink. This review discusses the possible role plants can play in cancer therapy by offering compounds that can complement PDT through their ability to act as PSs and NPs.

19.2 PLANTS IN CANCER THERAPY

Herbal medicines are used as adjuvant therapies in combination with chemotherapy and radiation therapy (Khan, 2014). The benefit of using herbal therapies as adjuvant therapy is to enhance the therapeutic effect and lessen post-treatment complications to the patients and to improve quality of life. Many cancer patients prefer complementary and alternative medicines along with mainline chemotherapy. Some of the plant-derived agents, when used along with the chemotherapy, show side effects as they interfere with the concurrent chemotherapy, however, some (example: ginger) did not show any overriding side effects or risk in treatment (Smith et al., 2014). The synergistic interactions within multicomponent drug preparations allow scientists to broaden the search for potential molecular targets in tumour cells. A combination approach is used by traditional Chinese medicine (TCM) where the effective doses of more than two compounds exert a synergistic effect that eventually induces cytotoxicity in cancer cells and elicits the tumour immune response. Alkaloids, saponins and flavonoids are the major bioactive secondary metabolites used in TCM and drug development (Rob, 1998). Vinblastine, vinorelbine, vincristine and vindesine are the important alkaloids used in cancer therapies. Trifolirhizin, curcumin and quercetin are important flavonoids with anticancer properties. The flavonoids showed various mechanisms in relation to cancer cell death, such as inhibition of proliferation and angiogenesis, cell cycle arrest at various phases of cell cycle, apoptosis induction, etc. Chikusetsusaponin IV, a butyl ester, polyphyllin D and diosgenin are examples of saponins with anticancer properties (Aung et al., 2017).

19.2.1 Anticancer Phytocompounds: Scope for Nanosynthesis?

Compounds derived from plants have been introduced as anticancer agents due to their ability to restrict proliferation and induce apoptosis through various mechanisms. These plant-derived agents act on cancer cells through antiproliferative properties (Yan et al., 2018), free radical inhibition (Lee et al., 2013), prevention of angiogenesis in tumour tissues (Lu et al., 2018), targeting kinases (Dou et al., 2018), transcriptional factors (Zhang et al., 2017b), cyclins and caspases (Yan et al., 2018), microRNAs (miRNAs) (Cojocneanu et al., 2015) and other mechanisms being discovered on a regular basis. More extensive research data on preliminary efficacy, toxicity, safety, pharmacokinetics and sustainability from preclinical research will render a strong platform for clinical trials.

6-Shogaol is a compound isolated from ginger (*Zingiber officinale*) that significantly inhibited the growth of lung cancer cells (NCI-H1650). The study showed a direct targeting of Akt1 and Akt2 and suppression of their actions (Saha et al., 2014). Another such compound is allicin from *Allium sativum*, a sulphur-containing compound responsible for inhibiting cholangiocarcinoma (CCA) (Chen et al., 2018). Reportedly, this compound reduced the STAT3 signalling pathway, resulting in declined levels of matrix metalloproteinase (MMP)-2 and -9. Andrographolide is a terpenoid isolated from *Andrographis paniculata* (*Acanthaceae*) that inhibits hypoxic conditions required for tumour growth (Li et al., 2015). Apigenin, found in fruits and vegetables, is flavonoid that possesses diverse anticancer properties including induction of apoptosis (Yan et al., 2018). This compound regulated the Bcl-2 protein and activated caspases that induces breast cancer cell-cycle arrest at G2/M (Yan et al., 2018). Baicalin and baicalein are anticancer compounds from *Scutellaria baicalensis* (*Lamiaceae*) that suppressed extracellular receptor kinase (ERK), mitogen-activated protein kinase (MAPK), and p38 signalling pathways to induce apoptosis in colon cancer cells (Dou et al., 2018). Plants from the *Cucurbitaceae* family produce well known alkaloid curcumin, which has been highly studied for its cell-death-inducing ability through autophagy, cell-cycle arrest, and down-regulation of the

TABLE 19.1
Anticancer Action of Phytocompounds

Plant Name	Compound	Action	Reference
Derris eriocarpa	Alpinumisoflavone	Suppresses Akt, increased ROS, DNA damage, apoptosis, and cell cycle arrest	Zhang et al., 2017a
Angelica gigas	Decursin and decursinol	Overexpressing the wild type androgen receptors	Wu et al., 2017
Melilotus officinalis	Dicumarol	Inhibits the kinase activity of pyruvate dehydrogenase kinase 1, higher level of ROS, attenuates the mitochondrial membrane potential (MMP), induced apoptosis	Aras et al., 2016
Rheum palmatum	Emodin	Activates ER stress and TRIB3/nuclear factor-kB signalling, increases T-cell activation and reduced tumour angiogenesis, increases in ERK and p38 phosphorylation and suppression of p-JNK expression, inhibits IRF4, STAT6, and C/EBPb signalling	Lin et al., 2016
Glycyrrhiza glabra	Glycyrrhizin	Suppressing TxA2 pathway, inhibition of JAK/STAT, signalling pathway	Deng et al., 2017; Wu et al., 2018.
Saussurea involucrata	Hispidulin	Increases the expression of PPARγ and phosphorylation levels of AMPK, JNK and ERK proteins	Han et al., 2018
Azadirachta indica	Nimbolide	Increases ROS generation, reduces PI3K/AKT/mTOR and ERK signalling	Pooladanda et al., 2018
Physalis pubescens	Physapubescin B	Reduces the expression levels of Ki-67, Cdc25C, and full-length PARP	Ding et al., 2015
Thymus vulgaris, Origanum vulgare	Thymol	Depolarizes mitochondrial membrane potential to induce apoptosis	De La Chapa et al., 2018
Withania somnifera	Withaferin A	Overexpresses AKT, blocks interleukin-6-induced activation of STAT3	Suman et al., 2016; Choi and Kim, 2015
Vitis vinifera	resveratrol	Cell cycle arrest and apoptosis	Kweon et al., 2010

essential cell-survival and cell-death pathways like AKT/PI3K/mTOR (Zhao et al., 2016). Other phytochemicals and their role in cancer prevention are mentioned in Table 19.1. Many of these and other phytochemicals do wonders in cancer chemotherapy in the clinical phase. Vinca alkaloids from *Vinca rosea*, paclitaxel from *Taxus* sp., camptothecin from *Campotheca acuminata* are a few notable plant-derived compounds that have received wide acceptance in research and cancer treatment.

With the previously mentioned activities against different cancer cell lines, an approach to merge the technology behind phytochemistry and nanotechnology would enhance the mechanism of action against cancer. Green synthesis of metallic NPs from plants, especially those synthesized as gold, silver, zinc and copper, are among the most reliable and flexible ions used in therapies. The active compounds like phenolics, terpenoids, flavonoids and alkaloids can act as one of the best reducing agents, or the NPs, thus synthesized, can act as a vehicle in transporting potential compounds of therapeutic interest to the site of action. However, this hypothesis needs further research and clarification.

19.3 GREEN NANOPARTICLES IN PHOTODYNAMIC THERAPY

Synthesis of metal NPs, carbon nanotubes (CNTs) and quantum dots (QDs) are of great interest in nanoscience and technology (Huang et al., 2006). Interestingly, modification of the size and shape of NPs is possible by modulating concentrations of chemicals used for synthesis and their reaction conditions like temperature and pH. However, when these NPs are put into application, they suffer stability, toxicity, biocompatibility and accumulation issues. Hence, it is important to tackle these issues to achieve the expected research outcome.

Nanotechnology and PDT are creating new ventures in drug delivery systems and improved treatment efficacy with minimal side effects. The properties of NPs make them a good candidate to facilitate transport of drug and PSs into target sites for better accumulation and action (Calixto et al., 2016). PS encapsulation approaches are changing with the identification of new and efficient carriers. Common approaches include oil-based dispersions (Chen et al., 2005), liposomes, micelles (van Nostrum, 2004) and biodegradable NPs (Stevens et al., 2004). Synthesis of NPs from natural sources is given prime importance due to its reliability, sustainability and ecofriendly approach. The synthesis of metal/metal oxide nanoparticles from plants is rather simple. The wide range of phytochemicals including ketones, aldehydes, flavones, amides, terpenoids, carboxylic acids, phenols and ascorbic acids present in plants make them a good candidate in reducing metal salts to metal NPs (Doble et al., 2007). Initially, gold and silver NPs were investigated in plant extracts. The common plants employed for this are *Aloe barbadensis*, *Citrus limon*, *Avena sativa*, *Azadirachta indica*, *Osimum sanctum*, *Coriandrum sativum*, *Brassica juncea*, *Medicago sativa* and *Cymbopogon flexuosus*. Other major metals like zinc, nickel, cobalt, and copper NPs from plants are gaining popularity among researchers (Marchiol, 2012).

The property of PSs to selectively target tumours and cause minimal damage to vital tissues makes them an excellent candidate for cancer therapy. However, based on the type of tumour, PSs used and conditions of exposure, limitations such as poor PS delivery, resistance to aggressive tumours and cancer stem cells are some of the challenges to overcome (Rajendrakumar et al., 2018). The introduction of NPs has exponentially helped researchers to surpass these limitations (Saeed et al., 2018). The conjugation or encapsulation of PSs in NPs could improve the stability and solubility of PSs thereby boosting efficacy of PDT, target tumours more specifically and improve the oxygen supply at tumour sites.

The ability of cancer cells to restrict certain PSs from localization promoted the use of target-specific molecules in combination with PSs. These molecules will facilitate PS localization through receptor-specific binding on the cell surface thereby increasing intracellular kinetics and cell death (Casas et al., 2011). Targeting molecules like antibodies can be conjugated to NPs and PSs to design a drug delivery system. This system of drug delivery has proved successful for the target-specific killing of cancer stem cells populations, which are responsible for cancer recurrence in most cancer cases (Chizenga et al., 2019). This multimodal targeting has been successfully achieved by many researchers including research done in our group (Chizenga et al., 2019). Interestingly, the use of therapeutic photosensitizing compounds and NPs from plants in PDT of cancer can offer more promising and innovative tools in cancer research (see Figure 19.1).

19.4 PHYTOCOMPOUNDS IN PHOTODYNAMIC THERAPY

Recently, the use of natural products and herbal drugs have been used for treatment instead of synthetic chemotherapeutic drugs since they are environmentally sustainable and minimise major side effects. The investigations have proved that herbal products, including tumour-targeting compounds, can be used in numerous cancer treatments. PDT is a novel treatment modality using PSs that can be employed for the treatment of various diseases including cancer. The combination of PDT with other treatment modalities such as chemotherapy, radiation and herbal medicine is a promising approach in cancer treatment. The synergistic combination approach enhances the treatment response and outcomes (Khdair et al., 2010). This will also be helpful in treating resistant and metastatic tumours. The efficacy of PDT is dependent on many factors and the choice of appropriate PS is one of the most important. Natural products from plants have been used as PSs for PDT applications. The main challenges preventing the treatment of patients with phototoxic agents extracted from plants and herbal products is their safety, regulatory approval requirements, and demonstration of equal effectiveness to synthetic PSs (Fu et al., 2009).

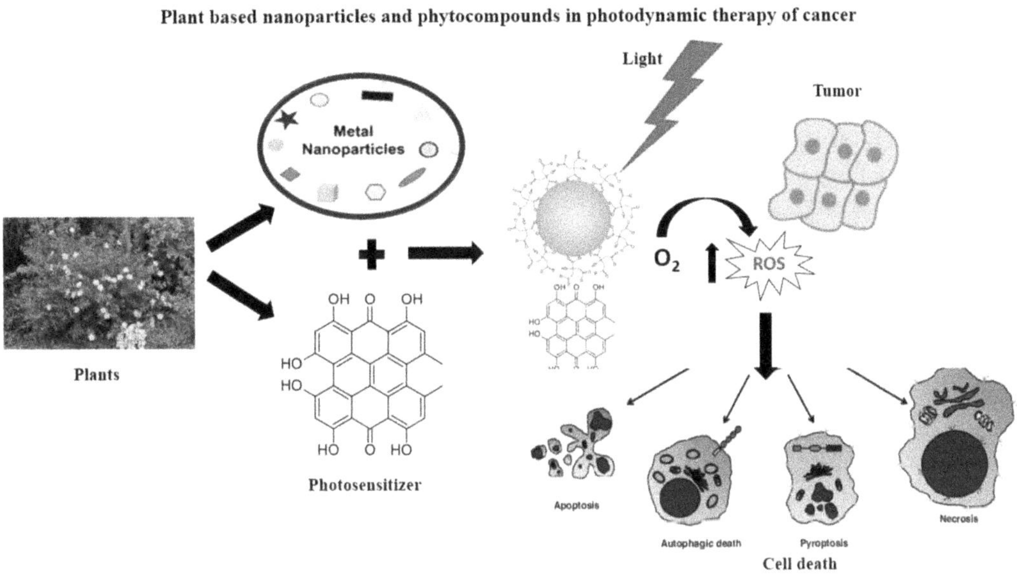

FIGURE 19.1 Plants are an excellent source of compounds that can be used to treat various diseases including cancer. This figure explains how plant-derived photosensitizers and nanoparticles can be used in the photodynamic therapy of cancer alone or in combination through various cell death mechanisms.

There are a few clinically approved plant-derived natural PSs available for cancer treatment. PSs contain several structural groups such as porphyrins. Porphyrin derivatives are one of the oldest natural PSs (Pandey and Zheng, 2000). Chlorins are another major class of PS that is originated from chlorophyll with a specific wavelength absorption range of 650–700 nm. The *Hypericum perforatum* or St John's Wort is a flowering plant well-known for its photosensitizing properties and used in various PDT applications. The compound hypericin is isolated from *Hypericum perforatum* and is used as a potential PS of first choice. There are studies showing the effect of hypericin on various tumours alone and in combination with other known natural or synthetic PSs including Chlorin(e6) that showed enhanced PDT efficacies. Hypericin-mediated PDT induces cell death via apoptosis by mitochondrial pathways activating various caspase apoptotic proteins. Studies showed that hypericin exhibited *in vitro* and *in vivo* phototoxic and antitumour effects on a range of cancers, with the absorption maxima of hypericin in the range of 595–610 nm (Skalkos et al., 2006). Several potential natural PSs with enhanced fluorescent properties have been isolated from Chinese medical plants such as *Cortex phellodendri* and *Rhizoma coptidis*. When cancer cells were treated with these plant extracts in combination with laser light irradiation, cell numbers were significantly reduced and showed the viability of cells decreased and induced the cell death (Liao et al., 1997). Wu (2013) reported that the bioactive compounds, extracts and active fractions obtained during the isolation of compounds from Chinese medicinal plants such as *Radix bupleuri*, *Rhizoma polygoni cuspidati*, *Rabdosia rubescens*, *Cortex magnolia officinalis*, and *Rhizoma chuanxiong*, in combination with exposure to light of a specific wavelength, induced cell death through autophagy (Wu, 2013). These studies reveal that photosensitizing photoactive compounds extracted from medicinal plants could be used as an alternative treatment for cancer using PDT. The major natural PS compounds are pheophorbide A, tolyporphin, chlorophyllin, curcumin, anthraquinones, hypericin, hypocrellin, cercosporin, thiophenes, 15(1)-hydroxypurpurin-7-lactone ethyl methyl diester and aloe-emodin. Some non-PS natural product compounds that sensitize the cancer cells for PDT applications are yet to be studies in detail.

Pheophorbide A, used in TCM is a chlorophyll product from green plants that has shown phototoxic effects on colorectal cancer cells. Curcumin from *Curcuma longa* showed phototoxicity under blue-light irradiation. Anthraquinone and thiophene compounds isolated from various medicinal plants have showed PDT effects on different type of cancer cells including skin, lung and breast cancers. The phototoxic effect of aloe-emodin from *Aloe vera* roots are due to the DNA and RNA damage to the cancer cells. Chlorophyllin PS localizes on the lysosomes and mitochondria of the malignant cells to induce cell death via apoptosis and autophagy after PDT. Hypocrellins from *Hypocrella bambusae* are used in deep-tissue PDT with a slight modification to the excitation wavelength from UV or blue light to red or near-infrared light. Derivatives of hypocrellins (H) such as HA, HB and DAHA (deacetylated hypocrellin A) also have been used in PDT applications. Compounds such as β-glucans, Indole-3-acetic acid and ascorbic acid are the well-known examples for such phytocompounds used in PDT (Mansoori et al., 2019). The natural PSs of plant origin may be a green approach to PDT of cancer. Low toxicity to normal and healthy cells and selective toxicity toward cancer cells are the major benefits of using natural PSs for PDT. The natural PSs, like synthetic PSs, promote cancer cell death via apoptosis, necrosis or autophagy induction. Since natural compounds are more readily available and accessible than synthetic compounds and lessen severe side effects, these compounds can be considered as promising PDT candidates in the future.

19.5 CONCLUSION

In conclusion, with emerging techniques in research and resistance to conventional treatment modalities, multidisciplinary therapeutic approaches are gaining interest. Alternative treatment strategies including PDT have shown significant success rates in the past few years. Many plants and bioactive compounds isolated from them are proven anticancer agents. However, their photoactive property and ability to synthesize NPs have boosted the scope of PDT in cancer research. This approach has led to the eradication of cancer with high efficiency and target specificity. This review promoted the effective role of plants and phytocompounds in multimodal therapy of many other diseases of serious concern.

CONFLICTS OF INTEREST

The authors declare no conflict of interest.

ACKNOWLEDGMENTS

This work is based on the research supported by the South African Research Chairs Initiative of the Department of Science and Technology and National Research Foundation of South Africa (Grant No 98337), as well as grants received from the University of Johannesburg (URC), the National Research Foundation (NRF), and the CSIR (Council for Scientific and industrial Research) – NLC (National Laser Centre) Laser Rental Pool Programme.

REFERENCES

Aras, Duru, Ozgur Cinar, Zeynep Cakar, Sinan Ozkavukcu, and Alp Can. "Can dicoumarol be used as a gonad-safe anticancer agent: An *in vitro* and *in vivo* experimental study." *MHR: Basic Science of Reproductive Medicine* 22, no. 1 (2016): 57–67.

Aung, Thazin I, Zhipeng Qu, R. Daniel Kortschak, and David L. Adelson. "Understanding the effectiveness of natural compound mixtures in cancer through their molecular mode of action." *International Journal of Molecular Sciences* 18, no. 3 (2017): 656.

Calixto, Giovana Maria Fioramonti, Jéssica Bernegossi, Laura Marise De Freitas, Carla Raquel Fontana, and Marlus Chorilli. "Nanotechnology-based drug delivery systems for photodynamic therapy of cancer: A review." *Molecules* 21, no. 3 (2016): 342.

Casas, Adriana, Gabriela Di Venosa, Tayyaba Hasan, and Al Batlle. "Mechanisms of resistance to photodynamic therapy." *Current Medicinal Chemistry* 18, no. 16 (2011): 2486–2515.

Chen, Bin, Brian Pogue, and Tayyaba Hasan. "Liposomal delivery of photosensitising agents." *Expert Opinion on Drug Delivery* 2, no. 3 (2005): 477–487.

Chen, Huinan, Biqiang Zhu, Lei Zhao, Yang Liu, Fuya Zhao, Jing Feng, Ye Jin, Jiayu Sun, Rui Geng, and Yunwei Wei. "Allicin inhibits proliferation and invasion *in vitro* and *in vivo* via SHP-1-mediated STAT3 signaling in cholangiocarcinoma." *Cellular Physiology and Biochemistry* 47, no. 2 (2018): 641–653.

Chizenga, Elvin Peter, Rahul Chandran, and Heidi Abrahamse. "Photodynamic therapy of cervical cancer by eradication of cervical cancer cells and cervical cancer stem cells." *Oncotarget* 10, no. 43 (2019): 4380.

Cho, Kwangjae, X. U. Wang, Shuming Nie, and Dong M. Shin. "Therapeutic nanoparticles for drug delivery in cancer." *Clinical Cancer Research* 14, no. 5 (2008): 1310–1316.

Choi, Bu Young, and Bong-Woo Kim. "Withaferin-A inhibits colon cancer cell growth by blocking STAT3 transcriptional activity." *Journal of Cancer Prevention* 20, no. 3 (2015): 185.

Cojocneanu, P. R., C. Braicu, L. Raduly, O. Zanoaga, N. Dragos, P. Monroig, D. Dumitrascu, and I. Berindan-Neagoe. "Phytochemicals modulate carcinogenic signaling pathways in breast and hormone-related cancers." *Onco Targets and Therapy* 8 (2015): 2053–2066.

De La Chapa, Jorge J., Prajjal Kanti Singha, Debbie R. Lee, and Cara B. Gonzales. "Thymol inhibits oral squamous cell carcinoma growth via mitochondria-mediated apoptosis." *Journal of Oral Pathology & Medicine* 47, no. 7 (2018): 674–682.

Deng, Qing-Ping, Mao-Jie Wang, Xing Zeng, George Gong Chen, and Run-Yue Huang. "Effects of glycyrrhizin in a mouse model of lung adenocarcinoma." *Cellular Physiology and Biochemistry* 41, no. 4 (2017): 1383–1392.

Ding, Wanjing, Zhijuan Hu, Zhewen Zhang, Qiaoqiao Ma, Huifang Tang, and Zhongjun Ma. "Physapubescin B exhibits potent activity against human prostate cancer *in vitro* and *in vivo*." *Journal of Agricultural and Food Chemistry* 63, no. 43 (2015): 9504–9512.

Doble, Mukesh, Rollins Ken, and Kumar Anil. *Green chemistry and engineering*. Cambridge, Academic Press, 2007.

Dou, Jie, Zhou Wang, Leon Ma, Bo Peng, Ke Mao, Chengqin Li, Mengqi Su, Changlin Zhou, and Guangyong Peng. "Baicalein and baicalin inhibit colon cancer using two distinct fashions of apoptosis and senescence." *Oncotarget* 9, no. 28 (2018): 20089.

Fu, Peter P., Hsiu-Mei Chiang, Qingsu Xia, T. A. O. Chen, Bai Hsiun Chen, Jue-Jie Yin, Kuo-Ching Wen, G. E. Lin, and Hongtao Yu. "Quality assurance and safety of herbal dietary supplements." *Journal of Environmental Science and Health Part C* 27, no. 2 (2009): 91–119.

Mei Han, Hui Gao, Jing Xie, Yin-Ping Yuan, Quan Yuan, Ming-Quan Gao, Kai-Li Liu, Xue-Hong Chen, Yan-Tao Han, and Zhi-Wu Han. "Hispidulin induces ER stress-mediated apoptosis in human hepatocellular carcinoma cells in vitro and in vivo by activating AMPK signaling pathway." *Acta Pharmacologica Sinca* 40, no. 5 (2019): 666–676.

Huang, Xiaohua, Ivan H. El-Sayed, Wei Qian, and Mostafa A. El-Sayed. "Cancer cell imaging and photothermal therapy in the near-infrared region by using gold nanorods." *Journal of the American Chemical Society* 128, no. 6 (2006): 2115–2120.

Khan, Haroon. "Medicinal plants in light of history: Recognized therapeutic modality." *Journal of Evidence-Based Complementary & Alternative Medicine* 19, no. 3 (2014): 216–219.

Khdair, Ayman, Di Chen, Yogesh Patil, Linan Ma, Q. Ping Dou, Malathy PV Shekhar, and Jayanth Panyam. "Nanoparticle-mediated combination chemotherapy and photodynamic therapy overcomes tumor drug resistance." *Journal of Controlled Release* 141, no. 2 (2010): 137–144.

Kweon, Sin Ho, Ju Han Song, and Tae Sung Kim. "Resveratrol-mediated reversal of doxorubicin resistance in acute myeloid leukemia cells via downregulation of MRP1 expression." *Biochemical and Biophysical Research Communications* 395, no. 1 (2010): 104–110.

Lee, Wai-Leng, Jing-Ying Huang, and Lie-Fen Shyur. "Phytoagents for cancer management: Regulation of nucleic acid oxidation, ROS, and related mechanisms." *Oxidative Medicine and Cellular Longevity* 2013 (2013).

Li, Jie, Chao Zhang, Hongchuan Jiang, and Jiao Cheng. "Andrographolide inhibits hypoxia-inducible factor-1 through phosphatidylinositol 3-kinase/AKT pathway and suppresses breast cancer growth." *OncoTargets and Therapy* 8, (2015): 427.

Liao, Jing, Ping-Ping Li, and C. J. Wu. "Screening new photosensitizers from Chinese medicinal herbs and searching for herbal photodynamic killing effects on human stomach cancer cells." *Chinese Journal of Integrated Traditional and Western Medicine* 17, no. 12 (1997): 726–729.

Lin, Wanfu, Maofeng Zhong, Huixia Yin, Yongan Chen, Qingxin Cao, Chen Wang, and Changquan Ling. "Emodin induces hepatocellular carcinoma cell apoptosis through MAPK and PI3K/AKT signaling pathways in vitro and *in vivo*." *Oncology Reports* 36, no. 2 (2016): 961–967.

Lu, Linwei, Zhengxiao Zhao, Lumei Liu, Weiyi Gong, and Jingcheng Dong. "Combination of baicalein and docetaxel additively inhibits the growth of non-small cell lung cancer *in vivo*." *Traditional Medicine and Modern Medicine* 1, no. 03 (2018): 213–218.

Malik, Parth, Ravi Shankar, Vibhuti Malik, Nitin Sharma, and Tapan Kumar Mukherjee. "Green chemistry based benign routes for nanoparticle synthesis." *Journal of Nanoparticles* 2014, (2014).

Mansoori, Behzad, Mohammadi Ali, Amin Doustvandi Mohammad, Mohammadnejad Fatemeh, Kamari Farzin, Gjerstorff F Morten, Baradaran Behzad, Hamblin R Michael "Photodynamic therapy for cancer: Role of natural products." *Photodiagnosis and Photodynamic Therapy* 26 (2019):395–404.

Marchiol, Luca. "Synthesis of metal nanoparticles in living plants." *Italian Journal of Agronomy* 7, no. 3 (2012): e37–e37.

Martinelli, Chiara, Carlotta Pucci, and Gianni Ciofani. "Nanostructured carriers as innovative tools for cancer diagnosis and therapy." *APL Bioengineering* 3, no. 1 (2019): 011502.

Pandey, Ravindra K., and Gang Zheng. "Porphyrins as photosensitizers in photodynamic therapy." *The Porphyrin Handbook* 6 (2000): 157–230.

Park, Y., Y. N. Hong, A. Weyers, Y. S. Kim, and R. J. Linhardt. "Polysaccharides and phytochemicals: A natural reservoir for the green synthesis of gold and silver nanoparticles." *IET Nanobiotechnology* 5, no. 3 (2011): 69–78.

Pooladanda, Venkatesh, Soumya Bandi, Sandhya Rani Mondi, Krishna Mohan Gottumukkala, and Chandraiah Godugu. "Nimbolide epigenetically regulates autophagy and apoptosis in breast cancer." *Toxicology In Vitro* 51 (2018): 114–128.

Rajendrakumar, Santhosh Kalash, Saji Uthaman, Chong-Su Cho, and In-Kyu Park. "Nanoparticle-based phototriggered cancer immunotherapy and its domino effect in the tumor microenvironment." *Biomacromolecules* 19, no. 6 (2018): 1869–1887.

Saeed, Madiha, Wenzhi Ren, and Aiguo Wu. "Therapeutic applications of iron oxide based nanoparticles in cancer: Basic concepts and recent advances." *Biomaterials Science* 6, no. 4 (2018): 708–725.

Saha, Achinto, Jorge Blando, Eric Silver, Linda Beltran, Jonathan Sessler, and John DiGiovanni. "6-Shogaol from dried ginger inhibits growth of prostate cancer cells both in vitro and in vivo through inhibition of STAT3 and NF-κB signaling." *Cancer Prevention Research* 7, no. 6 (2014): 627–638.

Skalkos, D., E. Gioti, C. D. Stalikas, H. Meyer, Th G. Papazoglou, and G. Filippidis. "Photophysical properties of *Hypericum perforatum* L. extracts–novel photosensitizers for PDT." *Journal of Photochemistry and Photobiology B: Biology* 82, no. 2 (2006): 146–151.

Smith, Peter J., Alexandra Clavarino, Jeremy Long, and Kathryn J. Steadman. "Why do some cancer patients receiving chemotherapy choose to take complementary and alternative medicines and what are the risks?" *Asia-Pacific Journal of Clinical Oncology* 10, no. 1 (2014): 1–10.

Stevens, Phillip J., Masaru Sekido, and Robert J. Lee. "Synthesis and evaluation of a hematoporphyrin derivative in a folate receptor-targeted solid-lipid nanoparticle formulation." *Anticancer Research* 24, no. 1 (2004): 161–166.

Suman, Suman, Trinath P. Das, Suman Sirimulla, Houda Alatassi, Murali K. Ankem, and Chendil Damodaran. "Withaferin—A suppress AKT induced tumor growth in colorectal cancer cells." *Oncotarget* 7, no. 12 (2016): 13854.

Thanki, Kaushik, Rahul P. Gangwal, Abhay T. Sangamwar, and Sanyog Jain. "Oral delivery of anticancer drugs: Challenges and opportunities." *Journal of Controlled Release* 170, no. 1 (2013): 15–40.

van Nostrum, Cornelus F. "Polymeric micelles to deliver photosensitizers for photodynamic therapy." *Advanced Drug Delivery Reviews* 56, no. 1 (2004): 9–16.

Verpoorte, Rob. "Exploration of nature's chemodiversity: The role of secondary metabolites as leads in drug development." *Drug Discovery Today* 3, no. 5 (1998): 232–238.

Wu, Jerry Yourui. "Chinese herbal composition for the treatment of macular degeneration and the process for manufacturing the same." U.S. Patent 8,372,450, issued February 12, 2013.

Wu, Wei, Su-Ni Tang, Yong Zhang, Manohar Puppala, Timothy K. Cooper, Chengguo Xing, Cheng Jiang, and Junxuan Lü. "Prostate cancer xenograft inhibitory activity and pharmacokinetics of decursinol, a metabolite of angelica gigas pyranocoumarins, in mouse models." *American Journal of Chinese Medicine* 45, no. 08 (2017): 1773–1792.

Wu, Xiaojin, Weitao Wang, Yuanyuan Chen, Xiangqun Liu, Jindong Wang, Xiaobin Qin, Dawei Yuan et al. "Glycyrrhizin suppresses the growth of human NSCLC cell line HCC827 by downregulating HMGB1 level." *BioMed Research International* 2018 (2018).

Yan, Xiao-Bo, Tao Xie, Sheng-Dong Wang, Zhan Wang, Heng-Yuan Li, and Zhao-Ming Ye. "Apigenin inhibits proliferation of human chondrosarcoma cells via cell cycle arrest and mitochondrial apoptosis induced by ROS generation-an *in vitro* and *in vivo* study." *International Journal Clinical and Experimental Medicine* 11, no. 3 (2018): 1615–1631.

Zhang, Bin, Xinglong Fan, Zhen Wang, Wenyong Zhu, and Jingbo Li. "Alpinumisoflavone radiosensitizes esophageal squamous cell carcinoma through inducing apoptosis and cell cycle arrest." *Biomedicine & Pharmacotherapy* 95 (2017a): 199–206.

Zhang, Wenjia, Jing Su, Huadan Xu, Shanshan Yu, Yanan Liu, Yong Zhang, Liankun Sun, Ying Yue, and Xiaoli Zhou. "Dicumarol inhibits PDK1 and targets multiple malignant behaviors of ovarian cancer cells." *PLoS One* 12, no. 6 (2017b): e0179672.

Zhao, Guangming, Xiaodong Han, Siwen Zheng, Zhen Li, Yang Sha, Jing Ni, Zhe Sun, Song Qiao, and Zhiqi Song. "Curcumin induces autophagy, inhibits proliferation and invasion by downregulating AKT/mTOR signaling pathway in human melanoma cells." *Oncology Reports* 35, no. 2 (2016): 1065–1074.

20 Quercetin Bioflavonoids Derived from Phytomedicinal Compounds for Targeted Drug Delivery and Their Antioxidant Properties

Selvaraj Rajesh Kumar and Nagamony Ponpandian

CONTENTS

20.1 WHAT IS QUERCETIN?

Quercetin is a natural pigment that belongs to a group of plant compounds called flavonoids. These flavonoids are present in many vegetables, grains, tea, wine and fruits (see Figure 20.1). Its name originates from the Latin word *quercetum*, meaning oak forest. Quercetin is usually found in various medicinal and edible plants. Quercetin is yellow and completely dissolves in alcohols/lipids, is sparingly soluble in hot water and insoluble in cold water. It is a universally well-known flavonoid derived from natural products with low toxicity. Quercetin was rich in several phytomedicinal plant families such as *Passifloraceae*, *Asteraceae*, *Solanaceae*, *Rhamnaceae*, etc. The flavonoid may contain several promising health benefits such as decreasing cancer risks, heart diseases, degenerative brain disorder, diabetes, etc. These favourable effects of quercetin originate from their ability to function with antioxidant and anti-inflammatory effects inside the human body. An average person consumes 10–100 mg of quercetin daily from different food sources. It is also frequently found in marketplaces where it is available in its isolated form.

DOI: 10.1201/9781003231745-20

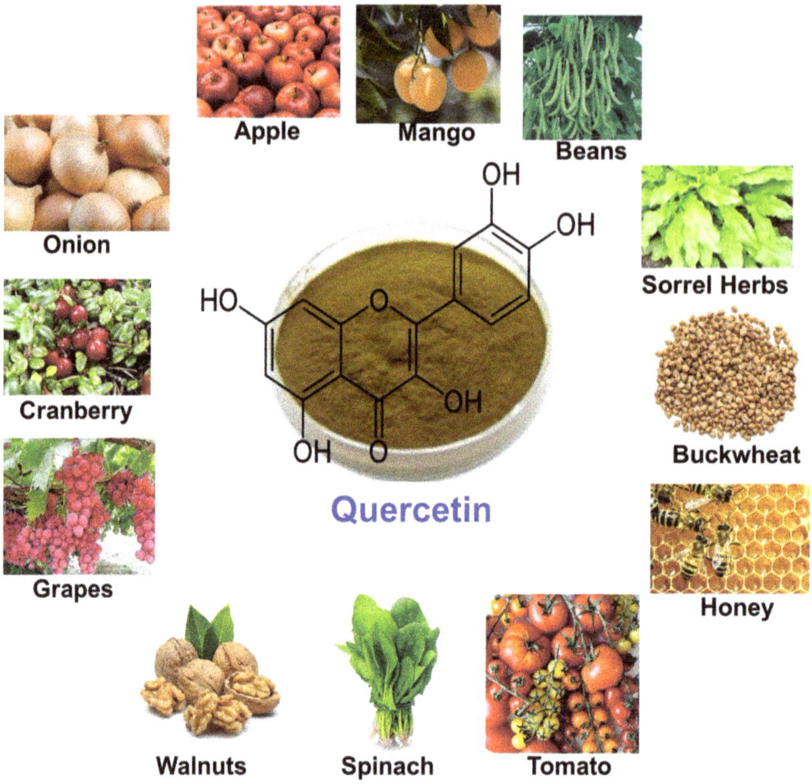

FIGURE 20.1 Derivation of quercetin flavonoids obtained from different phytomedicinal sources.

20.2 CHEMICAL STRUCTURES OF QUERCETIN

Quercetin (whose chemical name is 3,3′, 4′,5,7-pentahydroxyflavone) is mostly found in many natural phytomedicinal compounds. In plants, the availability of quercetin can be found in either bound or free (aglycone) form including methanol (quercetin methyl ethers), mostly with alcohols and carbohydrates (quercetin glycosides) and less frequently in sulphate and prenyl substituents. Quercetin contains two benzene rings connected with one pyrene ring and five hydroxyl (OH) groups (see Figure 20.1). The details of the main identifiers and their properties of quercetin (D'Andrea, 2015) are listed in Table 20.1. Usually, quercetin is found in its glycoside form, in which one or more OH molecules are replaced by diverse types of sugar molecules. Moreover, the wide range of quercetin's pharmacological and biochemical properties and their metabolites depend on the relative substitution of several functional groups on the flavonol molecule. The various sources, chemical and molecular structure of quercetin derivatives are shown in Table 20.2. The various polyphenolic compounds of quercetin have significant antioxidant ability to prevent free radical processes (de Oliveira et al., 2016).

20.3 QUERCETIN DERIVATIVES

Quercetin derivatives are the most abundant flavonoid molecules derived from many phytochemical plants, fruits, vegetables, etc. (see Table 20.3). Phytochemical investigations of several plant extracts have exposed that quercetin can be found in a free state (as an aglycone) and/or its derivative form of lipids (as prenylated quercetin), carbohydrates (as quercetin glycosides), a sulphate group (as quercetin sulphate) and alcohols (as quercetin ethers). Quercetin has multiple therapeutic

TABLE 20.1
Main Identifier and Its Properties of Quercetin

Identifiers	
IUPAC name	2-(3,4-dihydroxyphenyl)-3,5,7-trihydroxychromen-4-one
MeSH synonyms	1. Quercetin 2. 3,3′,4′,5,7-pentahydroxyflavone 3. Dikvertin
Molecular formula	$C_{15}H_{10}O_7$
Molecular weight	302.2357 g mol^{-1}
PubChem CID	5,280,343
Canonical SMILE	C1=CC(=C(C=C1C2=C(C(=O)C3=C(C=C(C=C3O2)O)O)O)O)O
Trivial chemical names	Meletin; Sophoretin; Xanthaurine
InChl key	REFJWTPEDVJJIY-UHFFFAOYSA-N
CAS	117–39–5
UN number	2811
EC number	204–187–1
Properties	
Colour	Yellow needles (dilute alcohol, + 2 water)
Density	1.799 g cm^{-3}
Melting point	316.5°C
Boiling point	Sublimes
Dissociation constants (in phenol)	$pKa_1 = 7.17$; $pKa_2 = 8.26$; $pKa_3 = 10.13$; $pKa_4 = 12.30$; $pKa_5 = 13.11$
Decomposition	During heating, emits irritating fumes and acrid smoke
Vapour pressure	2.81×10^{-14} mmHg at 25°C
Physical description	Yellow powder or yellow needles. Alcoholic solutions taste very bitter. Converts to anhydrous form at 203–207°F.
Spectral properties	Max absorption: 373 nm (log E = 4.32); 301 nm (log E = 3.89); 256 nm (log E = 4.32); Sadler Ref. Number: 594 (IR, PRISM)
Solubility	Very easily soluble in methanol and ether; normally soluble in acetone, ethanol, acetic acid and pyridine

Source: D'Andrea (2015).

properties including hepatoprotective, neuroprotective, antioxidative, antiproliferative, antiviral, antibacterial, anti-inflammatory and anticancer activities (Mehrbod et al., 2021). Depending on the family and its chemical structure, various pharmacological activities of quercetin (Batiha et al., 2020) are emphasized, as represented in Table 20.4.

20.4 STABILITY OF QUERCETIN

Quercetin's stability depends on pH, time, oxygen concentrations and temperature. Quercetin is highly unstable under alkaline conditions (pH >7) in the organic solution of methanol and acetonitrile. More importantly, the sample storage temperature and time could also influence quercetin's stability. In the case of *Phaseolus vulgaris* L. beans, 70% of quercetin loss was observed when it was boiling at 121°C for 10 minutes or atmospheric (100°C, 50 minutes) temperature. There was almost complete loss of quercetin (100%) in onion (*Allium cepa*) when stored at a low temperature of 4°C in the dark for 168 days. In similar temperature ranges for two months, the quercetin in strawberry juice was reduced to 46.1%.

TABLE 20.2
Chemical and Molecular Structures of Quercetin Derivatives

Molecular Formula (MW)	Common Name (Systematic Name)	Chemical Structure	Food Sources
$C_{15}H_{10}O_7$ (302.2)	Quercetin (3,5,7,3′,4′-pentahydroxyflavon)	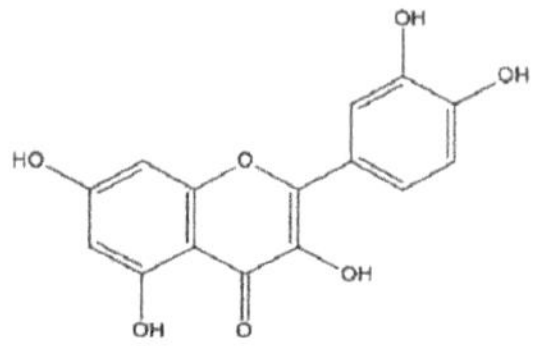	Herbs Plants Wine
$C_{16}H_{12}O_7$ (316.3)	(Quercetin 3′-methyl ether)	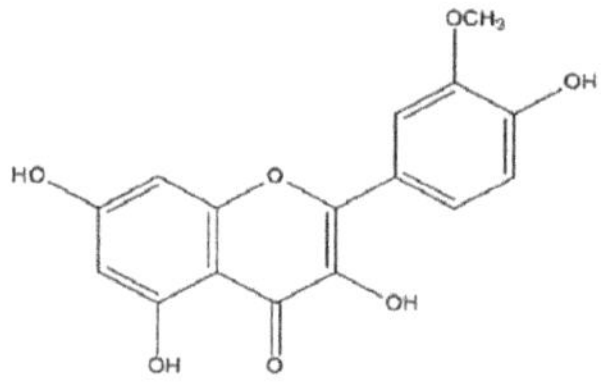	Beans
$C_{16}H_{12}O_7$ (316.2)	Rhamnetin (quercetin 7-O-methyl)	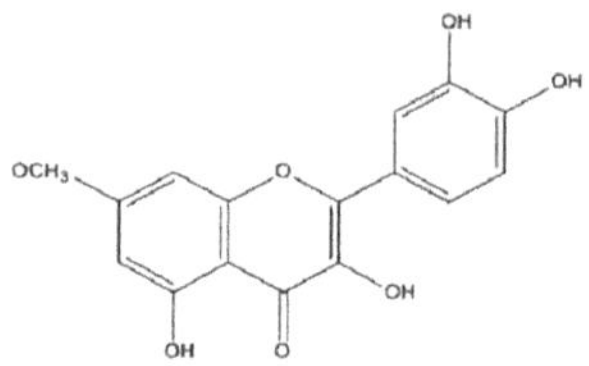	Rice Grape Walnuts
$C_{20}H_{18}O_{11}$ (434.3)	Quercetin 3-O-arabinoside	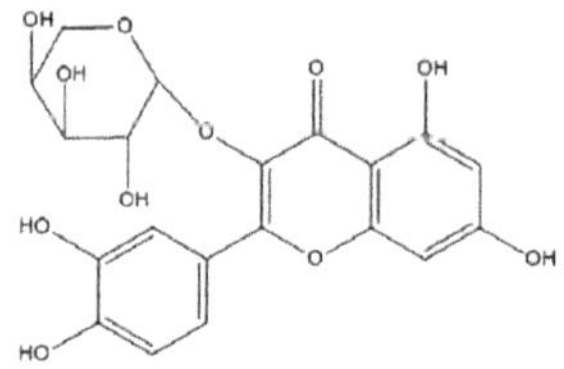	Walnuts Chokeberries Sorrels Lingonberries Apple
$C_{20}H_{18}O_{11}$ (434.3)	Reynoutrin (quercetin-3-O-xyloside)	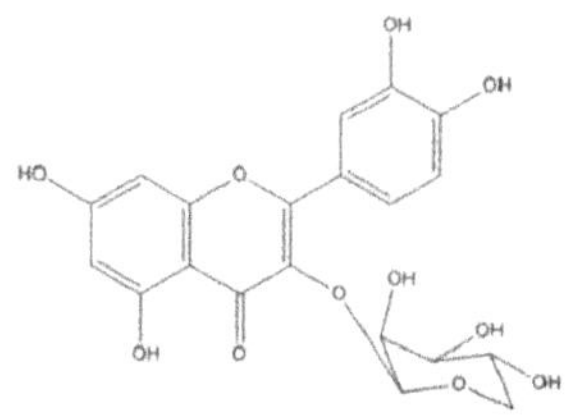	Grapes Apple Pear
$C_{21}H_{20}O_{12}$ (464.3)	Isoquercetin (quercetin-3-O-glucoside)	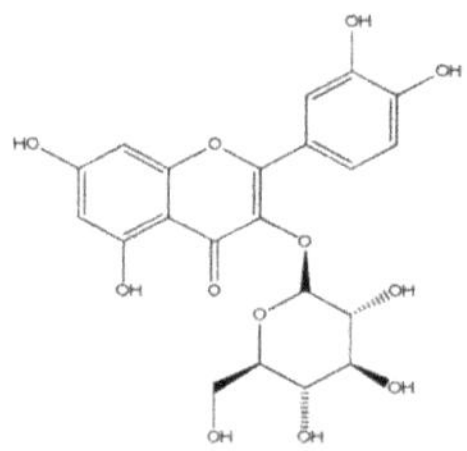	Plums Onion Beans Mango
$C_{21}H_{20}O_{12}$ (464.3)	Hyperoside (quercetin-3-O-galactoside)	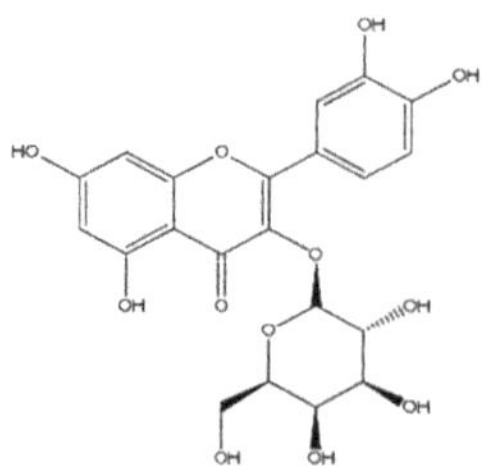	Blueberries Cranberries Mango Chokeberries

(Continued)

TABLE 20.2 ***(Continued)***

Molecular Formula (MW)	Common Name (Systematic Name)	Chemical Structure	Food Sources
$C_{21}H_{20}O_{11}$ (448.4)	Quercetrin (quercetin 3-O-rhamnoside)	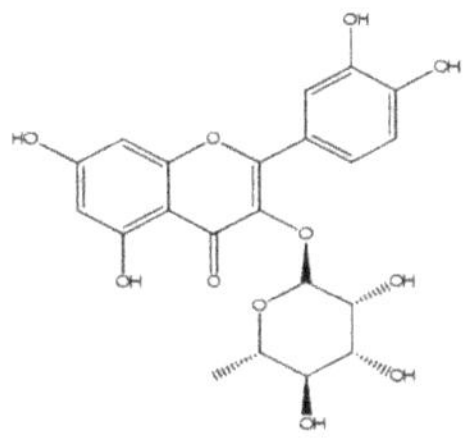	Spinach Mango
$C_{21}H_{20}O_{12}$ (464.3)	Quercimeritrin (quercetin-7-O-glucoside)	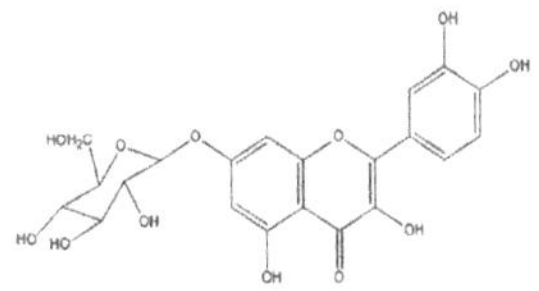	Beans
$C_{27}H_{30}O_{16}$ (610.5)	Rutin (quercetin-3-rutinoside)	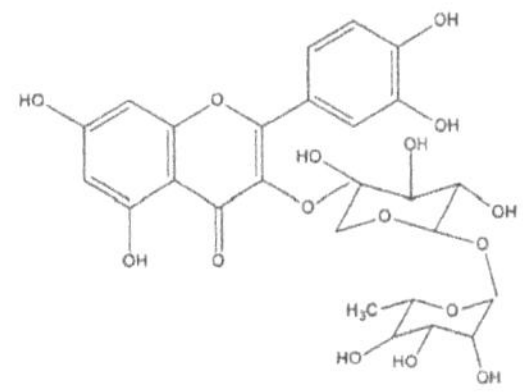	Cherries Plums Buckwheat Tomatoes
$C_{27}H_{30}O_{16}$ (610.5)	Quercetin-3-glucoside-7-rhamnoside	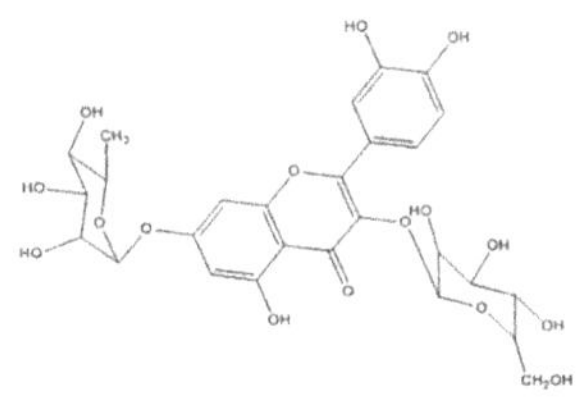	Honey Onions Pepper
$C_{27}H_{30}O_{17}$ (626.5)	Quercetin 3,7-diglucoside	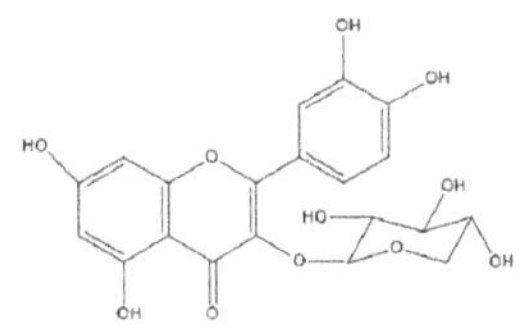	Onions Shallots

Sources: Gansukh et al. (2017), Mehrbod et al. (2021) and Zhao et al. (2019b).

20.5 VARIOUS DOSE LEVELS OF QUERCETIN

A high percentage of quercetin accumulates in various organs including liver, lungs, small intestines and kidneys, whereas lower levels are realized in the spleen, brain and heart. The dietary intake of quercetin in humans is in the range of 5–100 mg per day (Batiha et al., 2020). In the study, supplementary food containing quercetin compounds was fed to humans for two weeks. The results found that the intake of 50 mg of quercetin attained 178% higher serum levels, while 100 mg (359%) and 500 mg (570%) of quercetin intake shows highly increased serum levels. In the Western population, the level of daily intake of quercetin was found in the range of 1–30 mg, whereas some other countries may ingest a maximum of 200–1200 mg (D'Andrea, 2015).

The quercetin dose range of 10–50 mg/kg/bw can significantly reduce blood glucose levels, oxidation stress, inflammation and cell apoptosis levels in male diabetic rats and simultaneously improve cardiac injury marker stages. In addition, this flavonoid had promising results in reducing

TABLE 20.3
Quercetin Derivatives from Phytomedicinal Plants

Source	Botanical Name	Tissue	Concentration (mg kg^{-1})
Onion	*Allium cepa* red	Bulb	307
	Allium cepa	Bulb	221–347
	Allium cepa Patrick	Bulb	3830
Welsh onion	*Allium fistulosum*	Leaves	1498
Chokeberry	*Arontiamitschurinii* Viking	Fruit	89
Aloe vera	*Aloe barbadensis*	Leaves	95
Asparagus	*Asparagus officinalis* L.	Leaves	7–20
Broccoli	*Brassica oleracea*	Vegetables	30
Kale	*Brassica oleracea*	Vegetables	110
Cauliflower	*Brassica oleracea*	Vegetables	219
Red chili	*Capsicum annuum*	Fruit	800
Semambu	*Calamus scipronum*	Leaves	1188
Peepal	*Ficus religiosa*	Fruit	256
Apple	*Malus pumila*	Pomace	36–67
Mulberry	*Morus alba*	Fruit	359
Mango	*Mangifera indica* Ubá	Fruit peel	469
Sour cherry	*Prunus cerasus* Schattenmorelle	Fruit	25
Sweet cherry	*Prunus avium* Hartland	Fruit	28
French bean	*Phaseolus vulgaris*	Fruit	39
Apricot	*Prunus armeniaca*	Fruit	322
Tomato	*Solanum lycopersicum* Manisa C-33	Fruit	9.3
Wine grape	*Vitis vinifera*	Fruit	104
Bog bilberry	*Vaccinium uliginosum*	Fruit	158
Cranberry	*Vaccinium oxycoccos*	Fruit	83,121

Source: de Oliveira et al. (2016).

diabetic obstacles including neuropathy, retinopathy and nephropathy at various quercetin doses. The safety dose of quercetin is about 1000 mg m^{-2} and its foremost toxicological properties are hypokalaemia, hypertension and emesis (Vafadar et al., 2020). For instance, 945 mg m^{-2} of quercetin was a safe dose level for cancer patients through intravenous injection among the analysed concentration of 60–2000 mg m^{-2}. Above this concentration, it induces toxicity by decreasing serum potassium, vomiting, emesis, nephrotoxicity and hypertension. Moreover, the intravenous quercetin delivery half-life time was found to be 0.7–7.8 minutes, whereas the removal half-life time was 3.8–86 minutes.

In mice, the dietary feeding of 2 mg g^{-1} quercetin can increase the plasma concentration. In humans, the bioavailability of quercetin is slightly low; it fluctuates (0–50%) and rapidly vanishes with 60–120 minutes of half-life after intake through vegetables, fruits, rich foods or other supplements. Quercetin is possibly extracted through the respiratory tract, faecal and renal systems.

20.6 NANOMATERIAL CONJUGATION ON QUERCETIN

Understanding the quercetin interaction with nanomaterials is a very significant nanoformulation due to its sensitive applications. Quercetin was combined with various carriers to deliver to a particular target site such as nanoparticles, micelle complexes, lipid-based carriers and loaded/conjugation-based encapsulation to enhance bioavailability and chemical stability. Among various

TABLE 20.4
Pharmacological Properties of Quercetin Extracted from Different Phytomedicinal Plants

Family Name	Plant Name	Pharmacological Properties
Amaryllidaceae	*Allium cepa*	Cardioprotective, immunostimulant, and antioxidant properties
Apiaceae	*Centella asiatica*	Treatment for wound healing
Apiaceae	*Coriandrum sativum*	Reducing dyspepsia, cholesterol and blood pressure
Asparagaceae	*Asparagus officinalis*	Treatment for antitussive, antineoplastic and antiulcer
Asteraceae	*Lactuca sativa*	Improves osteoporosis laxative and iron deficiency anaemia
Brassicaceae	*Brassica oleracea* var. italica (broccoli)	Neuropathy, reduces glucose levels, lessens the risk of stroke
Brassicaceae	*Brassica oleracea* var. sabellica (Kale)	Possible to prevent fluid retention in cancer treatments
Capparaceae	*Capparis spinosa*	Effective antiatherosclerotic agent, disinfectants and vermifuges
Ericaceae	*Vaccinium oxycoccus*	Treatment for urinary tract infections
Hypericaceae	*Hypericum perforatum*	Effective antioxidant properties
Hypericaceae	*Hypericum hircinum*	Treatment of neurological effects and major depressive disorders
Moraceae	*Morus alba*	Diet treatment
Moringaceae	*Moringa oleifera*	Antibacterial, anti-inflammatory and antihypertensive
Rosaceae	*Malus domestica*	Decreasing the risk of cancer and cardiovascular diseases
Rosaceae	*Prunus avium*	Diuretic, tonic and astringent,
Solanaceae	*Solanum lycopersicum*	Food supplement and salads
Theaceae	*Camellia sinensis*	Antispasmodic, antidiabetic, analgesic, antiviral and bronchodilator

Source: Batiha et al. (2020).

pathways, loaded or conjugated quercetin with nanomaterials highly improved synergetic effects. The quercetin encapsulation principle with nanomaterials enhanced its solubility and maximized the stability with noteworthy bioavailability. Recently, quercetin was directly conjugated/loaded with different types of nanomaterials for promising drug delivery applications.

The schematic illustration of quercetin-conjugated magnetic nanoparticles through the nano-precipitation method is shown in Figure 20.2. In detail, the dextran was coated initially on the surface of Fe_3O_4 nanoparticles through the co-precipitation method. Then, the coupling agents of N-hydroxy succinimide and 1-ethyl-3(3-dimethylaminopropyl) carbodiimide hydrochloride were used to improve the magnetic nanoparticle surface functionalization and then conjugated with quercetin through the nanoprecipitation method (Kumar et al., 2014). The quercetin-conjugated nanoparticles induced apoptosis to many cancer cells including ovarian, breast, lung, blood and pancreatic (Parhi, Bharatiya, and Swain, 2020). Quercetin solubility was also enhanced through noncovalent bonding in the organic network of surfactant polymeric micelles in aqueous solution as compared with pure quercetin. For example, the quercetin-loaded spherical-shaped polymeric micelles in the size range of 140–300 nm improved the solubility and have MCF-7 cell inhibition

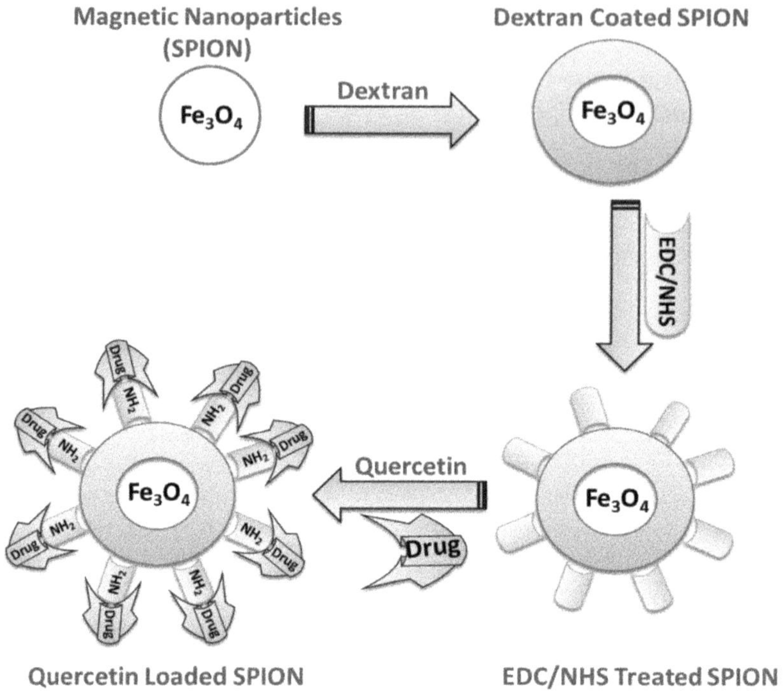

FIGURE 20.2 Quercetin conjugated on dextran-coated superparamagnetic Fe_3O_4 nanoparticles (SPIONs) through the nanoprecipitation method.

with hemocompatibility that indicates it as a good nanocarrier for cancer treatment. Moreover, 1–100 μM of quercetin-loaded or conjugated nanoparticles could also improve the solubility parameters and significantly reduce the cell mortality rate, fragmentation, growth factors and cell death of breast cancer cells (Kumar et al., 2014).

The quercetin-loaded silver nanoparticle shows an effective antioxidant property that reduced lipid oxidation and improved antibacterial activities. The quercetin-loaded poly(D, L-lactide-co-glycolide) (PLGA) nanoparticles with the size range of 100–150 nm were used as an antioxidant compound and enhanced antibacterial activity. Solid lipid nanoparticles loaded with quercetin protect retinal ganglion and cornea eye cells from oxidative damage by encapsulation that retains higher cytotoxicity and protects the eye from cancer attacks. Quercetin conjugated gold nanoparticles induce cell apoptosis in human breast cancer cells through epidermal growth factor receptor (EGFR)–mediated PI3K/Akt/GSK-3 β signalling by immunocytochemistry and immunoblotting. Quercetin loaded with the GO-PVP nanovehicle improves targeting efficiency, dissolution rate and induces more cell apoptosis in ovarian cells. Hence, the nanomaterials act as effective nanocarriers to quercetin for targeted drug delivery applications.

20.7 ANTIVIRAL ACTIVITIES OF QUERCETIN

Quercetin is an interesting natural compound for the antiviral treatment of a wide variety of viruses. Several research outcomes confirm that the quercetin derivatives obtained from phytomedicinal plants suppressed the influenza viral infection through neuraminidase inhibition pathways (Gansukh et al., 2017). Interestingly, quercetin compounds highly reduced the viral replications in the early stages of binding and fusion by inhibiting the reactive oxygen species (ROS) production of the viral host. It has been proven that hepatitis C and dengue type-2 viruses are highly suppressed when treated with quercetin compounds by destroying the nonstructural protein 3 protease activities. Moreover, quercetin has significant antiviral activities against the mosquito-borne flavivirus including Japanese encephalitis virus and human T-lymphotropic virus by reducing its replications (Coelho-dos-Reis et al., 2011). Quercetin could directly bind with RNA and inhibit the RNA-dependent polymerase to reduce the reproduction of viral RNA in the host cell.

To examine the influence of quercetin structural changes, different C3′, C3 and C5 hydroxyl molecule-substituted quercetin derivatives were prepared with different phenolic ester, aminoalkoxy, and alkoxy moieties. This modified quercetin improved their antiviral activities against A/swine/OH/511445/2007 (H1N1). Also, the quercetin-3-gallate and C3-analog were structurally associated with EGCG that represents favourable antiviral activity against the influenza virus in a MDCK cell culture and in reducing cytopathic effects. Thus, the modification of the functional group in the quercetin molecule might play an important role in understanding compound efficacy. The oral administration of quercetin flavonoid may not attain drug contact with the viral S protein due to its glycosylation for biotransformation in the gastrointestinal tract. Thus, using throat or nasal spray holding quercetin in an appropriate form was suggested to deliver a suitable volume to the active sites of the virus in its clinical trials. Hence, quercetin appears to be the most favourable natural antiviral compound against influenza viral studies.

20.8 ANTIMICROBIAL ACTIVITY OF QUERCETIN

Generally, traditional Indian and Chinese medicine use phytomedicinal herbs or plants for the treatment of oral bacterial infections through oral pathways. For example, quercetin was extracted from lotus leaf for the antibacterial treatment of mouth inflammation caused by various micro-organisms. The quercetin compound induced cell-wall damage, pore formations, membrane damage and bacterial aggregation effects in both Gram-negative and Gram-positive bacteria, leading to antibacterial activities. However, quercetin has more active antibacterial activities against Gram-positive than Gram-negative bacteria. Moreover, the antibacterial activities of quercetin were mainly dependent on the type of bacteria and concentration of quercetin. The minimal bactericidal concentration (MBC) and minimum inhibition concentration (MIC) of quercetin-treated *Staphylococcus aureus* (50 and 20 mcg mL^{-1}) and *Pseudomonas aeruginosa* (50 and 50 mcg mL^{-1}) have more effective antibacterial bacterial activities than those of *Escherichia coli* (>500 and 400 mcg mL^{-1}) and *Proteus vulgaris* (>500 and 300 mcg mL^{-1}), respectively. *Lactobacillus caseivarshirota* and *Shigella flexneri* do not have bactericidal activities even at >500 mcg mL^{-1} and the MIC/MBC values are not applicable. The PLGA-loaded quercetin nanoparticles improved the antibacterial activities of *Escherichia coli* as compared with *Micrococcus tetragenus* by distracting bacterial cell-wall integrity. Quercetin is also used for promising bactericidal properties that can be used for various food-packing materials. Quercetin derivatives could also influence strong antifungal activities against *Cryptococcus neoformans*, *Candida albicans* and *Aspergillus niger.*

20.9 ANTICANCER ACTIVITIES OF QUERCETIN

Quercetin is one of the promising flavonoids that exist in the diversity of phytomedicinal species including kale (110 mg kg^{-1}), broccoli (30 mg kg^{-1}), french beans (32–45 mg kg^{-1}), tomatoes (8 mg kg^{-1}), lettuce (14 mg kg^{-1}), onions (284–486 mg kg^{-1}), apples (21–72 mg kg^{-1}) and red grapes (15 mg kg^{-1}). Recently, quercetin derivatives showed effective anticancer activities in both *in vitro* and *in vivo* applications that involved many tumour cell lines and animal models. Quercetin has a strong chemo-protective activity that leads to apoptosis for several types of tumour cells including cervical, breast, liver, bladder, gastric, colon, lung, bone, blood, skin, eye, thyroid, ovarian and head and neck cancers. Also, quercetin enhances the mechanism of action in tumour cells such as good cellular signalling, inhibiting enzymes responsible for carcinogen activation, binding to targeted cellular receptors and proteins, cell propagation, Caspase 3 activation, mitochondria pathways, etc. (Batiha et al., 2020). Many literature reports specify that quercetin could control several cancer-related miRNAs such as

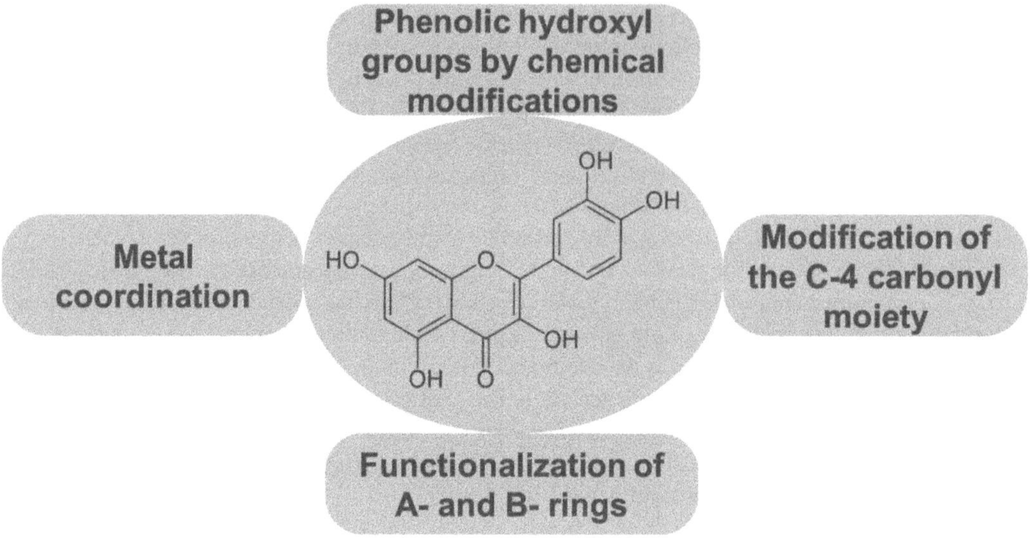

FIGURE 20.3 General methodology to modify quercetin with various compounds.

miR-21, miR-155, miR-146a and let-7, thus preventing cancer initiation and development (Kim et al., 2019). Most of the *in vitro* findings specify that the quercetin dose range of 1–100 μM could exhibit good antiproliferative effects. In cancer cells, quercetin could escalate the intracellular Ca^{2+} levels, thereby reducing the cell mitochondrial membrane potential that leads to cell death by stimulating apoptotic induction factor migration to the nucleus. In oral cancer, quercetin can block invasion and migration by regulating MMP-9/MMP-2 pathways, which confirms that quercetin demonstrates anticancer activities by regulating miR-16 (Zhao et al., 2019a). Thus, quercetin plays a promising role in various types of cancer cells with virtually no toxic side effects.

Quercetin with the required properties for potential anticancer activities has been attained through various chemical modification methods. These include the functionalization of A- and B-rings, chemical modification of phenolic hydroxyl groups, metal coordination, and in the combination with modifications at the C-4 carbonyl residue, as shown in Figure 20.3. It was also demonstrated that methoxy molecule insertion into a flavonoid resulted in metabolically high stable derivatives with improved bioavailability, solubility and anticancer activity. In quercetin, the hydroxyl groups on the side phenyl ring were fixed to significant amino-acid moieties at the dynamic spot of the two enzymes. Quercetin easily penetrates through cell membranes owing to its lipophilic properties and has a pleiotropic role in activating miscellaneous intracellular paths implicated in chemoprevention effects. Thus, the quercetin flavonoid can prevent various cancer growth through several mechanisms including pro-oxidation, cell-cycle arrest, antiproliferation and anti-inflammation.

20.10 ANTI-INFLAMMATORY AND ANTIOXIDANT PROPERTIES OF QUERCETIN

Quercetin's antioxidant properties are mostly based on its enzymatic activity, effect on glutathione, ROS and signal transduction pathways initiated by toxicological and environmental factors. In addition, quercetin displays robust antioxidant activity by keeping an oxidative balance. The oxidative damage induced through ONOO–,NO· and O_2– can produce harmful effects on tissues and cells in the human body and may cause several disorders such as diabetes cardiovascular diseases, cancers, etc. If a higher dose of quercetin is applied, the dynamic balance of glutathione was affected under the explosion of glutathione peroxidase. Thereby, glutathione is oxidized to glutathione disulphide and H_2O_2 is transformed to H_2O. Thus, the dynamic balance of glutathione at the level of

0.5% is the optimized range in rats. Quercetin can quench free radicals or ROS, thus supporting the inhibition of inflammation and several diseases caused by different pathogens.

Quercetin can protect neurons from oxidative damage despite decreasing lipid peroxidation. Numerous scientific reports have demonstrated quercetin as a treatment for neurodegenerative disorders including traumatic injury, ischemia, cognitive impairment and Huntington's and Parkinson's diseases. Quercetin also shields against mitochondrial dysfunction and improved dopaminergic neurodegeneration by stimulating MitoPark transgenic mouse models and the PKD1-Akt cell survival signalling axis of Parkinson's disease. Quercetin acts as a neuroprotective agent by antagonizing alternations in gene expression, including Casp7, Casp3, Park7, Park5 and Park2 genes. Therefore, the quercetin molecule can improve learning, memory and cognitive functions, all of which have been shown to be related to quercetin's antioxidant activities. However, the practical usage of quercetin for neurodegenerative treatment is inadequate due to its low bioavailability, water solubility and instability. Therefore, several investigations should be conducted to transform its basic structure to diminish its pharmacokinetic and other limitations for achieving improved and promising neuroprotective effects.

Quercetin has numerous effects on regulating signal pathways, including inhibiting, activating and down-regulating or up-regulating several molecules of the body, as shown in Figure 20.4. Quercetin has the distinct capability of scavenging highly reactive species including hydrogen peroxide, hydroxyl radicals and superoxide anions. Through these signal pathways, quercetin can enhance antioxidant activity, helping to heal injuries including atherosclerosis, spinal cord injury and cadmium or lead toxicity by preventing pathogen developments. It was reported that the shielding mechanism of quercetin on an acute spinal cord injury is facilitated by the up-regulation of superoxide dismutase (SOD) action, the down-regulation of malondialdehyde (MDA) levels and its inhibitory effect on the p38MAPK/iNOS signalling pathway to stimulate good antioxidant properties. The findings of this work are evidence of how quercetin acts as an anti-inflammatory and antioxidant factor without disturbing the endothelial cells from apoptosis by inhibiting the apoptotic process (Wu et al., 2014). Moreover, quercetin enhances the anti-inflammatory properties in human trials with coronary artery disease by reducing the transcriptional activity of the NF-κB. Hence, quercetin improved the antioxidant and anti-inflammatory properties for various biomedical fields.

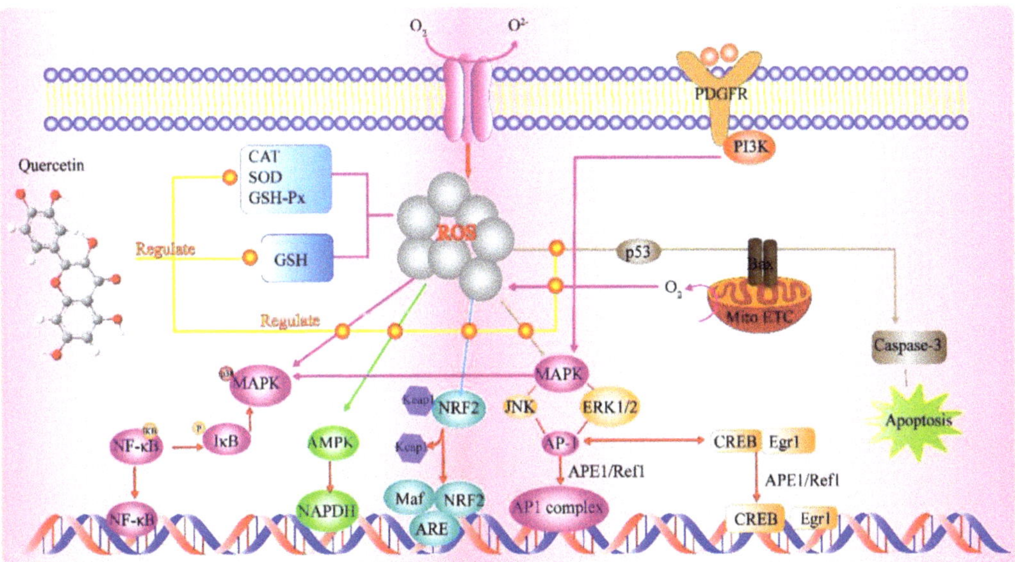

FIGURE 20.4 The antioxidant mechanism of quercetin by regulating signal pathways through reactive oxygen species (ROS) (Xu et al., 2019).

20.11 ANTIALLERGIC PROPERTIES OF QUERCETIN

In day-to-day life, various changes in our environmental factors including air pollution, outdoor and indoor allergens, micro-organisms, food habits and different unknown infections may contribute to serious allergic problems. Quercetin is a natural polyphenol that can be efficiently applied as a treatment of allergic rhinitis, late-late-phase bronchial asthma responses and restricted rheumatoid arthritis. Quercetin affects the inflammation and immune systems by performing mostly on leukocytes for targeting membrane proteins, several enzymes, intracellular signalling phosphatases and kinases. Especially, quercetin suppressed the allergic-mediator release from basophils and mast cells, which was another type of white blood cell induced in immune responses. This mast cell acts as a prominent immune system that is essential for the pathogenesis of autoimmune disorders and allergic reactions. Quercetin also suppressed the histamine, Ca^{2+} influx, leukotrienes release and protein kinase activation. Quercetin was extracted from an onion to treat smooth-muscle contraction and cytokines through an *in vitro* analysis for the murine model of asthma. The results found a release of tracheal rings, reduced inflammatory cytokine creation, and a decrease of cells in eosinophil peroxidase in the lungs.

20.12 QUERCETIN FOR THE TREATMENT OF COVID-19-INDUCED MORBIDITIES

To date, the COVID-19 pandemic continues to spread out around the world with significant mortality and morbidity, however, no precise drug or vaccine for fully effective and safe therapies are widely available for treatment of coronavirus infections. Quercetin derivatives have been used for various influenza virus treatments and its corresponding cellular life cycle is shown in Figure 20.5, including the impact of various stages of the viral life cycle such as binding, assembly, negative-strand synthesis, mRNA synthesis, and release. Also, the quercetin could indirectly affect the cytoplasm events that leads to apoptosis (Mehrbod et al., 2021). Quercetin can strongly bind with the spike protein (S) receptor of the virus with a stronger affinity for the main protease active site in

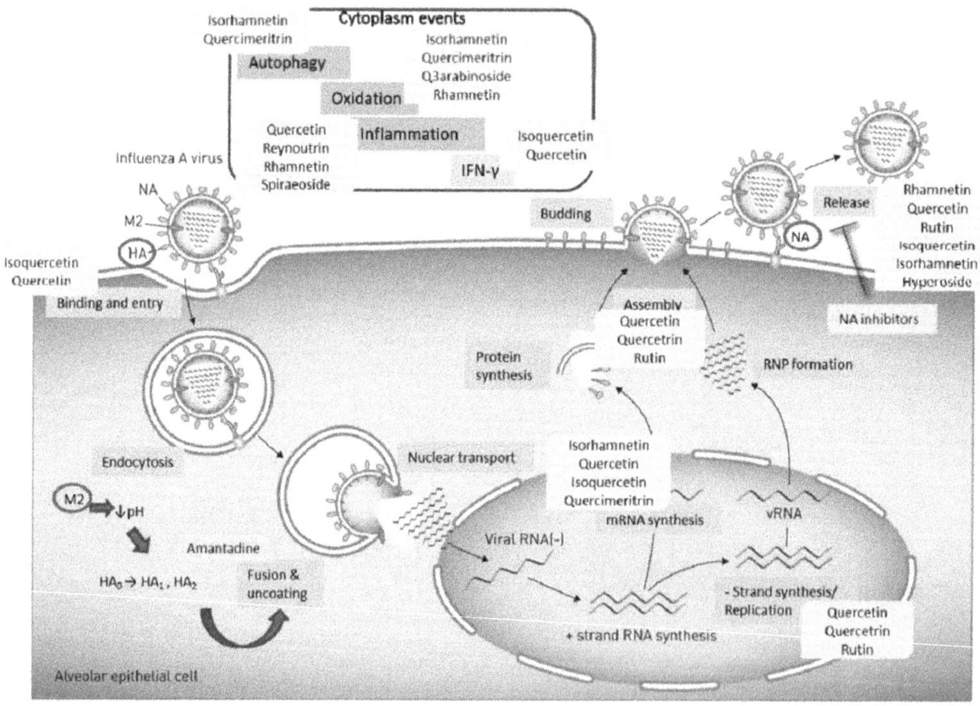

FIGURE 20.5 Quercetin derivatives against the influenza virus life-cycle pathways. (Mehrbod et al., 2021.)

COVID-19 than hydroxy-chloroquine. In earlier stages, these quercetin-derived compounds were also used for the promising treatment of MERS-CoV and SARS-CoV by inhibiting 3C-like protease and restricting both coronavirus replications. Theoretically, quercetin also can bind the S protein–human ACE2 receptor interface or isolate the viral S protein at its host receptor binding region to limit SARS-CoV-2 viral replication (Derosa et al., 2021).

The clinical trial of quercetin on the treatment and prevention of COVID-19 has been already registered at ClinicalTrials.gov. This was conducted by Iran (ID code: IRCT20200419047128N2) and Turkey (ID code: NCT04377789) to investigate whether the everyday usage of two 500 mg quercetin tablets can treat or prevent the high-risk form of COVID-19 and also investigates the inflammation-associated cytokines (Xia et al., 2020). Recently, the emerging pharmacological properties of quercetin were used for COVID-19 treatment, particularly for organ damage or injury from coronavirus (Diniz et al., 2020). For example, quercetin improved renal function and acute kidney injury, which are generally caused by COVID-19 or SARS-CoV. More recently, an experimental investigation was conducted at a hospital in Wuhan, China, showing that the conventional therapies of administering quercetin to COVID-19 patients are more effective. They used TCM therapies, including herbs with a high content of quercetin, and found that quercetin was medically safe, free from particular side effects and was able to help reduce COVID-19 symptoms (Luo et al., 2020). However, the interaction of COVID-19 with quercetin is still under investigation for practical clinical trials.

20.13 PROBLEMS ASSOCIATED WITH QUERCETIN

The implementation of pure quercetin in pharmaceutical production is still limited owing to its instability in physiological media, poor water solubility, fast enzymatic degradation and metabolism, short biological half-life and subsequent poor bioavailability. Quercetin was easily soluble in dimethyl sulphoxide (150 mg mL^{-1} at 25°C) and ethanol (4.0 mg mL^{-1} at 37°C) but poorly soluble in deionized water (0.01 mg mL^{-1} at 25°C). Thus, it is difficult to use quercetin directly via water-based dispersions rather than toxic solvents. Another concern of quercetin is its reduced bioactivity index due to the higher gastric pH in the gastrointestinal tract. Moreover, the low dissolution rate of highly concentrated quercetin could have negligible absorption in the gastrointestinal tract due to its high lipophilic properties. Its oral bioavailability has been proved to be less than 2% in humans and less than 17% in animals, thereby limiting its clinical usage in predictable dosage forms. For that reason, an improved quercetin oral design is required with better bioavailability and higher efficacy. Besides, quercetin also induces *in vitro* toxic effects on normal human cell lines, thus setting restrictions to its potential clinical and *in vivo* usages. On the other hand, quercetin also has drug-like compliance based on Lipinski's rule with its molecular weight of ≤500 Da, the compound has ≤5 hydrogen bond donors, its log P ≤5 more likely presents a high bioavailability and its ≤10 hydrogen bond acceptors. Quercetin is a multipurpose compound that can activate several responses to its active interaction once inside the human system. It was also reported that the oral route of quercetin would not completely attain drug interaction with a viral S protein due to its glycosylation for biotransformation in the gastrointestinal tract. Alternatively, throat or nasal sprays holding quercetin were recommended to deliver the proper volume to the virus-active sites (Zhao et al., 2019b) and be active in clinical trials. Therefore, before beginning a full-fledged biomedical routine for commercial usage, it is required that all the pros and cons and its *in vitro*/*in vivo* interactions are fully researched.

20.14 CONCLUSION

In summary, the quercetin derived from several phytomedicinal plants, fruits and vegetables can act as a promising drug for multidisciplinary pharmacological activities including antiulcer, antibacterial, anticancer, anti-inflammatory, antiallergy, antioxidant, antidiabetic and antiviral treatments. The synergistic link between flavonoid chemistry and nanoscience improves the epidemiological properties of quercetin. Moreover, quercetin is also a promising antiviral agent, particularly for the burden of the SARS-CoV-2 and COVID-19 pandemics. Thus, the effectiveness of quercetin in human clinical applications is remarkable, provided with improved safety and overall efficacy for suppressed toxicity and related side effects. However, more investigations should be made as the water solubility of quercetin is generally poor, which affects its bioavailability.

REFERENCES

Batiha, Gaber El-Saber, Amany Magdy Beshbishy, Muhammad Ikram, Zohair S. Mulla, Mohamed E. Abd El-Hack, Ayman E. Taha, Abdelazeem M. Algammal, and Yaser Hosny Ali Elewa. "The pharmacological activity, biochemical properties, and pharmacokinetics of the major natural polyphenolic flavonoid: Quercetin." *Foods* 9, no. 3 (2020): 374.

Coelho-dos-Reis, Jordana Grazziela A., Orlando Ab Gomes, Dener E. Bortolini, Marina L. Martins, Marcia R. Almeida, Camila S. Martins, Luciana D. Carvalho et al. "Evaluation of the effects of quercetin and Kaempherol on the surface of MT-2 cells visualized by atomic force microscopy." *Journal of Virological Methods* 174, no. 1–2 (2011): 47–52.

D'Andrea, Gabriele. "Quercetin: A flavonol with multifaceted therapeutic applications?" *Fitoterapia* 106 (2015): 256–271.

de Oliveira, Marcos Roberto, Seyed Mohammad Nabavi, Nady Braidy, William N. Setzer, Touqeer Ahmed, and Seyed Fazel Nabavi. "Quercetin and the mitochondria: A mechanistic view." *Biotechnology Advances* 34, no. 5 (2016): 532–549.

Derosa, Giuseppe, Pamela Maffioli, Angela D'Angelo, and Francesco Di Pierro. "A role for quercetin in coronavirus disease 2019 (COVID-19)." *Phytotherapy Research* 35, no. 3 (2021): 1230–1236.

Diniz, Lúcio Ricardo Leite, Marilia Trindade de Santana Souza, Allana Brunna Sucupira Duarte, and Damião Pergentino de Sousa. "Mechanistic aspects and therapeutic potential of quercetin against COVID-19-associated acute kidney injury." *Molecules* 25, no. 23 (2020): 5772.

Gansukh, Enkhtaivan, Manikandan Muthu, Diby Paul, Gopal Ethiraj, Sechul Chun, and Judy Gopal. "Nature nominee quercetin's anti-influenza combat strategy—Demonstrations and remonstrations." *Reviews in Medical Virology* 27, no. 3 (2017): e1930.

Kim, Doo Hwan, Haroon Khan, Hammad Ullah, Sherif TS Hassan, Karel Šmejkal, Thomas Efferth, Mohamad Fawzi Mahomoodally et al. "MicroRNA targeting by quercetin in cancer treatment and chemoprotection." *Pharmacological Research* 147 (2019): 104346.

Kumar, S. Rajesh, S. Priyatharshni, V. N. Babu, D. Mangalaraj, C. Viswanathan, S. Kannan, and N. Ponpandian. "Quercetin conjugated superparamagnetic magnetite nanoparticles for *in-vitro* analysis of breast cancer cell lines for chemotherapy applications." *Journal of Colloid and Interface Science* 436 (2014): 234–242.

Luo, Erdan, Daiyan Zhang, Hua Luo, Bowen Liu, Keming Zhao, Yonghua Zhao, Ying Bian, and Yitao Wang. "Treatment efficacy analysis of traditional Chinese medicine for novel coronavirus pneumonia (COVID-19): An empirical study from Wuhan, Hubei Province, China." *Chinese Medicine* 15, no. 1 (2020): 1–13.

Mehrbod, Parvaneh, Dorota Hudy, Divine Shyntum, Jarosław Markowski, Marek J. Łos, and Saeid Ghavami. "Quercetin as a natural therapeutic candidate for the treatment of influenza virus." *Biomolecules* 11, no. 1 (2021): 10.

Parhi, Biswajit, Debasrita Bharatiya, and Sarat K. Swain. "Application of quercetin flavonoid based hybrid nanocomposites: A review." *Saudi Pharmaceutical Journal: SPJ* 28, no. 12 (2020): 1719.

Vafadar, Asma, Zahra Shabaninejad, Ahmad Movahedpour, Farzaneh Fallahi, Mona Taghavipour, Younes Ghasemi, Maryam Akbari et al. "Quercetin and cancer: New insights into its therapeutic effects on ovarian cancer cells." *Cell & Bioscience* 10, no. 1 (2020): 1–17.

Wu, Jianjun, Xiaoshan Xu, Yi Li, Junping Kou, Fang Huang, Baolin Liu, and Kang Liu. "Quercetin, luteolin and epigallocatechin gallate alleviate TXNIP and NLRP3-mediated inflammation and apoptosis with regulation of AMPK in endothelial cells." *European Journal of Pharmacology* 745 (2014): 59–68.

Xia, Shuai, Qiaoshuai Lan, Shan Su, Xinling Wang, Wei Xu, Zezhong Liu, Yun Zhu, Qian Wang, Lu Lu, and Shibo Jiang. "The role of furin cleavage site in SARS-CoV-2 spike protein-mediated membrane fusion in the presence or absence of trypsin." *Signal Transduction and Targeted Therapy* 5, no. 1 (2020): 1–3.

Xu, Dong, Meng-Jiao Hu, Yan-Qiu Wang, and Yuan-Lu Cui. "Antioxidant activities of quercetin and its complexes for medicinal application." *Molecules* 24, no. 6 (2019): 1123.

Zhao, Junfang, Zheng Fang, Zhian Zha, Qiang Sun, Haibin Wang, Minglei Sun, and Bin Qiao. "Quercetin inhibits cell viability, migration and invasion by regulating miR-16/HOXA10 axis in oral cancer." *European Journal of Pharmacology* 847 (2019a): 11–18.

Zhao, Juanjuan, Jun Yang, and Yan Xie. "Improvement strategies for the oral bioavailability of poorly water-soluble flavonoids: An overview." *International Journal of Pharmaceutics* 570 (2019b): 118642.

21 Nanomaterials in Cardio Protection

Bhaarathy Velmurugan

CONTENTS

21.1 INTRODUCTION

A healthy heart is imperative for a steady flow of blood to carry oxygen and essential compounds required by numerous cells, tissues and organs throughout the body. The performance of the heart can be disturbed by diet, smoking, alcohol, infection and genes. Disease conditions where blood flow is obstructed are termed as ischemia. Ischemic diseases can affect different organs and tissues in the body. Ischemic heart disease, ischemic hepatitis, ischemic stroke and renal ischemia are some examples that occur in the heart, liver, brain and kidney, respectively. Cardiac diseases could occur as a single moiety or in combination with hypertension, drug-induced cardiac complications, diabetes, doxorubicin (DOX)-induced cardiomyopathy, age and many other factors.

21.1.1 Typical Classification of Heart Diseases

1. Coronary artery disease (CAD) — Narrowing of coronary arteries by accumulation of dietary fat and other substances leading to a minimal blood supply followed by ischemic injury
2. Myocardial infarction (MI) — Rapid or sudden obstruction of blood
3. Cardiomyopathy — Abnormally enlarged, thickened, or stiffened heart muscle
4. Congenital heart defects — Abnormal heart structure at birth caused by genetics or non-inherited risk factors

DOI: 10.1201/9781003231745-21

5. Others — Heart rhythm disorders (heart beats too fast or too slow), heart valve disorders (problems with valves controlling blood flow), pericarditis (inflamed pericardium), myocarditis (inflamed myocardium)

21.1.2 Cardio Protection

The characteristic events in ischemia are typically when a blockage occurs in the blood flow followed by reperfusion, which attempts to re-establish blood flow to the affected area. Ironically, reperfusion of blood causes reperfusion-induced cell death, leading ischemia reperfusion (IR) injury, a key topic to address. Fundamental events that are addressed in cardiac research are oxidative stress, calcium overload and intracellular acidosis inducing the opening of the mitochondrial permeability transition pore (mPTP) (Hausenloy and Yellon 2013). This causes the release of cytochrome C to the cytosol and activates the caspases, causing apoptosis. Downstream inflammatory responses include recruitment of neutrophils, monocytes and macrophages. Disease conditions, specifically acute myocardial infarction (AMI), are bound to experience IR injury. Alleviating any of these conditions would then be termed cardio protection.

21.2 NANOMATERIALS IN CARDIO PROTECTION

Mounting evidence in cardio protection research has flagged multiple ways to explore the use of nanostructures as image contrast agents, drug deliverers and as therapeutics in mitigating inflammatory, apoptotic and necrotic mechanisms in cardiac diseases.

Nanomaterials, in conjunction with the aforementioned diseases, are natural or synthesized particles in an unbound state, agglomerate or aggregate, where 50% or more of the particles have an external dimension in the range of 1–100 nm. In conditions where nanomaterials are loaded with drug/antioxidant or peptide conjugates, their size could grow over 100 nm into the micrometer range. Studies have taken into account empty nanovectors with and without conjugates. As nanomaterials exhibit a high surface-to-volume ratio, their surface modification and bioavailability can be exploited for use as image contrast assisting agents and carriers of therapeutics in diagnostics. Therapeutic nanocarriers were the first liposomal- and albumin-based nanodrugs approved by the U.S. Food and Drug Administration in the forms of Doxil (Barenholz 2012) and Abraxane (Kratz 2008) in cancer-related theragnostics. Like other disease complexities, the enhanced permeability and retention (EPR) effect serves as a crucial property in which multidimensional nanomaterials cause a slow and steady accumulation on the inflamed zones of infarcted myocardium.

The use of phytochemicals of plant origin has been increasingly considered as a natural source of supplements. Curcumin and ginseng have shown potential application in reducing oxidative stress by reducing the level of inflammatory markers. Nanomaterials in combination with plant-based medicine exhibit increased solubility of phytocompounds, fostering its clear bioavailability for action. It is also worth noting that natural compounds such as berberine have been highly tested for their efficient delivery of cardiovascular treatments (Feng et al. 2019) and cancer therapeutics (Šudomová et al. 2021) using polymer-based, micelle-, dendrimer- and lipid-based nanomaterials. However, alarming evidence from zebrafish embryos (Martini et al. 2020) shows a lethal and potent mutagenic effect. Such research on the zebrafish model sheds light onto possible alterations needed to the scale of nanophytomedicine applications.

21.3 MAJOR APPLICATIONS OF NANOMATERIALS IN CARDIO PROTECTION

Cardio protection using nanomaterials can be categorized in the following groups:

1. Drug carriers – Combined with stem cells or induced pluripotent stem cell (iPSC)-derived cardiomyocytes

2. Antioxidant carriers/conjugates
3. Image contract agent carriers
4. Drugs from cell-free approaches including modRNA, growth factors and exosomes

Distinctive types of nanomaterials that have thus far contributed to diverse applications are discussed with examples in the next sections.

21.4 ORGANIC NANOMATERIALS

21.4.1 Micelles

Micelles are characterized by their core-shell and self-assembled forms and are functionalized for various operative capacities. Polymeric micelles comprise a hydrophilic outer shell and hydrophobic inner core produced by spontaneous self-assembly. Mixed micelle synthesis was attempted by Aboud et al. (2020) with valsartan-encapsulated micelles. Studies analysed cisplatin-induced cardiac toxicity and noted its cardioprotective and pharmacokinetic effect in rats with reduced expression of markers in Mhrt/Nrf2 and Trx1 pathways. Although left ventricle diastolic dysfunction was improved, efficacy of the drug improving survival rate in the long run remains a challenge. Resveratrol and quercetin have also been loaded (in combination) in pluronic F127 micelles. This exhibited high solubility and bioavailability, showing significant protection on DOX-induced cardiotoxicity.

21.4.2 Polymersomes

Functional stability in extended circulation is a notable property that aims to produce polymersomes. Polymersomes are found to be superior in functional capacity compared to liposomes. Heat shock protein 70 (HSP70) epitope was found on the surface of the exosome and its interaction with TLR- 4 has been studied. To further this, polymersomes poly(oligo[ethylene glycol] methacrylate) and poly(2-[diisopropylamino]ethyl methacrylate) blocks have been functionalized with HSP70 peptides, resulting in cardioprotective effects noted *in vitro* IR injuries. Tunable properties such as pH-sensitive (Albuquerque et al. 2021) dual-responsive smart carriers holding thermoresponsive drugs (Oroojalian et al. 2018) are some of the novel features that polymersomes could dispense. While the efficiency of membrane permeability by polymersomes is sometimes considered a major challenge, researchers have clearly strategized the method of analyses using fluorescent dye release assay, osmotic swelling and influx assay based on the concentration gradient, pulse-field gradient NMR spectroscopy and enzyme-based activity (Miller et al. 2021).

21.4.3 PCL and PLGA Conjugates

Dimension-associated technology includes electrospinning and high-pitched developments in 3D bioprinting technology can be established. Different cell types, along with controlled capacity, craft exceptional treatment methods in cardio protection. Mesenchymal stem cells (MSCs) (Jang et al. 2017) and human induced pluripotent stem cell-derived cardiomyocytes (hiPSC-CMs) (Kumar et al. 2020) are perhaps the most widely tested cell moieties. Experiments on a 3D culture system have greatly improved the maturation of hiPSC-CMs. Notable models for cardiac toxicity and drug screening studies where hiPSC-CMs cultures on coaxial poly-Ɛ-caprolactone (PCL)-gelatin 3D scaffolds have clearly shown synchronous contractions and effective responses when tested with the antiarrhythmic drug E4031.

Poly(lactic-co-glycolic acid) (PLGA), a polymer, is one of the most common aliphatic polyesters to be tested for its potential therapeutic and diagnostic applications. PLGA is highly degradable into lactate, glycolate and hydrogen ions, allowing this particular class of polymer to be highly compatible with low cytotoxicity. Ichimura et al. (2016) has described the use of the EPR effect on the accumulation of pitavastatin-PLGA in the infracted myocardium of a porcine model. Other major studies have used isoproterenol stress-induced MI, a familiar model to assess stress-induced MI, and pharmacological drugs (Nakano et al. 2016, Tokutome et al. 2019) and phytochemical antioxidants (Sun et al. 2020, Salehi et al. 2020) and have largely assessed therapeutic efficiency.

21.4.4 Liposomes

Oxidative stress, reactive oxygen species (ROS) and DOX-induced cardiomyopathy are some areas where antioxidant-loaded or drug-loaded liposome carriers are under focused investigation. Liposomes, an acknowledged carrier of drugs, have been used as a multifactorial tool in combination with optosensitive and thermosensitive properties. Of the large number of studies, Bejarano et al. (2020) developed a novel nanocarrier system using near-infrared (NIR) light-induced release of angiotens in loaded LipoTherm gold where Ang (1–9) is released upon irradiation with NIR. After drug release, the complex system can be degraded by triton-X-100 into particles less than 15 nm in size, which allows them to be excreted normally. They are further confirmed by transmission electron microscopy and particle-size distribution using the dynamic light scattering (DLS) technique. *In vitro* studies using neonatal cardiomyocytes and an *ex vivo* Langendorff model have been used to identify cardioprotective effects. Results clearly showed an improved left ventricle developed pressure (LVDP). This work from Bejarano et al., is a step forward from using highly invasive osmotic pumps as drug deliverers to the previously described minimally invasive procedure.

Piperine, a major alkaloid in black pepper, has shown great advantages as a loaded component in lipid-based nanoparticles with applications in neurodegenerative diseases and epilepsy. The role of this bioactive compound as a major oxidative stress reducer has been studied using *in vitro* spectroscopy. Such compounds are expected to be further explored in combination with therapeutic (Smilkov et al. 2019) nanoconjugates to overcome the impact of oxidative stress in cardiovascular and other diseases.

21.4.5 Carbon Nanomaterials

Carbon nanotubes hold imminent possibilities in myocardial conduction (Behabtu et al. 2013). MI causing ischemia or scarring could potentially lead to ventricular tachycardia. Currently, treatments are given through defibrillators, which offer no sufficiency in reducing myocardial conduction and, when supplemented with antiarrhythmic drugs, cardiac toxicity profiles rise. To this end, McCauley and colleagues (2019) have synthesized carbon nanotube (CNT) fibres using high-throughput wet spinning. These CNT fibres, which are innate with high conductivity, stiffness and flexibility, are used to suture across the infarct zone.

Prolonged ischemia leading to AMI potentially causes cell death and makes it challenging for the heart to regenerate cells. Several studies have shown the relevance of apoptosis in AMI and heart failure (Teringova and Tousek 2017, Mishra et al. 2019). Caspase being the prime regulator of apoptosis has always been a target in gene-silencing studies to alleviate this condition. One such prominent attempt using single-walled carbon nanotubes (SWCNT) with a 1–3 nm diameter assisted siRNA knockdown of the caspase3 gene (Li et al. 2020). Phospholipid-PEGylated SWCNT *in vivo* studies on Sprague-Dawley rats have shown a clear reduction in infarct size and improved left ventricle ejection fraction.

21.5 INORGANIC NANOMATERIALS

Metallic and magnetic nanomaterials have gained attention for efforts in cell replacement therapy. One such effort has produced cardiac patches generated by decellularized extracellular matrix (ECM) from Sprague-Dawley rats. Decellularized ECM is then dissolved in a fresh silk fibroin (SF) solution and spray-dried to synthesize a smooth ECM/SF matrix. The notable click-chemistry method to incorporate gold nanoparticles onto the ECM/SF matrix was used. MSC adhesion was then initiated and subjected to the AMI animal model (Dong et al. 2020). Reduction in infarct size from 89% in control to 69% was noted by the authors and recommended as a potential cardiac patch with high retention of cardiomyocytes and initiation of neovascularization for angiogenesis.

Another common metal that has gained significant evidence in recent years for its anti-ageing and antioxidant properties is the copper nanoparticle (CuNP). A low concentration of CuNPs administered on an IR rat model along with exercise preconditioning helped reduce the oxidative stress response and inflammatory markers in infarcted myocardium (Sharma et al. 2018).

Similarly, superparamagnetic iron oxide nanomaterials (SPIONs) have gained immense recognition in magnetic resonance imaging for different disease pathologies. A recent approach tested SPION conjugated to human amniotic membrane-derived MSCs. Under a magnetic field, these nanostructures have shown improved cardiac function and reduced fibrosis tissue damage by reducing inflammatory responses (Naseroleslami, Aboutaleb, and Mokhtari 2021). Ginseng, a known traditional Chinese medicine, has been conjugated with SPIONs and studied on its anti-inflammatory effects on macrophage cells (Singh et al. 2018). These studies could be used to include possible natural agents to accompany existing inflammation triggers initiated by nanomaterial directly or as drug deliverers in cardio protection.

The use of gold-, silver- and silica-based nanomaterials has been tested in cardiac diseases; however, their harmful effects with increased cell cytotoxicity hamper their utilization. One such study where silver sulfide (Ag_2S) quantum dots surface functionalized with the angiotensin II peptide has proven to be a prominent NIR-emitting nanoprobe with low cytotoxicity and high resolution in *in vivo* imaging (Ortgies et al. 2019). Silver nanoparticles (AgNPs) have shown accumulation in the liver causing toxicity profiles to be noted. Studies using gingko biloba (Abd El-Maksoud et al. 2019) have shown reduced hepatotoxic effects of AgNPs and reduced mitochondrial abnormalities in Wistar rats. The ability of gingko biloba to contribute to mitochondrial biogenesis levels clearly suggests its ability as a complementary medicine in cardio protection.

21.6 ORGANIC-INORGANIC COMPOSITES

Cardiac remodelling and formation of scar tissue post-MI have been detrimental, capturing much attention in cardiac-tissue engineering. Mechanistic pathways have shown an over expression of matrix metalloproteinases (MMP) post-MI, triggering inflammatory and apoptotic pathways. Apoptosis was attempted in different studies using nanomaterials to sense overexpression and release a drug or peptide to alter the canonical pathways of inflammation.

Recent interventions reveal the use of dual-function MI-responsive smart hydrogels (Fan et al. 2019) to deliver drugs based on demand and promote angiogenesis. Here, the recombinant protein glutathione S transferase (GST), MMP2/9 cleavable peptide (TIMP) and basic fibroblast growth factor (bFGF) were conjugated with glutathione-modified collagen hydrogel (GST-TIMP-bFGF/Gel-GSH hydrogels). When released in the infract zone post-MI, MMP2 and MMP9 overexpression released bFGF from its conjugates and collagen hydrogel promoted the recovery of animal models, initiating neovascularization and higher contractility of the heart. Such smart nanomaterials deliver drugs only when the trigger of MMP2/MMP9 is sensed by the nanomaterial. In many disease conditions, such overexpression of a specific peptide is used as an alarm to trigger cell-replacement activities instead of allowing the natural downstream process, resulting in cell death. These conceivable ideas will be tangible when comprehensive studies in large animal and *ex vivo* studies complemented with live imaging are performed.

21.7 NANOMATERIALS IN DIAGNOSTIC AND THERANOSTIC APPLICATIONS

The potential of nanomaterials is investigated in diagnostic imaging and theranostic applications. Early intervention in diagnosing left ventricular dysfunction, cardiomyocyte hypertrophy and interstitial fibrosis that may eventually lead to heart failure and sudden death continues to remain a challenge. Patients with cardiac hypertrophy do not show typical clinical symptoms at early stages, which makes it difficult to diagnose. Long noncoding RNA (lncRNA), Chaer (Wang et al. 2016), Chast (Viereck et al. 2016), CHRF (Wang et al. 2014) and many other lncRNAs have been debated for their plausible role as emerging circulation biomarkers for cardiac hypertrophy and ischemic conditions (Greco et al. 2016). Nanomaterials could be the option to ensure specific bioavailability of early diagnosis and treatment options.

A recent innovative approach used estrogen, a known antihypertrophic, antifibrotic, 17β-estradiol (E2), as a potential drug for cardiac hypertrophy where the ejection fraction reached up to 53% *in vivo* (Iorga et al. 2016). However, a higher dosage of estrogen poses adverse nonspecific reactions in different organs. A study performed by Liwen Liu et al. (Zhao et al. 2018) used E2 bound with primary cardiomyocyte (PCM) peptide tagged to a nanoprobe with perfluorocarbon (PFC). As these conjugates are exposed to low-intensity focused ultrasound (LIFU), PCM-E2 is released at targets with minimal systemic toxicity in the transverse aortic constriction (TAC) model of Sprague-Dawley rats.

Current novel studies have also shown that oxygen delivery is possible through magnetic oxygen loaded nanobubbles (MOLNB) synthesized by adding SPIONs on the surface of polymeric nanobubbles to maximize oxygenation (Zullino et al. 2019, Ficiarà et al. 2020) at a site-specific zone of interest. Studies furthering such permutation, perhaps, could maximize the ability to use bubble inflation techniques with phase-change PFC nanobubbles (Brambila et al. 2020) for possible theragnostic viabilities in cardio protection.

21.8 CONCLUSION

Understanding the intricacies of cardiovascular disease and utilizing technical advancements to explore inventive formulations of nanomaterials is important to contribute to the research arena. Nanomaterials are highly appreciable for their dynamic nature, challenges exist as nanomaterial fabrication requires extreme precision to ensure highly reproducible results in both *in vitro* and *in vivo* studies. The feasibility in using a set of rules such as the Minimal Information Reporting in Bio-Nano Experimental Literature (MIRIBEL) (Faria et al. 2018) may ensure sustainability in results. This could eventually fill in the gaps in nanobioengineering research, recording plausible ideas with experimental and clinical value toward cardio protection.

REFERENCES

Abd El-Maksoud, Eman M., Mohamed A. Lebda, Aml E. Hashem, Nabil M. Taha, and Maher A. Kamel. "Correction to: Ginkgo biloba mitigates silver nanoparticles-induced hepatotoxicity in Wistar rats via improvement of mitochondrial biogenesis and antioxidant status." *Environmental Science and Pollution Research* 26, no. 30 (2019): 31552–31553.

Aboud, Heba M., Mohamed O. Mahmoud, Marwa Abdeltawab Mohammed, Mohammad Shafiq Awad, and Dina Sabry. "Preparation and appraisal of self-assembled valsartan-loaded amalgamated Pluronic F127/Tween 80 polymeric micelles: Boosted cardio protection via regulation of Mhrt/Nrf2 and Trx1 pathways in cisplatin-induced cardiotoxicity." *Journal of Drug Targeting* 28, no. 3 (2020): 282–299.

Albuquerque, Lindomar JC, Vladimir Sincari, Alessandro Jäger, Jan Kucka, Jana Humajova, Jan Pankrac, Petr Paral et al. "pH-responsive polymersome-mediated delivery of doxorubicin into tumor sites enhances the therapeutic efficacy and reduces cardiotoxic effects." *Journal of Controlled Release* 332 (2021): 529–538.

Barenholz, Yechezkel Chezy. "Doxil—the first FDA-approved nano-drug: Lessons learned." *Journal of Controlled Release* 160, no. 2 (2012): 117–134.

Behabtu, Natnael, Colin C. Young, Dmitri E. Tsentalovich, Olga Kleinerman, Xuan Wang, Anson WK Ma, E. Amram Bengio et al. "Strong, light, multifunctional fibers of carbon nanotubes with ultrahigh conductivity." *Science* 339, no. 6116 (2013): 182–186.

Bejarano, Julian, Aldo Rojas, Andrea Ramírez-Sagredo, Ana L. Riveros, Francisco Morales-Zavala, Yvo Flores, Jaime A. Riquelme et al. "Light-induced release of the cardioprotective peptide angiotensin-(1–9) from thermosensitive liposomes with gold nanoclusters." *Journal of Controlled Release* 328 (2020): 859–872.

Brambila, Carlos J., Jacques Lux, Robert F. Mattrey, Dustin Boyd, Mark A. Borden, and Caroline de Gracia Lux. "Bubble inflation using phase-change perfluorocarbon nanodroplets as a strategy for enhanced ultrasound imaging and therapy." *Langmuir* 36, no. 11 (2020): 2954–2965.

Dong, Yongda, Meiman Hong, Ruozhu Dai, Haiyun Wu, and Ping Zhu. "Engineered bioactive nanoparticles incorporated biofunctionalized ECM/silk proteins based cardiac patches combined with MSCs for the repair of myocardial infarction: *In vitro* and *in vivo* evaluations." *Science of the Total Environment* 707 (2020): 135976.

Fan, Caixia, Jiajia Shi, Yan Zhuang, Lulu Zhang, Lei Huang, Wen Yang, Bing Chen et al. "Myocardial-infarction-responsive smart hydrogels targeting matrix metalloproteinase for on-demand growth factor delivery." *Advanced Materials* 31, no. 40 (2019): 1902900.

Faria, Matthew, Mattias Björnmalm, Kristofer J. Thurecht, Stephen J. Kent, Robert G. Parton, Maria Kavallaris, Angus PR Johnston et al. "Minimum information reporting in bio–nano experimental literature." *Nature Nanotechnology* 13, no. 9 (2018): 777–785.

Feng, Xiaojun, Antoni Sureda, Samineh Jafari, Zahra Memariani, Devesh Tewari, Giuseppe Annunziata, Luigi Barrea et al. "Berberine in cardiovascular and metabolic diseases: From mechanisms to therapeutics." *Theranostics* 9, no. 7 (2019): 1923.

Ficiarà, Eleonora, Shoeb Anwar Ansari, Monica Argenziano, Luigi Cangemi, Chiara Monge, Roberta Cavalli, and Federico D'Agata. "Beyond oncological hyperthermia: Physically drivable magnetic nanobubbles as novel multipurpose theranostic carriers in the central nervous system." *Molecules* 25, no. 9 (2020): 2104.

Greco, Simona, Germana Zaccagnini, Alessandra Perfetti, Paola Fuschi, Rea Valaperta, Christine Voellenkle, Serenella Castelvecchio et al. "Long noncoding RNA dysregulation in ischemic heart failure." *Journal of Translational Medicine* 14, no. 1 (2016): 1–14.

Hausenloy, Derek J., and Derek M. Yellon. "Myocardial ischemia-reperfusion injury: A neglected therapeutic target." *Journal of Clinical Investigation* 123, no. 1 (2013): 92–100.

Ichimura, Kenzo, Tetsuya Matoba, Kaku Nakano, Masaki Tokutome, Katsuya Honda, Jun-ichiro Koga, and Kensuke Egashira. "A translational study of a new therapeutic approach for acute myocardial infarction: Nanoparticle-mediated delivery of pitavastatin into reperfused myocardium reduces ischemia-reperfusion injury in a preclinical porcine model." *PLoS One* 11, no. 9 (2016): e0162425.

Iorga, Andrea, Jingyuan Li, Salil Sharma, Soban Umar, Jean C. Bopassa, Rangarajan D. Nadadur, Alexander Centala et al. "Rescue of pressure overload-induced heart failure by estrogen therapy." *Journal of the American Heart Association* 5, no. 1 (2016): e002482.

Jang, Jinah, Hun-Jun Park, Seok-Won Kim, Heejin Kim, Ju Young Park, Soo Jin Na, Hyeon Ji Kim et al. "3D printed complex tissue construct using stem cell-laden decellularized extracellular matrix bioinks for cardiac repair." *Biomaterials* 112 (2017): 264–274.

Kratz, Felix. "Albumin as a drug carrier: Design of prodrugs, drug conjugates and nanoparticles." *Journal of Controlled Release* 132, no. 3 (2008): 171–183.

Kumar, Naresh, Divya Sridharan, Arunkumar Palaniappan, Julie A. Dougherty, Andras Czirok, Dona Greta Isai, Muhamad Mergaye, Mark G. Angelos, Heather M. Powell, and Mahmood Khan. "Scalable biomimetic coaxial aligned nanofiber cardiac patch: A potential model for 'clinical trials in a dish.'" *Frontiers in Bioengineering and Biotechnology* 8 (2020): 1045.

Li, Yi, Hong Yu, Liang Zhao, Yuting Zhu, Rui Bai, Zhigeng Jin, Zhenhong Fu et al. "Effects of carbon nanotube-mediated *Caspase3* gene silencing on cardiomyocyte apoptosis and cardiac function during early acute myocardial infarction." *Nanoscale* 12, no. 42 (2020): 21599–21604.

Martini, Davide, Cecilia Pucci, Chiara Gabellini, Mario Pellegrino, and Massimiliano Andreazzoli. "Exposure to the natural alkaloid Berberine affects cardiovascular system morphogenesis and functionality during zebrafish development." *Scientific Reports* 10, no. 1 (2020): 1–16.

McCauley, Mark D., Flavia Vitale, J. Stephen Yan, Colin C. Young, Brian Greet, Marco Orecchioni, Srikanth Perike et al. "*In vivo* restoration of myocardial conduction with carbon nanotube fibers." *Circulation: Arrhythmia and Electrophysiology* 12, no. 8 (2019): e007256.

Miller, Alisha J., Amanda K. Pearce, Jeffrey C. Foster, and Rachel K. O'Reilly. "Probing and tuning the permeability of polymersomes." *ACS Central Science* 7, no. 1 (2021): 30–38.
Mishra, P. K., A. Adameova, J. A. Hill, C. P. Baines, P. M. Kang, J. M. Downey, J. Narula et al. "Erratum: Guidelines for evaluating myocardial cell death (American journal of physiology-heart and circulatory physiology (2019) 317 (H891-H922." *American Journal of Physiology-Heart and Circulatory Physiology* 317, no. 6 (2019): H1390.
Nakano, Yasuhiro, Tetsuya Matoba, Masaki Tokutome, Daiki Funamoto, Shunsuke Katsuki, Gentaro Ikeda, Kazuhiro Nagaoka et al. "Nanoparticle-mediated delivery of irbesartan induces cardio protection from myocardial ischemia-reperfusion injury by antagonizing monocyte-mediated inflammation." *Scientific Reports* 6, no. 1 (2016): 1–14.
Naseroleslami, Maryam, Nahid Aboutaleb, and Behnaz Mokhtari. "Amniotic membrane mesenchymal stem cells labeled by iron oxide nanoparticles exert cardioprotective effects against isoproterenol (ISO)-induced myocardial damage by targeting inflammatory MAPK/NF-κB pathway." *Drug Delivery and Translational Research* 11 (2021): 242–254.
Online document: "<Official Journal of the European Union. Legislation. (English edition, L275) 2011 Oct 20;54 ISSN 1977-0677.pdf>."
Oroojalian, Fatemeh, Maryam Babaei, Seyed Mohammad Taghdisi, Khalil Abnous, Mohammad Ramezani, and Mona Alibolandi. "Encapsulation of thermo-responsive gel in pH-sensitive polymersomes as dual-responsive smart carriers for controlled release of doxorubicin." *Journal of Controlled Release* 288 (2018): 45–61.
Ortgies, Dirk H., Ángel Luis García-Villalón, Miriam Granado, Sara Amor, Emma Martín Rodríguez, Harrisson DA Santos, Jingke Yao, Jorge Rubio-Retama, and Daniel Jaque. "Infrared fluorescence imaging of infarcted hearts with Ag 2 S nanodots." *Nano Research* 12, no. 4 (2019): 749–757.
Salehi, Bahare, Del Prado-Audelo, L. María, Hernán Cortés, Gerardo Leyva-Gómez, Zorica Stojanović-Radić, Yengkhom Disco Singh et al. "Therapeutic applications of curcumin nanomedicine formulations in cardiovascular diseases." *Journal of Clinical Medicine* 9, no. 3 (2020): 746.
Sharma, Arun K., Ashish Kumar, Megha Sahu, Gunjan Sharma, Ashok K. Datusalia, and Satyendra K. Rajput. "Exercise preconditioning and low dose copper nanoparticles exhibits cardio protection through targeting GSK-3β phosphorylation in ischemia/reperfusion induced myocardial infarction." *Microvascular Research* 120 (2018): 59–66.
Singh, Hina, Juan Du, Priyanka Singh, Gafurjon Tom Mavlonov, and Tae Hoo Yi. "Development of superparamagnetic iron oxide nanoparticles via direct conjugation with ginsenosides and its *in-vitro* study." *Journal of Photochemistry and Photobiology B: Biology* 185 (2018): 100–110.
Smilkov, Katarina, Darinka G. Ackova, Aleksandar Cvetkovski, Tatjana Ruskovska, Bojana Vidovic, and Mustafa Atalay. "Piperine: Old spice and new nutraceutical?" *Current Pharmaceutical Design* 25, no. 15 (2019): 1729–1739.
Šudomová, Miroslava, Kateřina Berchová-Bímová, Stefania Marzocco, Alena Liskova, Peter Kubatka, and Sherif TS Hassan. "Berberine in human oncogenic herpesvirus infections and their linked cancers." *Viruses* 13, no. 6 (2021): 1014.
Sun, Liqiang, Yucai Hu, Anurag Mishra, Nagaraja Sreeharsha, Jeet B. Moktan, Piyush Kumar, and Lei Wang. "Protective role of poly (lactic-co-glycolic) acid nanoparticle loaded with resveratrol against isoproterenol-induced myocardial infarction." *Biofactors* 46, no. 3 (2020): 421–431.
Teringova, Elena, and Petr Tousek. "Apoptosis in ischemic heart disease." *Journal of Translational Medicine* 15, no. 1 (2017): 1–7.
Tokutome, Masaki, Tetsuya Matoba, Yasuhiro Nakano, Arihide Okahara, Masaki Fujiwara, Jun-Ichiro Koga, Kaku Nakano, Hiroyuki Tsutsui, and Kensuke Egashira. "Peroxisome proliferator-activated receptor-gamma targeting nanomedicine promotes cardiac healing after acute myocardial infarction by skewing monocyte/macrophage polarization in preclinical animal models." *Cardiovascular Research* 115, no. 2 (2019): 419–431.
Viereck, Janika, Regalla Kumarswamy, Ariana Foinquinos, Ke Xiao, Petros Avramopoulos, Meik Kunz, Marcus Dittrich et al. "Long noncoding RNA chast promotes cardiac remodeling." *Science Translational Medicine* 8, no. 326 (2016): 326ra22–326ra22.
Wang, Kun, Fang Liu, Lu-Yu Zhou, Bo Long, Shu-Min Yuan, Yin Wang, Cui-Yun Liu, Teng Sun, Xiao-Jie Zhang, and Pei-Feng Li. "The long noncoding RNA CHRF regulates cardiac hypertrophy by targeting miR-489." *Circulation Research* 114, no. 9 (2014): 1377–1388.
Wang, Zhihua, Xiao-Jing Zhang, Yan-Xiao Ji, Peng Zhang, Ke-Qiong Deng, Jun Gong, Shuxun Ren et al. "The long noncoding RNA Chaer defines an epigenetic checkpoint in cardiac hypertrophy." *Nature Medicine* 22, no. 10 (2016): 1131–1139.

Zhao, Xueli, Wen Luo, Jing Hu, Lei Zuo, Jing Wang, Rui Hu, Bo Wang et al. "Cardiomyocyte-targeted and 17β-estradiol-loaded acoustic nanoprobes as a theranostic platform for cardiac hypertrophy." *Journal of Nanobiotechnology* 16, no. 1 (2018): 1–14.

Zullino, Sara, Monica Argenziano, Shoeb Ansari, Roberta Ciprian, Lucia Nasi, Franca Albertini, Roberta Cavalli, and Caterina Guiot. "Superparamagnetic oxygen-loaded nanobubbles to enhance tumor oxygenation during hyperthermia." *Frontiers in Pharmacology* 10 (2019): 1001.

22 Thymoquinone

A Multifunctional Phytomedicine for Nanoencapsulated Drug Delivery for Cancer and Brain Disorder Treatments

D. Navadeepthy and N. Ponpandian

CONTENTS

22.1 AN INTRODUCTION TO PHYTOCHEMICALS AND THYMOQUINONE

Humans have always relied on plants and plant-based foods such as vegetables, fruits, herbs and spices as major sources of health and nutrition. Phytochemicals are bioactive constituents naturally present in food items that contain both medicinal values and therapeutic properties. These naturally occurring phytomedicines or plant medicines have been used for the treatment of various ailments for centuries. In Asian countries, medicinal plants are the most prominent source of healthcare. India is a rich source of plant biodiversity and medicinal plants have been used for therapeutic treatments since ancient times. Ayurveda is an ancient form of healthcare dating back approximately more than 3000 years. It mainly uses phytochemicals in its preparations and formulations. Very recently, these phytochemicals have been extracted and incorporated in contemporary synthetic drugs for the practice of medicine. The major phytochemicals been extracted for drug design are flavonoids, glycosides, tannins, phytophenols, resins, lectins, etoposides,

DOI: 10.1201/9781003231745-22

tanniposides, fatty acids, waxes and alkaloids (Gali-Muhtasib, Roessner and Schneider-Stock, 2006). These phytochemicals are known to possess antibacterial, antifungal, anticancerous, antidiabetic, antioxidant, anti-inflammatory, antimicrobial, immunodegenerative and neuroprotective properties in nature. Thus, they are being investigated to cure various chronic and degenerative diseases. Among the various plants and plant extracted phytochemicals, thymoquinone (TQ) is a promising phytochemical extracted from the volatile essential oil of black cumin seeds from *Nigella sativa* L. It is also called black caraway, nutmeg flower, kalonji, fennel flower and Roman coriander. The other important pharmacological bioactive constituents of *Nigella sativa* are dithymoquinone (DTQ), thymohydroquinone (THQ) and thymol (THY). Among the available phytochemicals of *Nigella sativa,* TQ holds a vast space in the treatment of chronic ailments because of the pharmacological properties such as anticancer, anti-inflammatory, antioxidant and promising biological activities both *in vivo* and *in vitro* experiments in various human health issues (Khader, Bresgen and Eckl, 2009). *Nigella sativa* is used to cure various ailments in the human body and a description on the effects of the seed is shown in Figure 22.1 (Tavakkoli *et al.*, 2017). Also, TQ is known to attenuate the side effects of diseases. The chemical formula

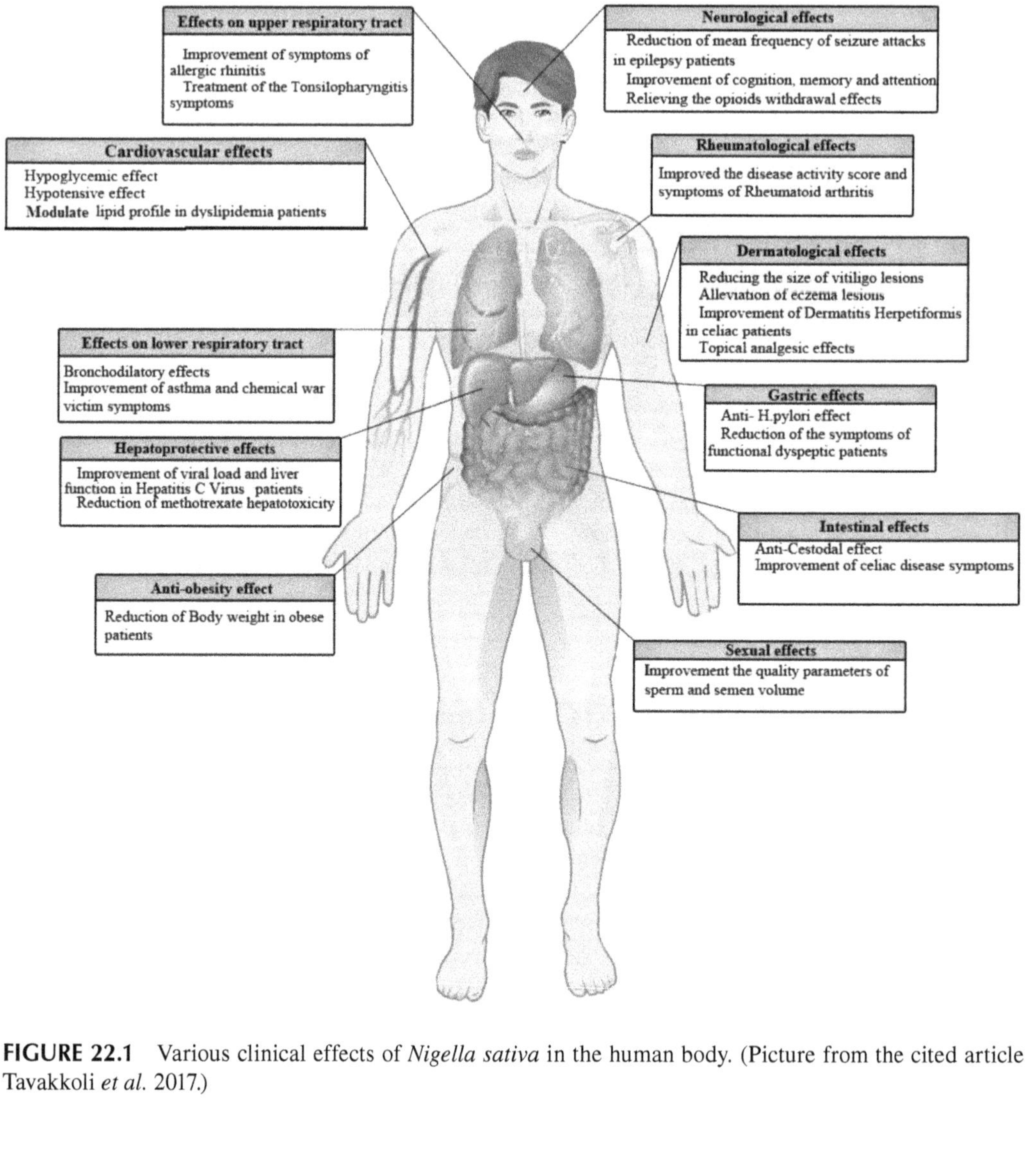

FIGURE 22.1 Various clinical effects of *Nigella sativa* in the human body. (Picture from the cited article Tavakkoli *et al.* 2017.)

of TQ(2-methyl-5-isopropyl-1,4-benzoquinone) is $C_{10}H_{12}O_2$ with a molecular weight of 164.2 g/mol. In this chapter, the structure and properties of TQ and their activity toward the treatment of various diseases are discussed in depth.

22.2 ORIGIN, STRUCTURE AND PROPERTIES OF THYMOQUINONE

TQ was first extracted and tested in 1963 by El-Dakhakhany and proven to uphold anti-inflammatory, antineoplastic and antioxidant effects in both *in vitro* and *in vivo* studies (El-Dakhakhany, 1963). TQ is extracted from *Nigella sativa* L., which comes under the category of flowering plants that are predominantly found in countries bordering the Mediterranean Sea, India, Pakistan and Iran. The seed is composed of other varying constituents like fixed oils, volatile oils, coumarins, minerals, fibres, alkaloids, saponins, etc. TQ is obtained by the oxidation of THY, a natural compound. The reaction takes place in the presence of manganese oxide in an acidic medium. The seed oil is made up on one-third TQ, which is why the oil is bright yellow in colour. Also, TQ can be extracted from other plants such as *Nepeta leucophylla*, *Tetraclinis articulate*, *Callitris quadrivalvis*, *Juniperus cedrus*, *Monarda fistulosa*, etc. (Ahmad, 2017). The concentration of TQ in the seed oil is found to be 18–25 μg/ml (Raschi *et al.*, 2010) and belongs to the monoterpenes class of natural compounds with keto-enol tautomerism (Samarghandian, Farkhondeh and Samini, 2018). The basic structure of TQ is $C_{10}H_{12}O_2$, where in the ten carbon structure, six carbons are involved in forming quinine ring moiety, and the C2 position holds the seventh methyl group carbon. The eighth, ninth and tenth propyl group carbons position at C5. The most common derivatives of TQ are DTQ and THQ. The structure of TQ is given in Figure 22.2. The powder X-ray diffraction (XRD) analysis of TQ revealed the triclinic structure. TQ as a major pharmacological herbal element has been known and explored for its attributed antioxidant, anti-inflammatory, antihistamic, antitumor, immunomodulatory and antihyperglycemic properties. Also, TQ exhibits neuroprotective, nephroprotective, gastroprotective and hepatoprotective activities.

22.3 DOSAGE AND TOXICITY

Dosage level and toxicity are important safety assessment procedures to validate a new drug before testing in humans. TQ is also tested by various *in vitro* and *in vivo* studies to study its toxicological properties. In this regard, the toxicity of TQ is tested on model animals and also in humans to check its effects and dosage. Since TQ is insoluble in water, the lethal dose (LD_{50}) of TQ depends on the route of administration and the carrier being used. On oral administration of TQ on mice, the LD_{50} was 870.9 mg/kg and in i.p. (intraperitoneal) injection, it became 104.7 mg/kg. In a similar manner in rats it was found to be 794.3 mg/kg on oral administration and 57.5 mg/kg in i.p. (Tubesha *et al.*, 2013). After oral administration, the side effects observed were difficulty in respiration and peritonitis in rats and mice. But it was found that long-term

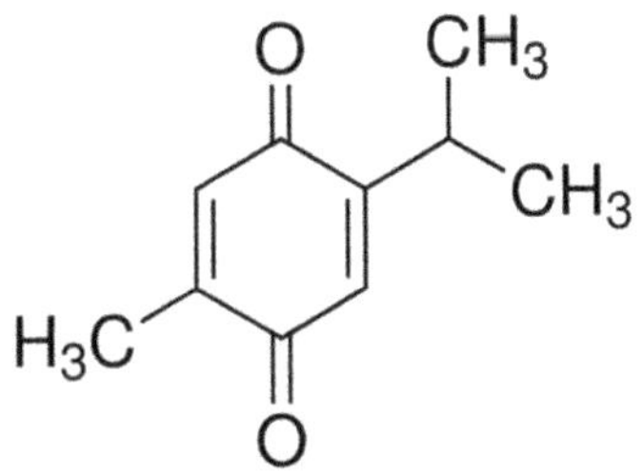

Thymoquinone
(2-Isopropyl-5-methylbenzo-1,4-quinone)

FIGURE 22.2 Structure of thymoquinone.

administration of TQ causes liver toxicity, but at tolerable dosage it does not affect any organs in humans (AbuKhader, 2013).

Bioavailability is also an important parameter in drug design and incorporation. From the site of administration, the rate and amount of a drug that reaches circulation is known as the bioavailability of that particular drug. Various parameters affect the bioavailability of a drug, such as dosage, solubility and formulation. Thus, while considering a drug for incorporation, bioavailability should be enhanced by modifying the parameters. Recently, nanoformulations were used to enhance the efficacy and bioavailability of the drug (Ahmad, 2017). Humans and animals are exposed to many environmental and biological hazards. Oxidant species and free radicals could become highly toxic when involved in organ and cellular dysfunction. DNA, lipid and protein damage can result from overproduction of these species. However, moderate or low reactive oxygen and nitrogen species can also engage in the alteration of normal physiological responses. The cell has many ways, including antioxidants and antioxidant enzymes, to deal with free radical generation. TQ is a persuasive antioxidant extracted from the seeds of *Nigella sativa*. It damages free radicals and prevents oxidants from damaging the cells.

22.4 IMPORTANT PHARMACOLOGICAL EFFECTS OF TQ

22.4.1 Antioxidative and Anti-Inflammatory Effects

The human body is exposed to many serious hazards both environmentally and biologically. On aerobic metabolism due to exogenous sources, reactive oxygen species (ROS) such as hydroxyl radical (HO•), superoxide anion radical (O2−•) and per-oxyl radical (ROO•) are continually generated in cells. Free radical formation and ROS production are deadly and cause organ or cell damage (Khalil *et al.*, 2020). Also DNA, lipid and protein damage can occur due to ROS excess production. Low or moderate ROS and nitrogen species can lead to not only excess production, but alteration in the normal physiological process. In human physiology, oxidative stress plays a crucial role in affecting the cell system. Various energy-dependent systems are there to prevent oxidative stress, where antioxidants and enzymes are involved in free radical regeneration. Inflammation is a biological response of human body against complex biological reactions. The stimuli of inflammation are the enzymes cyclooxygenase and lipoxygenase. They produce prostaglandins and leukotrienes, respectively. TQ exhibited anti-inflammatory effects on various experimental models as well as by different mechanisms in each study that will be discussed in upcoming sections (Khader and Eckl, 2014). Some of the major activities of TQ are shown as descriptive chart in Figure 22.3.

22.4.2 Anticancer Properties

Cancer was once considered incurable, but the invention of new medicines and drug delivery through targeting cancer cells has brought more insight to cancer treatment. Carcinogens such as chemicals, radiation or viruses, natural or manmade, are the main reason for cancer in humans. Natural phytochemicals are exceptional sources with anticancer properties. Compared to the synthetic drugs, natural phytochemicals are effective and are not highly toxic to healthy tissue. TQ is one of the anticancer agents widely used for treating various types of cancer (Chiuh Woo *et al.*, 2011). TQ exerts two major roles either as an antioxidant or a pro-oxidant. It begins its function generally by ROS generation in the tumour cells (pro-oxidant behaviour). It interacts with the thiol or amino groups and undergoes a series of oxidoreduction reactions and is then metabolized to either semiquinone or THQ, leading to the production of ROS (Ballout and Gali-Muhtasib, 2021). TQ has been employed to cure various types of human cancer cells such as breast cancer cells, human cervical squamous, human ovarian adenocarcinoma, osteosarcoma, glioblasta cells, colorectal carcinoma, pancreatic cancer cells, lung cancer cells and skin tumour genesis. Both *in vivo* and *in vitro* studies were carried out (Darakhshan *et al.*, 2015).

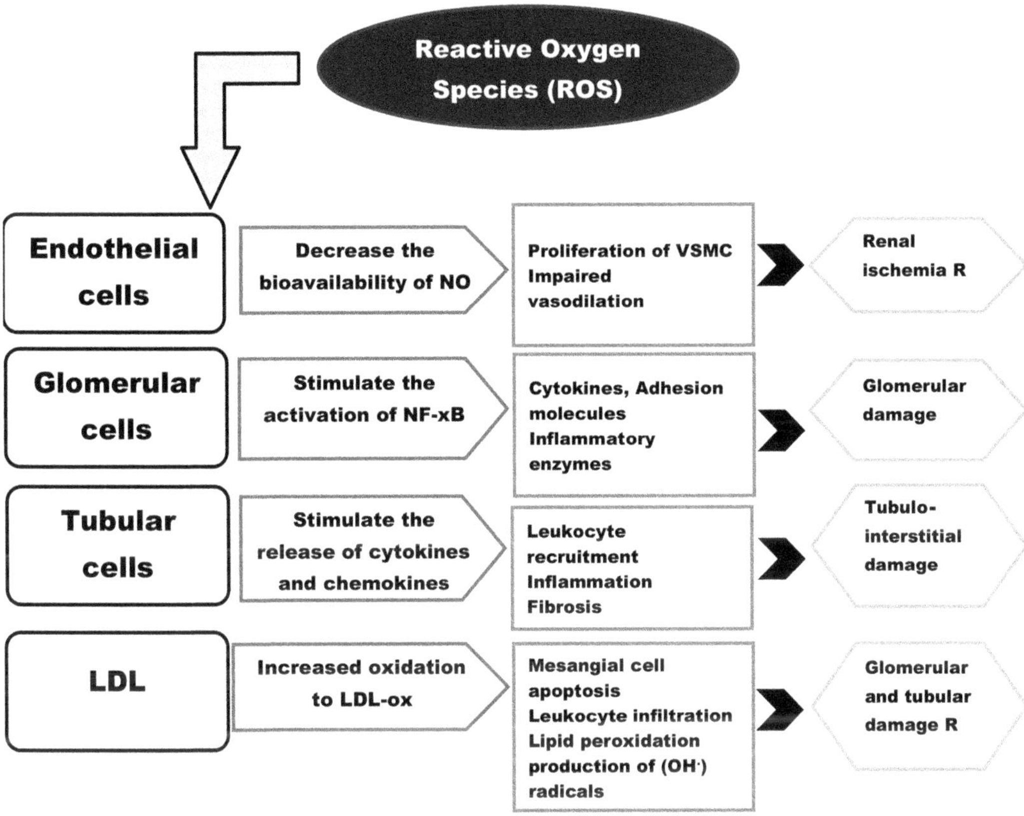

FIGURE 22.3 The increase in ROS may have serious effects. This figure shows the impact on the cells, the effects and the damage it causes. (Listed from Khalil and Collins, 2010.)

22.4.3 Antimicrobial Effects

Over the past few decades, plants and plant-based antimicrobial drugs have been used as natural forms of treatment. They are prominently used more than the commercial antibiotics and antibacterial chemotherapeutic drugs since they lag behind in resistance and the side effects they cause. TQ is also an important compound with antibacterial properties against few bacterial strains such as *Escherichia coli*, *Bacillus subtilis*, *Streptococcus faecalis*, *Pseudomonas aeruginosa* and *Staphylococcus aureus* (Novy *et al.*, 2014). *S. aureus* is one of the most serious causes of food poisoning and severe infection — and is sensitive to TQ. TQ also shows good bactericidal activity toward Gram-positive cocci and avoided biofilm formation in human pathogens. It prevented cell adhesion in glass slides and was found to be a potential growth inhibitor even at low concentrations of Gram-positive bacteria. TQ has an adverse effect on cell metabolism in biofilms. TQ antibacterial activity was also demonstrated for other bacteria such as *S. aureus*, *S. salivarius* and *S. oralis*. In addition, TQ shows its bactericidal activity toward oral bacterial strains. It has been reported that TQ had synergistic effects in combination with antibacterial agents (Kouidhi *et al.*, 2011).

22.4.4 Effects on Metabolic Disorders

TQ is effective in antihyperlipidaemic and antidiabetic activities and is involved in treating other metabolic disorders. According to study results, the possible mechanism observed is that TQ is responsible for the decrease in the expression of gluconeogenic enzymes viz. fructose-1,6-biphosphatase, glucose 6-phosphatase, hepatic glucose and production. Additionally it enhances the uptake of low-density lipoprotein cholesterol (LDL-C) by up-regulation of LDL-C hepatic receptors (Al Wafai, 2013; Al-Rasheed *et al.*, 2014).

22.4.5 Antidiabetic Effects

Diabetes is one of the most prevalent diseases around the world. It is a metabolic disorder caused by an elevation in blood glucose level. The variation in the glucose level develops either from a deficiency of insulin or an over-secretion of insulin, which is due to the destruction or inactivity of β-cells. Diabetes may further lead to various complications in the cardiovascular system, neuropathy and nephropathy (Al-Rasheed *et al.*, 2014; Farkhondeh, Samarghandian and Borji, 2017). Many medicinal plants are being used as drugs for diabetes. Among them, TQ has also been tested to treat diabetes in rats and mice. One study observed that treatment of diabetic mice during pregnancy and lactation periods with a dosage of 20 mg/kg body weight of TQ protected the offspring from diabetes and other related complications. The incorporation of TQ has decreased blood glucose levels and lipids and restored the number of circulating lymphocytes, aberrant AKT phosphorylation plasma pro-inflammatory cytokines (IL-1β, IL-6 and TNF-a), T- cell proliferation and free radical production (Al Wafai, 2013; Umeno *et al.*, 2016; Kumar, 2019).

22.4.6 Neuroprotective Agent

Oxidative stress initiates various problems in humans. Mainly, it can lead to sensory or motor function disorder or autonomic nervous system dysfunction, including arrhythmias, gastroparesis, incontinence and sexual dysfunction. There are reports that demonstrated several neuropharmacological properties of TQ (Ullah *et al.*, 2012). TQ is used to prevent cell death induced by MPP+ and rotenone in dopamineic neurons in the cell cultures of patient with Parkinson's disease. It protected the immunoreactive cells from the damage caused by MPP+ and rotenone toxicities. The antioxidant properties of TQ played a vital role in protecting the cells from further damage (Khaled *et al.*, 2009). In streptozotocin-induced diabetic rats, TQ inhibited oxidative stress and neuropathy. TQ reduced norepinephrine and dopamine levels and increased serotonin levels (Hamdy and Taha, 2009). TQ caused the development of morphine tolerance and dependence in mice and also suppressed morphine-induced brain oxidative stress.

22.5 TQ IN DRUG DELIVERY

As discussed in the previous sections, TQ has been used extensively in therapeutic processes as antioxidant, anti-inflammatory and anticancer agents and for the treatment of various diseases. But to improve the targeting capacity and bioavailability and to avoid unspecific binding, different types of nanoformulations are introduced. Because of this, targeted drug delivery with high efficiency has been achieved. The nanotechnological strategy for targeted drug delivery is one of the growing techniques for effective therapeutic treatments (El-Far *et al.*, 2018). To overcome the limitations of TQ, encapsulation or binding with nanocarriers could improve the efficiency of TQ delivery. Encapsulation of TQ with various nanocarriers enhances the *in vivo* solubility, stability, bioavailability and also protects it from binding with other cells. Due to its superior properties encapsulated TQ gains more attention over pristine TQ in the field of drug designing.

The hydrophobic nature of TQ and its poor solubility in low pH are the major disadvantages of pure TQ because these, in turn, reduce its bioavailability, possibly requiring higher dosage levels. TQ also is reported to be less stable in chemical and physiological environments and is highly unstable when exposed to light and temperature. Thus, effort has been made to improve its bioavailability by nanocarrier formulations. Nanocarriers are chosen for their nonbiodegradable and highly biocompatible properties. Polymers and lipids are typically employed as nanocarriers (Odeh *et al.*, 2012; Nallamuthu, Parthasarathi and Khanum, 2013). The TQ lipid nanocarriers are developed using liposomes, solid lipid nanoparticles (SLNs), lipospheres, nanoemulsions, niosomes and proniosomes. Similarly, polymeric carriers are formed using poly lactic-co-glycolic acid (PLGA), polyethylene glycol (PEG), cyclodextrin and the natural polymer chitosan. Each nanocarrier substance and its combination with TQ will be clearly discussed in upcoming sections.

22.6 DIFFERENT TYPES OF NANOCARRIERS FOR TQ LOADING AND DRUG DELIVERY

22.6.1 Lipid-Based TQ Nanoparticles

As discussed, the major challenges faced by TQ are its solubility and bioavailability. Lipids are one of the biocompatible nanostructures used as nanocarriers. Lipid-based nanoformulations provide a novel approach to improve the efficiency of TQ, enhance its stability and further more advantages. Also, the lipid matrix system moderates drug release and the dosage level with systematic circulation and reduces the side effects caused by oral delivery. Targeted drug delivery by lipid nanoparticles is a successful way to deliver drugs specifically to each organ based on the particular disease. Lipid nanocarriers are classified into two major categories: particulate and nonparticulate (Almajali *et al.*, 2021). The SLNs, nanostructured lipid carriers (NLCs), niosomes, liposomes, dendrimers, nanogels, microparticles, microspheres and proniosomes are the particulate type of nanocarriers. Self-nanoemulsifying drug delivery systems (SNEDDS), nanogels and nanoemulsions fall under the nonparticulate type (Rathore *et al.*, 2020). Lipid-based nanocarriers are composed of phospholipid bilayer vesicles such as phosphatidylcholine and phosphatidylethanolamine, the most popular phospholipids found in nature, with other membrane bilayer components, such as cholesterol and hydrophilic polymers around each liposomal vesicle (Rathore *et al.*, 2020; Ballout *et al.*, 2018).

22.6.2 Polymer TQ Nanocarriers

Another important nanoformulation of TQ is the use of polymeric nanostructures. Either synthetic polymers like polyvinylchloride, PLGA, PEG, polyhydroxyalkanoate and cyclodextrin are used or natural polymers like chitosan. Like lipids, polymeric nanoparticles are also favourable for drug delivery due to its tunable particle size and shape, biodegradability, biocompatibility with tissues and cells and controlled- or sustained-release profile. Polymer nanocarriers are generally nontoxic and stable in the bloodstream. They do not possess immunogenicity and thrombogenicity. These polymer nanocarriers are mainly used for targeted drug delivery to the brain and central nervous system (CNS) since they can easily cross the blood-brain barrier. The polymeric encapsulation of TQ can protect it from harsh environments and surface modifications. The polymeric nanocarriers could effectively modulate the pharmacokinetics and diffusion kinetics of TQ into various organs by crossing the barrier. The encapsulation of TQ by polymeric substances can be done by various approaches (Rathore *et al.*, 2020). Based on the drugs, encapsulation with polymeric nanoparticles can be done by various formulations. There are three main preparation methods on preparing TQ-loaded polymeric nanocarriers: nanoprecipitation, solvent-evaporation and ionic-gelation.

22.7 NANOFORMULATED TQ FOR THE TREATMENT OF VARIOUS CANCER TYPES IN HUMANS

Cancer is one among the deadly diseases once thought to be incurable. Along with the development of phytochemicals and synthetic drugs derived from natural plant extracts, cancer has become a treatable disease if properly diagnosed. Thus, phytochemicals play a vital role in cancer treatment. TQ can help in treating various types of human cancer cell lines. Due to the problems associated with bare TQ, nanoformulations are being developed for targeted delivery of TQ for cancer cells (Almajali *et al.*, 2021). Various studies were reported on the successful delivery of TQ to the cancer cells without damaging the noncancerous cells. Based on the reports, the various nanocarriers used and their activity toward proliferation of cancer cells are discussed in detail below.

22.7.1 Anticancer Activity of Various TQ-Loaded Nanoformulations

Docetaxel and TQ were co-encapsulated within mPEG-DSPE-vitamin E TPGS-lipid nanocapsules for human breast cancer treatment by Zafar et al. The group achieved a desirable amount of drug release tested in normal and triple-negative breast cancer and achieved superior anticancer properties in dual drug-loaded nanocapsules. In addition, the dual drug-loaded nanocapsules reduced the toxicities associated with liver, kidney and oxidative stress. A haemolysis assay revealed the biocompatibility and stability in pH with protein resistance. As a whole, the work states the effective chemotherapeutic behaviour of the nanocapsules toward breast cancer treatments with minimal side effects and reduced drug dosage (Zafar *et al.*, 2020). TQ-loaded cubosomes were formulated to study their effects in treating breast cancer. The nanoformulation was done by emulsification homogenization and the physicochemical characteristics were studied. Further, the *in vitro* antitumour activity was tested on breast cancer cell lines (MCF-7 and MDA-MB-231) and compared to a nontumorigenic cell line (MCF-10A). The results showed that the TQ-cubosomal nanoformulation had a narrow unimodal distribution. The cell viability tests revealed the antitumor activity of the TQ-cubosome combination. Localization and trafficking studies revealed that the cubosomes accumulate in the cytoplasm of the studied breast cancer cells (Sarieddine and Alwattar, 2020). The TQ-Phospholipon nanocarrier was formulated by the antisolvent precipitation method for the lung cancer A549 cell line. A remarkable increase in dose-dependent cytotoxicity was observed. The TQ-phytosomes increased apoptotic induction and cell necrosis, which was revealed by the annexin V staining technique via activation of caspase-3. In ROS analysis, TQ-phytosomes significantly increased ROS generation. Thus, TQ-phytosomes effectively proved the anticancer efficiency in lung cancer cell lines (Alhakamy *et al.*, 2020). Chitosan-loaded TQ with lipid nanostructures were formulated to study the anticancer activity toward human breast cancer. The chitosan functionalization of NLCs improved the uptake and endosomal escape effect of the drug and led to the improved cytotoxic effects toward MCF-7 and triple-negative (MDA-MB-231) breast cancer cells. TQ-PLGA nanocarriers were prepared and characterized for size, zeta potential, encapsulation efficiency, and release profile for anticancer activities in melanoma cancer. Due to the sustained release and small size of the carrier, the drug can easily accumulate in cancer tissue along with enhanced permeability and retention effects.

22.7.2 Nanoformulated TQ for the Treatment of Brain Disorders in Humans

As discussed in the previous sections and the knowledge from studies of TQ being used for neuropharmacological treatments, it is evident that TQ is an efficient drug for neurological disorders (Moradi *et al.*, 2020). Hence, TQ nanoformulation and proper delivery is quite necessary. The hydrophobic chemical nature of TQ affects its water solubility and is its major disadvantage in drug formulation. Increasing bioavailability and the absorption rate are essential for pharmacological activities. The various formulations in the treatment of neurological disorders are discussed next (Manek, Darvas and Petroianu, 2020).

Alam and colleagues formulated SLN-loaded TQ for the treatment of stress and depression in animal models. The nanoformulated drug delivery to brain tissue effectively decreased depression in the animal model and also had an anti-inflammatory effect. This, in turn, was involved in the improvement of brain metabolism in the animals (Alam *et al.*, 2020). Alzheimer's disease (AD) is a neurodegenerative disorder in humans that presents with effects such as pathological dementia and multiple cognitive deficits also accompanied by behavioural changes and abnormal mood fluctuations. Yusuf et al. formulated the polysorbate-80-coated TQ PLGA (P-80-TQN) nanoparticles for targeting the brain cells to treat AD.

22.8 CONCLUSION

This chapter presented a deep understanding and investigation on thymoquinone, a phytochemical obtained from the seed oil of *Nigella sativa*. Being a phytochemical, TQ has efficacy in treating various chronic disorders in humans due to its antioxidant, anti-inflammatory, antimicrobial and anticancer properties. As a drug, TQ is unstable under low pH and extreme biological environments, which prompted the researchers to work on the nanoformulation of TQ. Thus, the nanoencapsulation of TQ was a better solution to effectively deliver the drug to the infected area with enhanced bioavailability. This chapter details nanoformulation with lipids and polymers and their drug delivery properties for curing various types of cancers and brain disorders. This review on the drug delivery properties and research cited could prompt researchers to develop new methods of encapsulation for drug delivery and phytochemical loading.

REFERENCES

AbuKhader, Majed M. "Thymoquinone in the clinical treatment of cancer: Fact or fiction?" *Pharmacognosy Reviews* 7, no. 14 (2013): 117.

Ahmad, Niyaz. "Rasagiline-encapsulated chitosan-coated PLGA nanoparticles targeted to the brain in the treatment of Parkinson's disease." *Journal of Liquid Chromatography & Related Technologies* 40, no. 13 (2017): 677–690.

Al Wafai, Rana J. "*Nigella sativa* and thymoquinone suppress cyclooxygenase-2 and oxidative stress in pancreatic tissue of streptozotocin-induced diabetic rats." *Pancreas* 42, no. 5 (2013): 841–849.

Alam, Mahtab, Saima Zameer, Abul Kalam Najmi, Farhan Jalees Ahmad, Syed Sarim Imam, and Mohd Akhtar. "Thymoquinone loaded solid lipid nanoparticles demonstrated antidepressant-like activity in rats via indoleamine 2, 3-dioxygenase pathway." *Drug Research* 70, no. 05 (2020): 206–213.

Alhakamy, Nabil A., Shaimaa M. Badr-Eldin, Usama A Fahmy, Nabil K. Alruwaili, Zuhier A. Awan, Giuseppe Caruso, Mohamed A. Alfaleh et al. "Thymoquinone-loaded soy-phospholipid-based phytosomes exhibit anticancer potential against human lung cancer cells." *Pharmaceutics* 12, no. 8 (2020): 761.

Almajali, Belal, Hamid Ali Nagi Al-Jamal, Wan Rohani Wan Taib, Imilia Ismail, Muhammad Farid Johan, Abd Almonem Doolaanea, and Wisam Nabeel Ibrahim. "Thymoquinone, as a novel therapeutic candidate of cancers." *Pharmaceuticals* 14, no. 4 (2021): 369.

Al-Rasheed, Nawal, Nouf Al-Rasheed, Yieldez Bassiouni, Laila Faddah, and Azza M. Mohamad. "Potential protective effects of *Nigella sativa* and *Allium sativum* against fructose-induced metabolic syndrome in rats." *Journal of Oleo Science* 63, no. 8 (2014): 839–848.

Ballout, Farah Rabih, and Hala Gali-Muhtasib. "Thymoquinone: A potential therapy against cancer stem cells." *Pharmacognosy Reviews* 14, no. 28 (2021).

Ballout, Farah, Zeina Habli, Omar Nasser Rahal, Maamoun Fatfat, and Hala Gali-Muhtasib. "Thymoquinone-based nanotechnology for cancer therapy: Promises and challenges." *Drug Discovery Today* 23, no. 5 (2018): 1089–1098.

Chiuh Woo, Chern, Ser Yue Loo, Veronica Gee, Chun Wei Yap, Gautam Sethi, Alan Prem Kumar, and Kwong Huat Benny Tan. "Anticancer activity of thymoquinone in breast cancer cells: Possible involvement of PPAR-γ pathway." *Biochemical Pharmacology* 82, no. 5 (2011): 464–475.

Darakhshan, Sara, Ali Bidmeshki, Pour Abasalt, Hosseinzadeh Colagar, Sajjad Sisakhtnezha. "Thymoquinone and its therapeutic potentials." *Pharmacological Research* 93 (2015): 138–158.

El-Dakhakhany, "Studies on the chemical constitution of Egyptian *Nigella sativa* L. Oil." *Planta Medica* 11 (2) (1963): 134–138.

El-Far, Ali H., Soad K. Al Jaouni, Weikun Li, and Shaker A. Mousa. "Protective roles of thymoquinone nanoformulations: Potential nanonutraceuticals in human diseases." *Nutrients* 10, no. 10 (2018): 1369.

Farkhondeh, Tahereh, Saeed Samarghandian, and Abasalt Borji. "An overview on cardioprotective and anti-diabetic effects of thymoquinone." *Asian Pacific Journal of Tropical Medicine* 10, no. 9 (2017): 849–854.

Gali-Muhtasib, H., A. Roessner, and R. Schneider-Stock. "Thymoquinone: A promising anti-cancer drug from natural sources." *International Journal of Biochemistry and Cell Biology*, 38, no. 8 (2006): 1249–1253.

Hamdy, Nadia M., and Ragia A. Taha. "Effects of *Nigella sativa* oil and thymoquinone on oxidative stress and neuropathy in streptozotocin-induced diabetic rats." *Pharmacology* 84, no. 3 (2009): 127–134.

Khader, Mohannad, Nikolaus Bresgen, and P. M. Eckl. "*In vitro* toxicological properties of thymoquinone." *Food and Chemical Toxicology* 47, no. 1 (2009): 129–133.

Khader, Mohannad, and Peter M. Eckl. "Thymoquinone: An emerging natural drug with a wide range of medical applications." *Iranian Journal of Basic Medical Sciences* 17, no. 12 (2014): 950.

Khaled, Radad, Rudolf Moldzio, Mokhtar Taha, Wolf-Dieter Rausch *et al.* "Thymoquinone protects dopaminergic neurons against MPP+ and rotenone." *Phytotherapy Research* 23, no. 5 (2009): 696–700.

Khalil, Ahmad S., and James J. Collins, "Synthestic biology: Applications come of age." *Nature Review Genetics* 11 (2010): 367–379.

Khalil, Palwasha, Sara Masood, Attiq Ur Rehman, Faiza Khalil, and Javeria Nawaf. "Preventive role of thymoquinone against certain chronic health issues: A review." *International Journal of Nutrition Sciences* 5 (2020): 151–158.

Kouidhi, Bochra, Tarek Zmantar, Hanene Jrah, Yosra Souiden, Kamel Chaieb, Kacem Mahdouani, and Amina Bakhrouf. "Antibacterial and resistance-modifying activities of thymoquinone against oral pathogens." *Annals of Clinical Microbiology and Antimicrobials* 10, no. 1 (2011): 1–7.

Kumar, S. "Antidiabetic activity enhancement in streptozotocin + nicotinamide–induced diabetic rats through combinational polymeric nanoformulation." *International Journal of Nanomedicine* 14 (2019): 4383–4395.

Manek, Eniko, Ferenc Darvas, and Georg A. Petroianu. "Use of biodegradable, chitosan-based nanoparticles in the treatment of Alzheimer's disease." *Molecules* 25, no. 20 (2020): 4866.

Moradi, Seyed Zachariah, Saeideh Momtaz, Zahra Bayrami, Mohammad Hosein Farzaei, and Mohammad Abdollahi. "Nanoformulations of herbal extracts in treatment of neurodegenerative disorders." *Frontiers in Bioengineering and Biotechnology* 8 (2020): 238.

Nallamuthu, I., A. Parthasarathi, and F. Khanum. "Thymoquinone-loaded PLGA nanoparticles: Antioxidant and anti-microbial properties." *International Current Pharmaceutical Journal* 2, no. 12 (2013): 202–207.

Novy, Pavel, Pavel Kloucek, Johana Rondevaldova, Jaroslav Havlik, Lenka Kourimska, and Ladislav Kokoska. "Thymoquinone vapor significantly affects the results of *Staphylococcus aureus* sensitivity tests using the standard broth microdilution method." *Fitoterapia* 94 (2014): 102–107.

Odeh, Fadwa, Said I. Ismail, Rana Abu-Dahab, Ismail S. Mahmoud, and Abeer Al Bawab. "Thymoquinone in liposomes: A study of loading efficiency and biological activity towards breast cancer." *Drug Delivery* 19, no. 8 (2012): 371–377.

Raschi, A. B., E. Romano, A. M. Benavente, A. Ben Altabef, and M. E. Tuttolomondo. "Structural and vibrational analysis of thymoquinone." *Spectrochimica Acta Part A: Molecular and Biomolecular Spectroscopy* 77, no. 2 (2010): 497–505.

Rathore, Charul, Michael J. Rathbone, Dinesh K. Chellappan, Murtaza M. Tambuwala, Terezinha De Jesus A. Pinto, Harish Dureja, Chetna Hemrajani, Gaurav Gupta, Kamal Dua, and Poonam Negi. "Nanocarriers: More than tour de force for thymoquinone." *Expert Opinion on Drug Delivery* 17, no. 4 (2020): 479–494.

Samarghandian, Saeed, Tahereh Farkhondeh, and Fariborz Samini. "A review on possible therapeutic effect of *Nigella sativa* and thymoquinone in neurodegenerative diseases." *CNS & Neurological Disorders-Drug Targets (Formerly Current Drug Targets-CNS & Neurological Disorders)* 17, no. 6 (2018): 412–420.

Sarieddine, R., and Alwattar, J. K. "Anticancer activity of thymoquinone cubic phase nanoparticles against human breast cancer: Formulation, cytotoxicity and subcellular localization." *International Journal of Nanomedicine* 15 (2020): 9557.

Tavakkoli, Alireza, Vahid Mahdian, Bibi Marjan Razavi, and Hossein Hosseinzadeh. "Review on clinical trials of black seed (*Nigella sativa*) and its active constituent, thymoquinone." *Journal of Pharmacopuncture* 20, no. 3 (2017): 179.

Tubesha, Zaki, Mustapha Umar Imam, Rozi Mahmud, and Maznah Ismail. "Study on the potential toxicity of a thymoquinone-rich fraction nanoemulsion in Sprague Dawley rats." *Molecules* 18, no. 7 (2013): 7460–7472.

Ullah, Ikram, Najeeb Ullah, Muhammad Imran Naseer, Hae Young Lee, and Myeong Ok Kim. "Neuroprotection with metformin and thymoquinone against ethanol-induced apoptotic neurodegeneration in prenatal rat cortical neurons." *BMC Neuroscience* 13, no. 1 (2012): 1–11.

Umeno, Aya, Masanori Horie, Kazutoshi Murotomi, Yoshihiro Nakajima, and Yasukazu Yoshida. "Antioxidative and antidiabetic effects of natural polyphenols and isoflavones." *Molecules* 21, no. 6 (2016): 708.

Zafar, Sobiya, Sohail Akhter, Nupur Garg, Angamuthu Selvapandiyan, Gaurav Kumar Jain, and Farhan Jalees Ahmad. "Co-encapsulation of docetaxel and thymoquinone in mPEG-DSPE-vitamin E TPGS-lipid nanocapsules for breast cancer therapy: Formulation optimization and implications on cellular and *in vivo* toxicity." *European Journal of Pharmaceutics and Biopharmaceutics* 148 (2020): 10–26.

23 Regulatory Aspects of Herbal Nanoformulations

Lavanya Mude, Gowthamarajan Kuppusamy, Veera Venkata Satyanarayana Reddy Karri, Arun R., Nandha Kumar and Muhammad Thanzim

CONTENTS

23.1 INTRODUCTION: REGULATORY AFFAIRS

Regulatory affair is a specialized field in the pharmaceutical and biotechnology industry with the unique blend of science and management to achieve a commercially significant aim inside a drug development organization. It covers all aspects of drug development, from the first nonclinical investigations through regular manufacturing and commercialization.

Regulatory affairs departments work with federal, state and local regulatory agencies around the world, including the Food and Drug Administration (FDA, United States), Therapeutic Goods Administration (TGA, Australia), Medicines and Healthcare products Regulatory Agency (MHRA, United Kingdom), Medicines Control Council (MCC, South Africa) and others to ensure that the registration, licensing, developing, manufacturing, marketing, and labelling of pharmaceutical and medical products is done following all applicable rules.

DOI: 10.1201/9781003231745-23

23.1.1 Herbal Formulations

Herbal formulations (herbal medicine) are a type of traditional medicine that has become the most popular form of healthcare globally because of the belief that they are safer with fewer side effects and lower cost. Herbal medicine uses herbs and other forms of traditional medicine to treat various diseases, including skin-related problems (1). These medicines are prepared by subjecting herbal substances to multiple treatments such as extraction, distillation, expression, fractionation, purification and fermentation. Nowadays, herbal medicines are available in herbal powders, tinctures, extracts, essential oils, etc. Herbal formulations are of different characters with unique action, application, and therapeutic roles of treating or curing a disease (2). Herbal formulations can be phytomedicines, nutraceuticals, cosmetics, etc.

23.1.1.1 Phytomedicines

Phytomedicine is derived from a whole plant or parts of the plant used for drug discovery. In ancient texts, there were about 20,000 registered plant species in India. Among them, 500 plants are recognized for medical purposes, and about 800 plants are used for indigenous medical systems (3). Plants make primary metabolites like sugars, lipids and other secondary metabolites with therapeutic effects, which produce drugs as phytomedicine. Phytomedicine from natural extracts consists of active chemicals that possess several biological activities and functions. Some phytochemicals that can be beneficial as plant-based medications are alkaloids, glycosides, phenols, terpenoids, flavonoids, etc. They are manufactured as purified pharmaceutical formulations such as tinctures, syrups, medicinal oils (liquid dosage forms) and tablets, granules and capsules (solid dosage forms) with therapeutic and healing properties (4). Some of the marketed phytomedicines are as follows:

- Anticancer: Vinblastine, vincristine, paclitaxel, podophyllotoxin
- Antimalarial: Quinine, artemisinin
- Digestive system: Pepticare
- Central nervous system (CNS) stimulant: Cocaine, caffeine
- Cardiotonic: Digoxin
- Nervous system: Bacopamine
- Arthritis: Triphala, codeine, morphine

According to the World Health Organization (WHO), although synthetic drugs are widely used, most people still practice traditional phytomedicines in developing countries for healthcare (4).

23.1.1.2 Nutraceuticals

The term nutraceutical combines the words "nutrition" and "pharmaceuticals." Nutraceuticals are dietary components that are naturally found in foods and that have a health or medicinal advantage. The potential nutritional, safety and therapeutic impacts of nutraceuticals have shown great promise. Some studies demonstrated successful pathological complications such as cardiovascular diseases, neurological disorders, diabetes and cancer. Nutraceuticals are categorized as traditional, nontraditional and commercial nutraceuticals (5, 6).

23.1.1.2.1 Traditional Nutraceuticals

Traditional nutraceuticals are obtained directly from nature without any changes in their natural form. Traditional nutraceuticals are classified into chemical constituents (nutrients, herbals and phytochemicals), probiotic micro-organisms and nutraceutical enzymes. Nutrients such as amino acids, various vitamins and fatty acids have many health benefits to treat various diseases. Phytochemicals are secondary metabolites of plant products that include alkaloids, terpenoids, phenolic acids, etc. Probiotic micro-organisms are the live microbial food ingredients that are advantageous to the human gastrointestinal tract and improve absorption and metabolism by removing toxic substances (7). Nutraceutical enzymes are structurally proteinous, produced from plant, animal and microbial sources, and serve as a biocatalyst. Enzyme therapy has now been utilized for various uncommon disorders, including Gaucher disease and Hunter syndrome (8).

23.1.1.2.2 Nontraditional Nutraceuticals

Nontraditional nutraceuticals are produced biotechnologically to boost the immune system. These are divided into fortified nutraceuticals and recombinant nutraceuticals. Fortified nutraceuticals are bred to add nutrients to the main ingredient. Examples include milk fortified with cholecalciferol, which is used for D vitamin deficiency. Recombinant nutraceuticals are produced by extracting bioactive components through a fermentation process. Examples include yogurt, bread, alcohol, cheese, etc. (9).

23.1.1.3 Herbal Cosmetics

Cosmetics comes from the Greek word *kosmtikos*, which means power and having skill in decorating. In India, according to the Drugs and Cosmetics Act (C&D Act) 1940, cosmetics are applied to the human skin by rubbing, pouring or sprinkling on to beautify and promote the attractiveness of the skin (10). Herbal cosmetics are formulated purely by using one or more herbal ingredients to form a base to enhance the health and beauty of skin and hair. Compared to synthetic cosmetics, these are free from harmful chemicals with no or relatively few side effects and are compatible with all skin types (11).

Some of the examples for herbal cosmetics are as follows:

- Dry skin treatment: Aloe vera, jojoba oil, coconut oil, olive oil, etc.
- Skin protection: Turmeric, green tea, calendula, etc.
- Antiageing treatment: Ginkgo, carrot, rhodiola rosea
- Dandruff treatment: Henna, shikakai, neem
- Hair care: Coconut oil, olive oil, amla, almond oil, etc.

23.2 NANO HERBAL FORMULATIONS AND THEIR ROLE IN TREATING VARIOUS DISEASES OR DISORDERS

Nano herbal formulation is an emerging trend that has already yielded some exciting results in developing novel phytochemical delivery systems. Compared to traditional herbal drug preparations, nano herbal formulation has many benefits, including improved component solubility, increased bioavailability, improved absorption, a reduced dose of herbal medicinal products and achieving steady-state therapeutic levels of the drug over a longer period of time (12).

23.2.1 Nano Herbal Formulation for the Treatment of Neurodegenerative Disorders

Neurodegenerative disorders (NDs) are described as a range of functional or structural abnormalities of the nervous system or neuron that cause disability in movement, thinking, and memory. Alzheimer's disease (AD), Parkinson's disease (PD), and multiple sclerosis (MS) are some examples of NDs. AD is a chronic, untreatable disorder associated with behavioural, social and work impairments caused by damaged brain cells that eventually leads to death. PD is a chronic ND that affects muscle mobility and control by the constant loss of dopamine in substantia nigra. Nerve-cell damage in the brain and spinal cord occurs as a result of MS (13).

Some of the nano herbal formulations for the treatment of NDs are curcumin (CUR), quercetin (QC), piperine (PIP), resveratrol (RSV), gallic acid (GA) and ferulic acid (FA). CUR modulates the inflammation process and is effective in aggregating protein by making it advantageous for treating NDs. However, CUR showed limited absorption, rapid metabolism by the liver and poor bioavailability. To overcome such defects, CUR was encapsulated as nanoparticles, nanoliposomes, nanocapsules, and nanomicelles. QC is a bioflavonoid known for neuroprotective and anti-inflammatory activities. PIP is an alkaloid that shows its effectiveness in CNS by acting on acetylcholine (14).

23.2.2 Nano Herbal Formulation for the Treatment of Colorectal Cancer

Cancer is a devasting disease and a foremost cause of death worldwide. The risk of developing colorectal cancer (CRC) is identified in people above 50 years of age. Generally, CRC begins as a polyp localized in the proximal colon, rectum and distal colon. The polyp continues to invade nearby tissues through the lymph and circulatory systems. Cancer treatments include radiotherapy, chemotherapy and surgery, which in turn are associated with severe side effects. To overcome these challenges, herbal medicines such as curcumin, RSV, lycopene, gingerol and folate have gained attention as an alternative treatment. CUR possesses anticancer, antibacterial, anti-inflammatory and antibiotic properties. However, CUR has some drawbacks such as faster elimination from the body, low water solubility and poor absorption. Therefore, CUR nanoformulations were developed, such as nanogels, nanoemulsions, nanoparticles, cyclodextrins, etc. In CRC, CUR nanoformulations promote apoptosis by up-regulating the p53 suppressor gene, which leads to cell death (15).

23.2.3 Nano Herbal Formulation for the Treatment of Diabetes

Diabetes is a metabolic disorder that causes higher levels of blood sugar due to insulin deficiency. Some of the natural products used for diabetes are CUR, capsicum oleoresin, quercetin, emodin, scutellarin, silybin, etc. CUR exhibits antitumour, anti-inflammatory, antioxidant and hypolipidemic properties. The poor water solubility of CUR is resolved by encapsulated CUR with nanoparticles. Capsicum olcoresin is an organic solvent that possesses antioxidant, anti-inflammatory and anticancer activities. Nanoemulsions of capsicum oleoresin decrease higher glucose levels. QC nanoparticles increase insulin secretion by reducing nitric oxide (NO) synthase activation. Nanoformulation of scutellarin promotes neovascularization by enhancing vascular permeability, which further improves diabetic retinopathy (16).

23.3 REGULATORY ASPECTS OF NANOFORMULATIONS

Regulatory aspects are essential in nanomedicine and nanocosmeceuticals. Regulatory aspects are in place to safely develop pharmaceuticals containing nanoparticles and methods of drug delivery systems, which are regulated by the FDA.

23.3.1 Nanomedicine

Nanomedicine is a nanotechnological application to healthcare with enhanced physicochemical and biological properties using nanoscale structures for treatment and diagnosis. Nanomedicines have a vast potential to overcome failures of traditional medicine that cannot be adequately formulated by variables such as inadequate water solubility, toxicity, limited bioavailability or lack of target specificity (17, 18). Innovative nanomedicines have shown a potential impact on global healthcare by focusing on the targeted drug delivery systems with polymeric nanoparticles, liposomes, nanoemulsions, nanotubes, quantum dots, etc. Nanomedicine features microscopic dimensions with a large surface area to penetrate living tissues (19). However, despite their many advantages, nanomaterials used in nanomedicines have some limitations, which are shown in Table 23.1.

As per the Nanotechnology Characterization Laboratory (NCL), nanotechnological products are subjected to a proper assessment in which the particle properties are evaluated in six key categories shown in Table 23.2.

TABLE 23.1
Advantages and Limitations of Nanomaterials Used in Nanomedicine (20)

Nanomaterials	Particle Size Range	Advantages	Limitations
Polymeric nanoparticles	1–100 nm	• Biocompatibility • Nonimmunogenicity • Enhance the stability of any volatile dosage forms • Can be quickly produced by several methods in large quantities	• Low drug-loading capacity • Shorter half-life • Expensive • Cause burst release of the drug
Solid lipid nanoparticles	1–100 nm	• Less toxicity and good storage stability • Easy to prepare for large-scale production • Drug releases in a controlled manner and improved bioavailability • Bypasses the first-pass metabolism • Increases patient compliance	• May cause burst release of the drug • Limited drug-loading capacity • Drug release profile changes because of lipase degradation
Carbon nanotubes	0.4–40 nm	• Accessible to synthesis and encapsulate different types of bioactive agents • Protects entrapped drug molecules • Sustained release	• Extremely toxic • Nonbiodegradable • Insoluble in water • Accumulates in internal organs
Quantum dots	2–10 nm	• Highly effective in converting shorter wavelength to longer wavelength • Theranostic applications • Stable and easy to handle	• Potential material toxicity • Cannot diffuse across the cellular membrane due to their large physical size • Accumulates in the reticuloendothelial system
Dendrimers	10–100 nm	• Uniformity in size and shape • More biocompatible with low cytotoxicity • High biopermeability • Highly soluble and stable • High drug-encapsulation efficiency • Suitable for incorporating lipophilic drugs	• Cationic dendrimers cause cytotoxicity • Rapid clearance • Poor drug-release profile
Nanoemulsions	20–500 nm	• Drugs that are hydrophilic and lipophilic can be encapsulated easily • Added surfactants significantly improve intestinal penetration • Highly stable, biodegradable/biocompatible	• Poor GI stability • Chances of premature antigens • A large quantity of surfactant is required to produce nanometric-sized oil droplets.
Liposomes	25–50 nm	• Ability to entrap both hydrophilic and hydrophobic drugs • Biodegradable and biocompatible • Protects the drug from metabolic degradation	• Limited antigen loading • Naked liposomes have poor GI stability • Difficulties in preparation and sterilization • Weaker mucus penetration • Cationic lipids cause toxicity • Low stability

TABLE 23.2
Physicochemical Parameters of Nanomaterials According to the Nanotechnology Characterization Laboratory (NCL)

Parameters	Importance	Analytical Methods Used
Particle size/size distribution	• The size and area of nanomaterials are essential material properties from the standpoint of toxicological and biological applications. • Particle-size distribution (PSD) is significant because nanomaterials are generally polydisperse, which means they are usually composed of various particle sizes.	• Transmission electron microscopy (TEM) • Laser diffraction studies • Dynamic light scattering (DLS) • Multi-angle light scattering (MALS) • Single-particle inductively coupled plasma mass
Composition and purity	• In determining the toxicity of nanomaterials, chemical composition is a significant component in addition to particle size and shape. • The existence of residuals, foreign materials and chemical agents is responsible for toxicity issues. So, the purity of any nanomaterial is also an essential factor for biomedical applications.	• Reversed-phase high-performance liquid chromatography (RP-HPLC) • Liquid chromatography-mass spectrometry (LC-MS) • Gas chromatography-mass spectrometry (GC-MS) • Thermogravimetric analysis (TGA) • Size exclusion chromatography multi-angle light scattering (SEC-MALS)
Surface charge	• Surface charge is also significant in nanomatcrial toxicity since it determines how they interact with biological systems. • The surface charge influences the organism by changing nanomaterial size and shape through aggregate formation.	• Zeta potential • Asymmetric flow fluid flow fractionation (AF4) • LC-MS • TGA • Quartz Crystal Microbalance with Dissipation Monitoring (QCM-D)
Stability	• Pharmaceutical stability is a retention of the same characteristics over a period of time after the production of the pharmaceuticals.	• Shelf life • Temperature • pH and light sensitivity • Excipients and solvents • Centrifugation • Filtration • Lyophilization
Lot-to-lot variation (LTLV)	• LTLV is a significant cause for analytical inaccuracy. • It is considered by the manufacturers to check whether a lot is suitable for use and to detect any variations in lots of reagents.	• DLS • Zeta potential • LC-MS • AF4 • TGA • SEC-MALS
Starting material characterization	• Starting material characterization plays a vital role in controlling the quality of product development. • Before formulation, the starting material should be tested for purity, product degradation, stability, toxicity and other characteristics.	• Drugs: chemical structure, product degradation, impurities • Polymers: molecular mass, polydispersity, purity, degree of functionalization • Lipids: free fatty acid and its distribution, lysophospholipids • Solvents: chemical and biological impurities

23.3.1.1 Regulatory Aspects

Nanomedicine is considered a medicine in one nation and a medical device in another depending on their classification. No specific standards for preclinical and clinical testing of nanomedicine have been established by regulatory agencies like the European Medicines Agency (EMA) and the FDA. The manufacturing and regulatory aspects of nanopharmaceuticals remain challenges. Regulatory elements play an essential role in nanomedicine to ensure the quality, safety and efficacy of finished products. The guidelines apply to nanomaterials and encourage their inventions by increasing their benefit-to-risk ratio (21).

23.3.1.2 EMA

The first international scientific workshop on nanomedicine was conducted by the EMA in London on 2 September 2010, to share regulatory experiences and address the issues related to nanomedicine quality, safety and efficacy. Twenty-seven countries, including India, participated in the EMA workshop. The discussion was mainly on medicines and medical devices containing nanoparticles such as liposomes, nanoshells, quantum dots and dendrimers, which have shown potential advantages over conventional drug delivery systems. The benefits include increased absorption, metabolism and drug residence time in the body to treat diseases (22). The EMA has so far reviewed 18 marketing authorization applications for nanomedicines used as anti-cancer, anti-infective and immune-modulating agents. The applications were assessed under the European Union (EU) regulatory framework established to analyse benefits and risks as well as evaluate nanomedicinal products. Some of the approved nanomedicines by the European Commission (EC) are as follows (23):

- Nanoparticles: Nab-paclitaxel, Abraxane, Zevalin
- Liposomes: Doxyrubicin hydrochloride, amphotericin B, cytarabine
- Nanoemulsions: Cyclosporine, Oncaspar
- Nanocrystals: Fenofibrate, Lipidil

When the products are based on viable cells or tissues, these products will fall under advanced therapy medicinal products (ATMPs). As per regulation (EC) 1394/2007, ATMPs must fulfil the established criteria such as safety, purity, potency and clinical efficacy before the product can be marketed. According to the EU directive, human medicines are amended as 2001/83/E.C. and medicinal devices as 93/42/EEC.

23.3.1.3 U.S. FDA

The FDA interacts with worldwide stakeholders, including regulatory agencies, to communicate and share information on nanotechnological products. The FDA provides guidance on food, cosmetics and animals for nanomaterials. However, no specific guidance on nanomedicine has been issued. The FDA regulates products such as biological products, dietary supplements, medical devices, drugs, etc. In 2006, the FDA launched an initiative to expand its scientific expertise in nanotechnology to ensure that FDA scientists were ready to meet the regulatory challenges that nanotechnology products could offer. The FDA established a Nanotechnology Task Force (NTF) to identify regulatory measures that promote innovative, safe and effective FDA-regulated products to prevent public health issues. In 2007, the NTF issued a report on scientific and regulatory matters concerning nanotechnology to develop safe and effective products to protect and promote public health. The FDA has issued guidance to the industry, including advice on determining the regulatory status and safety of nanotechnology products. The NTF focused on two scientific problems to better understand nanoscale material interaction with biological systems and four regulatory policy issues regarding evaluating the safety and effectiveness of nanomaterials. Some of the approved nanomedicines by the FDA are as follows (24):

- Liposomes: Doxil, Myocet, DaunoXome
- Lipid nanoparticles: Abelcet, Amphotec
- Nanoemulsion: Diprivan, Durezol, Restasis

23.3.1.4 Central Drugs Standard Control Organisation in India

The guidelines for the evaluation of nanomedicines were developed by the Department of Biotechnology (DBT), Central Drugs Standard Control Organisation (CDSCO) and the Indian Council of Medical Research (ICMR) under the Ministry of Health and Family Welfare (MoHFW). The CDSCO is responsible for approving and setting standards for various medicines, but the implementation occurs in state and union territories. In 2019, CDSCO issued the new drug and clinical trial rules to follow the general guidelines. The main aim of the guidelines is to assess the quality, safety testing and effectiveness of nanopharmaceutical products. In India, the guidelines for the evaluation of nanopharmaceticals and nanomedicine were considerations from Schedule Y of the Drugs and Cosmetics Act (D&C Act) in 1945, as well as from the New Drugs and Clinical Trials Rules in 2019. The guidelines state that the particles should be in nanoscale range in all testing conditions, possible advantages and disadvantages of nanomaterials compared to conventional material should be stated clearly, the detailed manufacturing process should be provided, finished products containing nanomaterials should be safe and effective and intended to treat various diseases and disorders (25).

23.3.2 Nanocosmeceuticals

The term cosmeceutical was coined in 1962 by Raymond Reed, a U.S. cosmetic chemist who received a medal award for his studies in cosmetic field. Cosmeceuticals are personal-care products that have an active medicinal substance with increased therapeutic efficiency. In 1986, the term nanocosmetics was coined by Christian Dior, a cosmetic chemist, and describes cosmetics preparations that contain nanoparticles. Nanocosmeceutical preparations include solid lipid nanoparticles, niosomes, nanoliposomes, nanocapsules, nanoemulsions, dendrimers, etc. (26).

23.3.2.1 Regulatory Aspects

In the past, there have been many challenges in regulating nanoparticulate systems related to the safety of nanomaterials in cosmetics. On 7 November 2018, the EC issued the tenth amended edition of "Scientific Committee on Consumer Safety" (SCCS) to test the safety assessment of cosmetic ingredients. This guidance also addresses nanomaterial safety and efficacy (27).

23.3.2.2 U.S. FDA

The FDA issued guidance for the industry to monitor the safety and efficacy of nanomaterials in cosmetic products. According to the FD&C, no premarket approval is required to market cosmetics. Under Section 301(a) of the FD&C Act (21 USC 331[a]), adulterated or misbranded cosmetics are forbidden in the interstate market. The ingredients used to prepare cosmetics products should not make products adulterated under Section 601 of the FD&C Act (21 U.S.C.361) or misbranded under Section 602 of the FD&C Act (21 USC 362). The Federal Register of 3 March 1975 (40 FR 8912 at 8916) advised that "the safety of a cosmetic product is sufficiently justified by nanomaterial characterization, toxicological considerations, and safety testing." Nanomaterial considerations include evaluating physicochemical properties and impurities of cosmetic ingredients to determine the safety of the cosmetic product (28). While formulating the nanocosmetic product, manufacturers should consider each ingredient's physical and chemical properties, stability conditions, bioavailability and toxicity. The FDA organized a public conference on 8 September 2008 to examine facts and information associated with scientific and regulatory challenges in nanotechnology. The International Cooperation on Cosmetics Regulations (ICCR) provided information on current nanotechnological advancements and the safety of nanomaterials. Some of the marketed nanocosmeceutical products are as follows (29):

- Liposomes: Sesderma C vit liposomal serum, emergin C Hyper-vitalizer face cream
- Niosomes: Antiageing response cream, MayuNiosome base cream
- Solid lipid nanoparticles: Cutanova cream Nanovital Q10

23.4 CONCLUSION

It is well known that traditional medicines are widely used and that plants contain many natural chemicals that could develop novel drugs. Although traditional herbal medicines have a long and rich history, they face many challenges regarding their regulatory issues, safety and efficacy assessments, quality control and a lack of or limited knowledge about them. Nanomedicines have a vast potential to overcome the failures of traditional medicine. Today, the nanopharmaceuticals market is experiencing rapid growth. It is expected to grow in the coming decades due to the development of innovative nanoparticle-based drug delivery systems and the emergence of new therapeutic classes such as smart delivery systems that function biologically. In conclusion, for preclinical and clinical testing of nanoparticle-based drug delivery systems, there are currently no specific regulatory requirements and there are only a handful of papers that provide guidelines from the world's leading regulatory agencies like the FDA and EMA regarding nanomedicine. Therefore, the development of nanomedicine must be expedited to clinical trial stages, which is a major undertaking.

REFERENCES

1. Kunle, Oluyemisi Folashade, Henry Omoregie Egharevba, and Peter Ochogu Ahmadu. "Standardization of herbal medicines—A review." *International Journal of Biodiversity and Conservation* 4, no. 3 (2012): 101–112.
2. Sarojini K, Lakshminarayanan Arivarasu, Smiline Girija A S. "Herbal formulation: Review of efficacy, safety, and regulations." *International Journal of Research in Pharmaceutical Sciences* 11, no. 3 (2020): 1506–1510.
3. Kamboj, Ved P. "Herbal medicine." *Current Science* 78, no. 1 (2000): 35–39
4. Mukeshwar, P., Mousumi Debnath, Shobit Gupta, and Surender K. Chikara. "Phytomedicine: An ancient approach turning into future potential source of therapeutics." *Journal of Pharmacognosy and Phytotherapy* 3, no. 2 (2011): 27–37
5. Kalra, Ekta K. "Nutraceutical-definition and introduction." *Aaps Pharmsci* 5, no. 3 (2003): 27–28.
6. Sapkale, Anita P., Mangesh S. Thorat, Prasad R. Vir, and Meera C. Singh. "Nutraceuticals—global status and applications: A review." *International Journal of Pharmaceutical and Chemical Sciences* 1, no. 3 (2012): 1166–1181.
7. Ruchi, Sharma. "Role of nutraceuticals in health care: A review." *International Journal of Green Pharmacy (IJGP)* 11, no. 03 (2017).
8. Chintale Ashwini, G., S. Kadam Vaishali, S. Sakhare Ram, O. Birajdar Ganesh, and N. Nalwad Digambar. "Role of nutraceuticals in various diseases: A comprehensive review." *International Journal of Research in Pharmacy and Chemistry* 3 (2013): 290–299.
9. Shirwaikar, Annie, Vinit Parmar, and Saleemulla Khan. "The changing face of nutraceuticals—An overview." *International Journal of Pharmacy & Life Sciences* 2, no. 7 (2011).
10. Pastrana, Homero, Alba Avila, and Candace SJ Tsai. "Nanomaterials in cosmetic products: The challenges with regard to current legal frameworks and consumer exposure." *Nanoethics* 12, no. 2 (2018): 123–137.
11. Joshi, Laxmi S., and Harshal A. Pawar. "Herbal cosmetics and cosmeceuticals: An overview." *Natural Products Chemistry & Research* 3, no. 2 (2015): 170.
12. Patil, Rinku Y., Shubhangi A. Patil, Niranjan D. Chivate, and Yogesh N. Patil. "Herbal drug nanoparticles: Advancements in herbal treatment." *Research Journal of Pharmacy and Technology* 11, no. 1 (2018): 421–426.

13. Przedborski, Serge, Miquel Vila, and Vernice Jackson-Lewis. "Series Introduction: Neurodegeneration: What is it and where are we?" *Journal of Clinical Investigation* 111, no. 1 (2003): 3–10.
14. Moradi, S. Z., Momtaz, S., Bayrami, Z., Farzaei, M. H., & Abdollahi, M. "Nanoformulations of herbal extracts in treatment of neurodegenerative disorders." *Frontiers in Bioengineering and Biotechnology*, *8*, (2020): 238.
15. Carrouel, Florence, Stephane Viennot, Livia Ottolenghi, Cedric Gaillard, and Denis Bourgeois. "Nanoparticles as anti-microbial, anti-inflammatory, and remineralizing agents in oral care cosmetics: A review of the current situation." *Nanomaterials* 10, no. 1 (2020): 140.
16. Viswanathan, Vijay, Rajesh Kesavan, K. V. Kavitha, and Satyavani Kumpatla. "A pilot study on the effects of a polyherbal formulation cream on diabetic foot ulcers." *Indian Journal of Medical Research* 134, no. 2 (2011): 168.
17. Sanchez, Florence, and Konstantin Sobolev. "Nanotechnology in concrete—A review." *Construction and Building Materials* 24, no. 11 (2010): 2060–2071.
18. Navalakhe, Rajshri M., and Tarala D. Nandedkar. "Application of nanotechnology in biomedicine." *Indian journal of Experimental Biology* 42, no. 2 (2007): 160–165.
19. Rathor, Sandeep, D. C. Bhatt, Sheikh Aamir, S. K. Singh, and Vikash Kumar. "A comprehensive review on role of nanoparticles in therapeutic delivery of medicine." *Pharmaceutical Nanotechnology* 5, no. 4 (2017): 263–275.
20. Das, Malay K., and Yashwant V. Pathak, eds. "Nano Medicine and Nano Safety" *Recent Trends and Clinical Evidences* (2020).
21. Raghvendra, G., Narendra Kumar Pandey, Bimlesh Kumar, Sachin Singh, Parth Sharma, and Yadav Sarvi. "Investigation of influence of shape on drug loading and entrapment efficiency of nanoparticles." *Journal of Pharmacy Research* 11, no. 7 (2017): 850–855.
22. Dorbeck-Jung, Bärbel R., and Nupur Chowdhury. "Is the European Medical Products Authorisation Regulation equipped to cope with the challenges of nanomedicines?" *Law & Policy* 33, no. 2 (2011): 276–303.
23. Hafner, Anita, Jasmina Lovrić, Gorana Perina Lakoš, and Ivan Pepić. "Nanotherapeutics in the EU: an overview on current state and future directions." *International Journal of Nanomedicine* 9 (2014): 1005.
24. Tinkle, Sally, Scott E. McNeil, Stefan Mühlebach, Raj Bawa, Gerrit Borchard, Yechezkel Barenholz, Lawrence Tamarkin, and Neil Desai. "Nanomedicines: Addressing the scientific and regulatory gap." *Annals of the New York Academy of Sciences* 1313, no. 1 (2014): 35–56.
25. Draft Guidance for Industry: Considering Whether an FDA Regulated Product Involves the Application of Nanotechnology, June 2011. Available online at: http://www.fda.gov/ RegulatoryInformation/ Guidances/ucm257698.htm
26 Mihranyan, Albert, Natalia Ferraz, and Maria Strømme. "Current status and future prospects of nanotechnology in cosmetics." *Progress in Materials Science* 57, no. 5 (2012): 875–910.
27 Bawa, Raj, Srikumaran Melethil, William J. Simmons, and Drew Harris. "Nanopharmaceuticals: Patenting issues and FDA regulatory challenges." *SciTech Lawyer* 5, no. 2 (2008): 10–15.
28. Draft Guidance for Industry: Safety of Nanomaterials in Cosmetic Products, April 2012. Available online at: http://www.fda.gov/Cosmetics/GuidanceComplianceRegulatory Information/ GuidanceDocuments/ucm300886.htm
29. Mansour, Heidi M., Chun Woong Park, and Raj Bawa. "Design and development of approved nano-pharmaceutical products." In *Handbook of Clinical Nanomedicine: Nanoparticles, Imaging, Therapy and Clinical Applications*, pp. 233–272. Pan Stanford Publishing Pte. Ltd., 2016.

24 Pharmacological Properties of Green Synthesized *Rubus fairholmianus* Gardner Zinc Oxide Nanoparticles

Gayathri Nataraj, Gayathri Jagadeesan, Sandhya Saminathan, Ashwini Lydia Manoharan, Kasipandi Muniyandi, Dipshika Motwani and Parimelazhagan Thangaraj

CONTENTS

DOI: 10.1201/9781003231745-24

24.1 INTRODUCTION

Nanoparticles (NPs) exist at the nanoscale size, which ranges from 1 to 100 nm, and have been used recently for molecular and submolecular functions, mainly electrical, optical and magnetic activity as well as structural behaviour [Sanguansri and Augustin, 2006]. NPs can be synthesized by both chemical and physical methods such as simple solution-based methods, chemical precipitation, sol-gel, solvothermal/hydrothermal, electrochemical and photochemical reduction [Sangeetha et al., 2011]. Sometimes, the size of the NPs will be maintained by using chemicals that may produce non-ecofriendly toxic by-products [Galdiero et al., 2011]. The "green synthesis of nanoparticles" method emerges as it uses biological compounds (plant samples) to produce nontoxic products [Iravani, 2019]. Metallic NPs have immense applications in the pharmaceutical industry, namely gold, silver, alloy, magnetic, platinum, zirconium, palladium, cadmium, zinc, etc. [Hanus and Harris, 2013]. Antioxidants play an important role in the functioning of all biosystems, where free radicals are generated due to the interaction of biomolecules with molecular oxygen [Rea et al., 1985]. Free radicals are responsible for the degradation of biomolecules while oxidation is also accountable for nutritional quality deterioration and discoloration of food. Consumption of oxidized foods generates lipid peroxides and low-molecular-weight compounds that cause serious diseases like hepatomegaly or necrosis of epithelial tissues [Winston et al., 1998, Regoli et al., 2000]. Hence, the antioxidant plays an important role in scavenging these toxic free radicals. Recently, many natural and synthetic antioxidants have been investigated to inhibit the oxidation reactions. The evaluation of the antioxidant activity of nanomaterials has become one of the significant basic studies in pharmaceutical science as well as in nanoscience and technology [Baldwin, 1968].

Inflammation is a complex immune response in living organisms against injury, insect bite or irritant. Inflammatory mediators such as nitric oxide (NO), inducible nitric oxide synthase (iNOS) and cyclooxygenase (COX)-2 modify the functions of tissues and organs [Wellen and Hotamisligil, 2005]. Understanding the cascade of inflammatory mediators from arachidonic acid, prostanoids have empowered researchers to adopt newer methods [Rho et al., 2020]. Therefore, the study was designed to observe the anti-inflammatory potential of green synthesized NPs from *Rubus fairholmianus*. The genus of *Rubus* that comes under the family Rosaceae is found worldwide and comprises 750 species in 12 subgenera [Dowidar et al., 2003]. The Koch-Rajbangsi and Rangia tribal community use the *R. fairholmianus* leaf decoction for early morning consumption to relieve headache and also leaves are used as an insecticide. The edible part of the fruits is used as stimulant. The ethnobotanical survey reported on the therapeutic potential of *R. fairholmianus* previously called *Rubus moluccanus* against anxiety, pain, diabetes, burn wounds, haemorrhoids, inflammation and bacterial infections [Matkowski et al., 2013]. Though it possesses many traditional medicinal uses, only a few scientific investigations were reported on this plant. Hence the study was designed to evaluate the polyphenols and the antioxidant potential of green synthesized NPs of *R. fairholmianus* against inflammation in the *in vitro* erythrocyte membrane lysis method. The results from this study will pave the way for product development using the phytocompounds with targeted delivery NPs.

24.2 MATERIAL METHODS

24.2.1 CHEMICALS

The chemicals 2,2-diphenyl-1-picrylhydrazyl (DPPH), 2,2′-azinobis (3-ethyl-benzothiozoline)-6-sulfonic acid (ABTS), trolox (6-hydroxy-2,5,7,8-tetramethylchroman-2-carboxylic acid), hydrogen peroxide, ethylene diamine tetra-acetic acid (EDTA), 2,4,6-Tris(2-pyridyl)-s-triazine (TPTZ), polyvinylpolypyrrolidone (PVPP), nitro blue tetrazolium salt (NBT), Folin-Ciocalteau phenol reagent, sodium nitroprusside and zinc nitrate were purchased from Himedia Laboratories (Mumbai, India); Sisco Research Laboratories (Mumbai, India); Sigma Aldrich (USA) and Merck (Bengaluru, India). All the chemicals and solvents used were of analytical grade.

24.2.2 Plant Collection and Extraction

The roots of *Rubus fairholmianus* were collected in November 2019 from Nilgiris, Tamil Nadu, India. The taxonomic identity of the plant was confirmed and certified by the Botanical Survey of India, Southern Regional Centre, Coimbatore, Tamil Nadu, India. The root of the plant was separated and washed under running tap water; the root was then dried and made into a fine powder by a mechanical grinder and stored at –20°C for further analysis.

24.2.3 Preparation of Plant Extract

The root powder was extracted using water; about 20 g of root powder was mixed with 200 ml of deionized water and heated at 80°C using a hot plate magnetic stirrer for 10 minutes. Then the extract was cooled, filtered and stored for the synthesis of zinc oxide nanoparticles (ZnO NPs)

24.2.4 Synthesis of Zinc Oxide Nanoparticles

About 50 mL of 0.05 M $Zn(NO_3)_2 \cdot 6H_2O$ was prepared in Millipore water and 10 mL of *R. fairholmianus* root aqueous extract was added in the solution. The mixture was stirred for two hours and adjusted to pH 12 using 0.02 M NaOH; the mixture was then centrifuged at 10,000 rpm for 10 minutes. The precipitate was purified by several redispersions in deionized water under centrifugation. The final precipitate was dried overnight in a hot oven at 60°C and calcinated in a muffle furnace at 450°C [Fakhari et al., 2019].

24.2.5 Characterization of Zinc Oxide Nanoparticles

The ultraviolet-visible (UV-Vis) spectra were recorded over a 200–800 nm range. Fourier-transform infrared (FTIR) analysis of the ZnO NPs was carried out using the KBr pellet method and the presence of functional groups was observed by the various vibrational modes. The phase structure and material identification of ZnO NPs were studied by an X-ray diffractometer. The scanning electron microscopy (SEM) technique was employed to visualize the size and morphology of the particles. Energy-dispersive X-ray spectroscopy (EDX) was employed to analyze the elemental composition of the synthesized particles.

24.2.6 Estimation of Secondary Metabolites

24.2.6.1 Total Phenolics, Tannins and Flavonoids

The secondary metabolites of the phenolic content of ZnO NPs were followed by Siddhuraju and Becker (2003), according to the Folin-Ciocalteau phenol method. The nontannin phenolic content was analyzed by the PVPP extracting method. The amount of tannins was calculated by subtracting the nontannin phenolics from the total phenolics. The flavonoid content of ZnO NPs was estimated by using the method from Zhishen et al. (1999).

24.2.7 *In Vitro* Antioxidant Assays

The *in vitro* antioxidant assays were followed by Shanmugam et al. (2018).

24.2.7.1 Phosphomolybdenum Assay

The antioxidant activity of the ZnO NPs was evaluated by the phosphomolybdenum method. We added 1 mL of reagent solution (0.6 M sulphuric acid, 28 mM sodium phosphate and 4 mM ammonium molybdate) to the test tubes and incubated in a water bath at 95°C for 90 minutes after being measured at 765 nm against a blank. The results were reported in ascorbic acid equivalents (AAE)/g extract.

24.2.7.2 Ferric Reducing Antioxidant Power Assay

The antioxidant capacities of green synthesized ZnO NPs were estimated according to the ferric reducing antioxidant power (FRAP) procedure. We incubated 900 μL of freshly prepared FRAP reagent and 100 μL of ZnO NPs at 37°C for 30 minutes, which measured at 593 nm against the reagent blank. The results were expressed in mg Fe(II) equivalent/g extract.

24.2.7.3 Metal Ion Chelating Activity

The antioxidant assays were followed and 100 μL of ZnO NPs and butylated hydroxytoluene (BHT) (standard) were added to 100 μL solution of 2 mM $FeCl_2$. With the addition of 400 μL of 5 mM ferrozine, the mixture was made up to 3 mL using deionized water, incubated at room temperature for 10 minutes, then measured at 562 nm against the blank. The results were expressed as mg EDTA equivalents/g extract.

24.2.7.4 DPPH Radical Scavenging Activity

We added 1 mL of ZnO NPs to 5 mL of a 0.1 mM methanolic solution of $DPPH^{\cdot}$ and placed it in the dark for 20 minutes at room temperature. The absorbance of the sample was measured at 517 nm against the blank. The results were calculated by percentage inhibition.

$$\%\ inhibition = [Control\ OD - Sample\ OD]/control\ OD \times 100$$

24.2.7.5 Superoxide Radical Scavenging Activity

About 3 mL of reaction mixtures (50 mM sodium phosphate buffer [pH 7.6], 20 μg riboflavin, 12 mM EDTA and 0.1 mg NBT) with 100 μL ZnO NPs in each test tube were illuminated for 90 seconds. The illuminated reaction mixture served as negative control while the mixture without extract in the dark was taken as blank. Immediately after illumination, the absorbance was measured at 590 nm. The activity was compared to rutin. The percentage inhibition of superoxide anion generation was calculated using the following formula:

$$\%\ inhibition = [Control\ OD - Sample\ OD]/control\ OD \times 100$$

24.2.7.6 Nitric Oxide Radical Scavenging Activity

We added 3 mL of sodium nitroprusside (5 mM) in 0.2 M phosphate-buffered saline of pH-7.4 to the 100 μL of ZnO NPs and incubated at room temperature for 150 minutes. The addition of 3 mL of Griess reagent (1% sulphanilamide, 2% H_3PO_4 and 0.1% N-[1-naphthyl]ethylenediamine dihydrochloride) was added and measured at 546 nm against the blank. The scavenging activity (%) was calculated as: [(Control OD – Sample OD)/Control OD] × 100.

24.2.8 *In Vitro* Anti-Inflammatory Activity by Membrane Stabilization Method

Following Shinde et al. (1999), the Alsever's solution was prepared by dissolving 2% dextrose, 0.8% sodium citrate, 0.05% citric acid and 0.42% sodium chloride in distilled water; the solution was then sterilized. Goat blood was collected and mixed with an equal volume of sterilized

Alsever's solution. The blood was centrifuged at 3000 rpm for 10 minutes and packed cells were washed three times with isosaline (0.9%, pH 7.2); the red blood cell (RBC) suspension (10%v/v) was made with isosaline solution. The reaction mixture (4.5 mL) contained 1 mL phosphate buffer (pH 7.4), 2 mL hyposaline (0.45%), 1 mL ZnO NPs (1 mg/mL) and 0.5 mL RBC suspension and was incubated at 37°C for 30 minutes and centrifuged. Diclofenac sodium was used as the reference drug. A reaction mixture without a plant sample was used as a negative control and the phosphate buffer served as a blank. The haemoglobin content in the supernatant solution was estimated spectrophotometrically at 560 nm and percent membrane stabilizing activity was calculated by using the formula: % inhibition = [(Control OD – Sample OD)/Control OD] × 100

24.2.9 Statistical Analyses

The results were expressed as mean ± standard deviation (SD). The data were statistically analyzed using SPSS version 20.0 by means of one-way ANOVA followed by the Duncan's test for antioxidant studies and Dunnett's t-test. Mean values were considered statistically significant when $p < 0.05$.

24.3 RESULTS

24.3.1 Synthesis of Zinc Oxide Nanoparticles

R. fairholmianus root extract was used for the synthesis of ZnO NPs. The synthesis was indicated by the root extract's change in colour from green to light yellow and, eventually, to dark yellow when mixed with ZnO at room temperature within two hours of incubation, after which the pure white-coloured powder was obtained (see Figure 24.1).

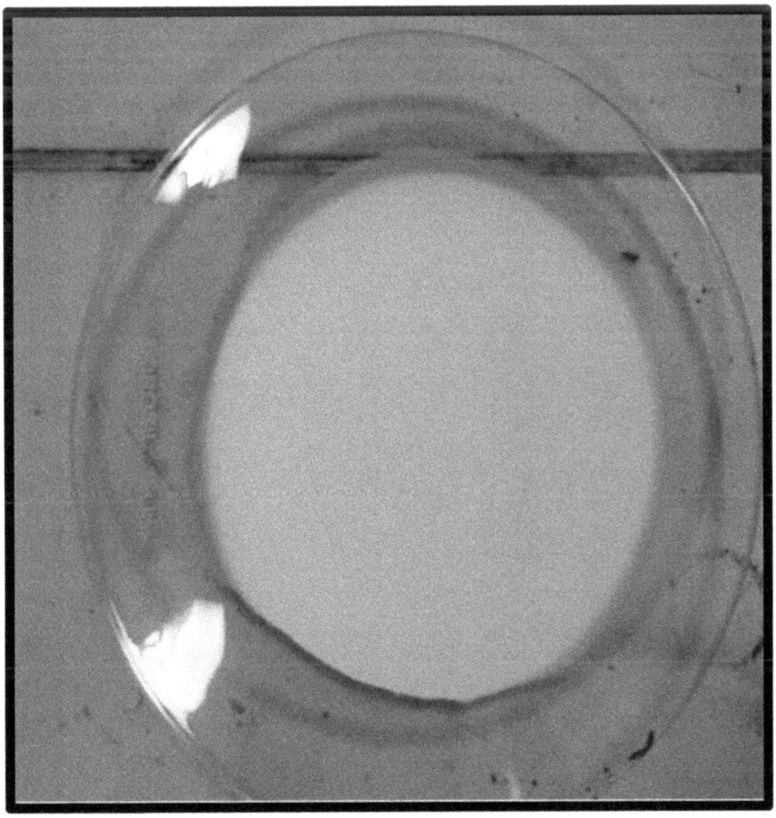

FIGURE 24.1 Synthesis of ZnO NP.

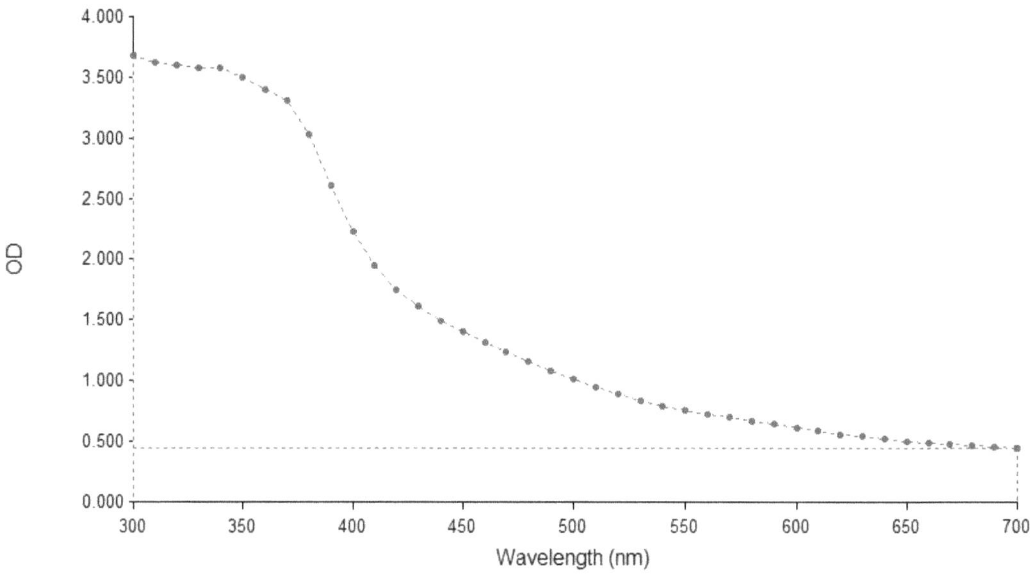

FIGURE 24.2 UV-visible spectrum.

24.3.2 Characterization of Zinc Oxide Nanoparticles

24.3.2.1 Ultraviolet-Visible Spectroscopy

UV-Vis spectroscopy analysis indicating a characteristic absorption peak of ZnO is shown in Figure 24.2. The ZnO absorption curve wavelength of 380 nm can be assigned to the intrinsic band-gap absorption of ZnO due to the electron transitions from the valence band to the conduction band, which was specific for ZnO NP. Similarly, several researchers have also observed the absorption spectrum at 380 nm.

24.3.2.2 FTIR Analysis

Infrared studies were carried out to ascertain the purity and nature of the NPs (see Figure 24.3). The spectrum of zinc nitrate NPs has also shown bands at 3386.39 cm^{-1}, 3262.97 cm^{-1}, 2933.2 cm^{-1},

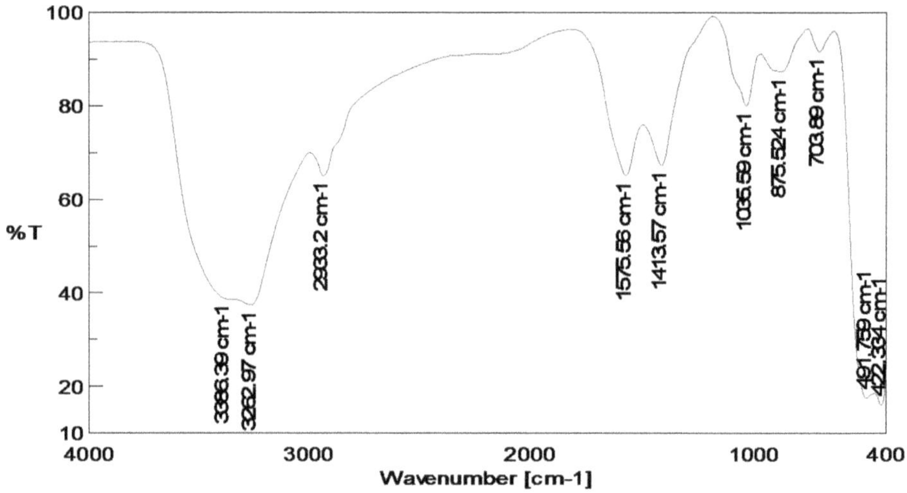

FIGURE 24.3 FTIR spectrum of ZnO NP.

FIGURE 24.4 Different magnification of SEM micrographs of ZnO NPs.

1575.56 cm^{-1}, 1413.57 cm^{-1}, 1035.59 cm^{-1}, 875.524 cm^{-1}, 703.89 cm^{-1}, 491.759 cm^{-1} and 422.334 cm^{-1}, respectively. The broad peak around 3386.39–2933.2 cm^{-1} indicates the presence of N-H strands and asymmetrical C-H stretching vibrations. The absorption peaks in the regions between 600–400 cm^{-1} confirm stretching vibrations of ZnO. The Zn–O frequencies observed for the synthesized ZnO NPs are in accordance with literature values.

24.3.2.3 Scanning Electron Microscopy

The shape, structure and size of the synthesized ZnO NPs were determined by SEM analyses (see Figure 24.4). The micrographs of ZnO NPs proved that they have a nanosized range, spherical shape and uniform distribution. The SEM results illustrated that different precursors affected the size and shape of the NPs. As it is seen when using zinc nitrate as a precursor, spherical ZnO NPs are formed and the NPs grow and accumulate to form flower-shaped bundles.

24.3.2.4 Energy-Dispersive X-ray Spectroscopy (EDX)

The elemental composition analyses of the ZnO NPs were examined from the EDX plot of the SEM images. The vertical axis displays the number of X-ray counts while the horizontal axis displays energy in KeV (see Figure 24.5). The weight percentage of zinc and oxygen in ZnO NPs has been reported to contain 34.13% zinc and 65.87% oxygen when synthesized.

24.3.2.5 XRD Analysis

The crystalline size and structural properties of the ZnO NPs are revealed by using powder X-ray diffraction (XRD). The XRD is carried out with Cu K-α radiation (k = 0.1540 nm) and 2 theta (2θ) ranges from 200 to 800. The XRD pattern of biosynthesized ZnO NPs from *R. fairholmianus* is given in Figure 24.6. XRD peaks obtained at 2θ values 31.32, 34.36, 42.58 and

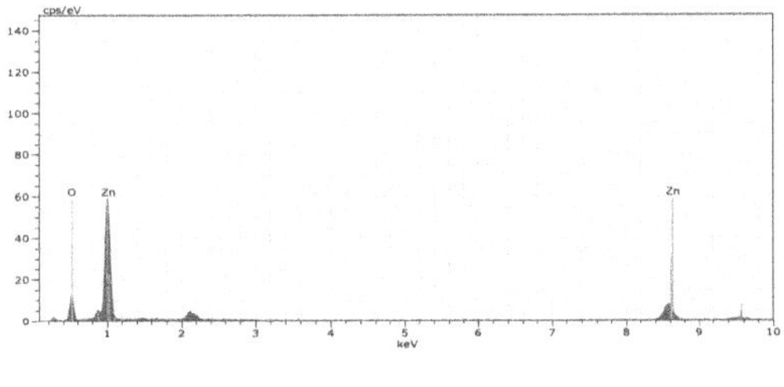

Elements	Weight %	Atom %
Zn	40.99	34.13
O	19.36	65.87

FIGURE 24.5 EDX analysis of ZnO NPs.

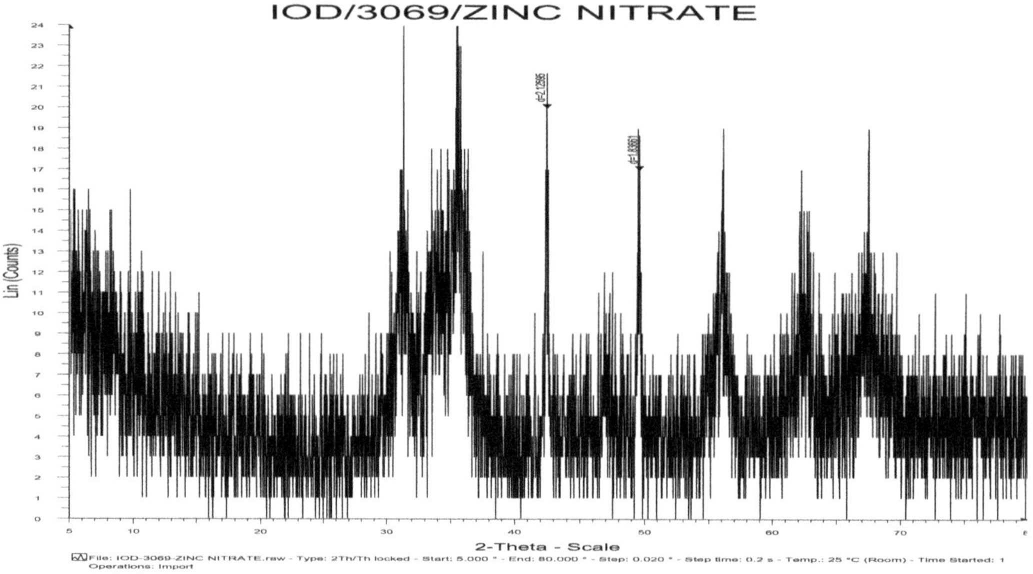

FIGURE 24.6 XRD images of ZnO NP.

62.28 corresponded to lattice plane (100), (002), (102) and (110), respectively. The sharp and narrow diffraction peaks revealed that the product is well-crystalline in nature and confirms the ZnO-hexagonal phase.

24.3.3 Estimation of Secondary Metabolites

24.3.3.1 Total Phenolics, Tannins and Flavonoids

Phenolics in food can be particularly important for consumers because of their beneficial health properties. The amount of total phenolic content of *R. fairholmianus* ZnO NPs was analyzed and is presented in Table 24.1. The total phenolic content in ZnO NPs was found to be 412.82 ± 0.27 mg GAE/g. Samples were analyzed for their tannin content and the results were expressed in Gallic acid equivalents. The ZnO NPs showed 25.06 ± 1.71 mg GAE/g of high tannin content. The flavonoid contents in the samples of *R. fairholmianus* ZnO NPs were analyzed using rutin as reference and the results are presented in Table 24.1. The ZnO NPs showed significant flavonoid content (99.38 ± 7.77 mg RE/g).

24.3.4 *In Vitro* Antioxidant Assays

The phosphomolybdenum assay is a quantitative method used to measure the reduction degree of Mo (VI) to Mo (V) by the plant extract possessing antioxidant compounds. The ZnO NPs showed

TABLE 24.1
Total Phenol, Flavonoid and Tannin Content

Sample	Total Phenol Content (mg GAE/g sample)	Total Flavonoid Content (mgRE/g sample)	Total Tannin Content (mg GAE/g sample)
ZnO NP	412.82 ± 9.27	99.38 ± 7.77	25.06 ± 1.71

Note: Mean values were considered statistically significant when $p < 0.05$.

TABLE 24.2
Phosphomolybdenum Assay, FRAP Assay and Metal Chelating Assay

Sample	Phosphomolybdenum Assay (mg AAE/g)	FRAP Assay (mg Fe (II)/g)	Metal Chelating Assay (mg EDTA/g)
ZnO NP	9.48 ± 0.36	94.33 ± 4.93	415 ± 54.08

Note: Mean values were considered statistically significant when $p < 0.05$.

9.48 ± 0.36 mg AAE/g. The ferric reducing ability of samples was estimated based on their ability to reduce TPTZ-Fe(III) complex to TPTZ-Fe(II) and *R. fairholmianus* ZnO NPs (94.33 ± 4.93 mg Fe [II]/g) showed a higher power of ferric reduction. A chelation power assay was carried out to assess the chelation capacity of the samples and illustrated that the ZnO NPs possessed remarkable chelation power of 415 ± 54.08 mg EDTA/g. The results of antioxidant activity are shown in Table 24.2.

The DPPH radical has been widely used to test the antioxidant activities of foods. ZnO NPs of *R. fairholmianus* showed 77.60% inhibition against the DPPH radical, which was considerably high in comparison with the gallic acid standard. The NO radical scavenging activities of ZnO NPs of *R. fairholmianus* were analyzed and 43.14% inhibitions were obtained. The results of superoxide anion scavenging activities of ZnO NPs revealed that the samples were found to be an efficient scavenger of superoxide radicals generated in a riboflavin-NBT-light system *in vitro*. The scavenging activity of ZnO NPs showed 60.75% inhibition compared to rutin. The results are presented in Table 24.3.

24.3.5 *In Vitro* Anti-Inflammatory Activity by Membrane Stabilization Method

The membrane-stabilizing activity of *R. fairholmianus* ZnO NPs were tested on the erythrocyte membrane. The lysosomal enzymes released during inflammation produce a variety of disorders. RBC membranes are similar to lysosomal membrane components; the prevention of hypotonicity induced RBC membrane lysis and was taken as a measure of the anti-inflammatory activity of the drugs. The results are shown in Table 24.4. The ZnO NPs showed 80.17% inhibition, which would protect the erythrocyte membrane against hypotonic induced lysis. The activity was compared to the standard diclofenac.

24.4 DISCUSSION

The synthesis of NPs using plant mediators has recently become more popular as these biosynthesized NPs are uniform and stable in nature. However, NP size, shape and formation differ among different plant species due to the reaction with metal ions and different biomolecules that are

TABLE 24.3
DPPH, Nitric Oxide Radical Scavenging and Superoxide Anion Scavenging Activities

Sample	DPPH	Nitric Oxide Radical Scavenging Activity	Superoxide Anion Scavenging Activity
ZnO NP	77.60%	43.14%	60.75%
Gallic acid	69.26%	—	—
Rutin		56.32%	74.78%

TABLE 24.4
***In Vitro* Anti-Inflammatory Activity**

Sample	Anti-Inflammatory	Diclofenac
ZnO NP	80.17%	75.16%

Note: Mean values were considered statistically significant when $p < 0.05$.

generally found in the plant extract. The high blue shift absorption for synthesized ZnO NPs in comparison with the bulk ZnO can be due to a high decrease in particle size [Bhuyan et al., 2015]. Infrared studies were carried out to ascertain the purity and nature of the NPs and also to find the presence of phytochemicals in the extract. The phytocompounds in the plant such as alcohols, phenols, amines, carboxylic acids and so on can also be capable of interacting with the zinc surface and aid in the stabilization of ZnO NPs [Singh et al., 2018]. This agglomeration process may be due to the polarity and electrostatic attraction of ZnO NPs [Fakhari et al., 2019].

The EDX spectra obtained from the analysis revealed that the required phase of Zn and O is present in the samples and confirmed as high purity for the synthesized ZnO NPs [Alamdari et al., 2020]. The presence of polyphenolics in food products is significant mainly because of their oxidative stability, contribution toward antioxidant potential and antimicrobial activity [Shahidi and Ambigaipalan, 2015]. The mechanism of plant phenolic components is capable of highly imparting antioxidant properties due to its free-radical neutralizing power and also decomposition of hydroperoxide free radicals [Lee et al., 1998]. The antioxidant activities of plant polyphenolic compounds are mainly attributed to their redox properties, which allow them to act as reducing agents, hydrogen donors, quenchers of singlet oxygen as well as possess metal chelating properties [Kasote et al., 2015]. The properties are used to develop drugs in pharmaceutical science, nanoscience and nanotechnology. Nanostructured materials are considered to have better antioxidant activity [Valgimigli et al., 2018]. DPPH is a stable nitrogen-entered free radical commonly used for analysis of compound or plant extracts to detect free radical scavenging capacity. The stable DPPH radical is reduced by accepting hydrogen or electron [Bjelakovic and Gluud, 2007]. NO is a vital bioregulatory molecule in the nervous, immune and cardiovascular systems, muscle sclerosis, arthritis and ulcerative colitis [Rees et al., 1989]. Hydrogen peroxide is a nonradical oxygen species, but occasionally may become toxic and produce hydroxyl (OH^-) radicals that can cause severe damage to the cells. Therefore, the removal of the oxides through natural antioxidants is important for the defence system of the human body [Kumar et al., 2012]. Hydrogen peroxide inside a cell at a low dose can accelerate the dissolution of NPs and produce much stronger oxidative stress [He et al., 2013]. Biosynthesised NPs can produce greater amounts of hydrogen peroxide, which may produce damage to the tissues and induce greater inflammations mainly because they can cause stronger leakage of cathepsins from impaired lysosomes and an efflux of K+ ions that may contribute to the production of superoxide and hydrogen peroxide in the membranes of mitochondria [Yang et al., 2012]. Various studies have reported that the membrane stabilization method for studying *in vitro* anti-inflammatory activity is most common and shows stabilization between erythrocyte membrane and lysosomal membrane using natural phytocompounds; the results imply that the ZnO_2-NPs may well stabilize lysosomal membranes [Schimmer and Parker, 2005].

24.5 CONCLUSION

This study is an ecofriendly approach to the economic biological process developed to synthesize spherical-shaped ZnO NPs through a green synthesis method using *Rubus fairholmianus* root extract. The characterized ZnO NPs act as an effective agent in possessing polyphenolic content and antioxidant activities. These results demonstrate that ZnO NPs can be used as effective therapeutics with multipurpose activity in biomedical applications.

ACKNOWLEDGMENT

All the authors acknowledge the RUSA 2.0 BEICH (Ref No: BU/RUSA/2019/2019/299-14) and the Department of Science and Technology, Government of India — FIST (Ref No: SR/FST/LS-II/2017/106[C]) in the Department of Botany, Bharathiar University, Coimbatore, for granting the instrument facilities during the research work.

REFERENCES

Alamdari, Sanaz, Morteza Sasani Ghamsari, Chan Lee, Wooje Han, Hyung-Ho Park, Majid Jafar Tafreshi, Hosein Afarideh, and Mohammad Hosein Majles Ara. "Preparation and characterization of zinc oxide nanoparticles using leaf extract of *Sambucus ebulus*." *Applied Sciences* 10, no. 10 (2020): 3620.

Baldwin, R. L. "Estimation of theoretical calorific relationships as a teaching technique. A review." *Journal of Dairy Science* 51, no. 1 (1968): 104–111.

Bhuyan, Tamanna, Kavita Mishra, Manika Khanuja, Ram Prasad, and Ajit Varma. "Biosynthesis of zinc oxide nanoparticles from *Azadirachta indica* for antibacterial and photocatalytic applications." *Materials Science in Semiconductor Processing* 32 (2015): 55–61.

Bjelakovic, Goran, and Christian Gluud. "Surviving antioxidant supplements." *Journal of the National Cancer Institute* 99, no. 10 (2007): 742–743.

Dowidar, Amin E., M. A. Loutfy, Ehab A. Kamel, A. H. M. Ahmad, and Hanan HL Hafez. "Studies on the Rosaceae I-seed and/or achene macro and micromorphology." *Pakistan Journal of Biological Sciences (Pakistan)* 6, no. 20 (2003): 1178–1791.

Fakhari, Shabnam, Mina Jamzad, and Hassan Kabiri Fard. "Green synthesis of zinc oxide nanoparticles: A comparison." *Green Chemistry Letters and Reviews* 12, no. 1 (2019): 19–24.

Galdiero, Stefania, Annarita Falanga, Mariateresa Vitiello, Marco Cantisani, Veronica Marra, and Massimiliano Galdiero. "Silver nanoparticles as potential antiviral agents." *Molecules* 16, no. 10 (2011): 8894–8918.

George, Blassan P., Thangaraj Parimelazhagan, and Rahul Chandran. "Anti-inflammatory and wound healing properties of *Rubus fairholmianus* Gard. root—An *in vivo* study." *Industrial Crops and Products* 54 (2014): 216–225.

Hanus, Monica J., and Andrew T. Harris. "Nanotechnology innovations for the construction industry." *Progress in Materials Science* 58, no. 7 (2013): 1056–1102.

He, Weiwei, Yu-Ting Zhou, Wayne G. Wamer, Xiaona Hu, Xiaochun Wu, Zhi Zheng, Mary D. Boudreau, and Jun-Jie Yin. "Intrinsic catalytic activity of Au nanoparticles with respect to hydrogen peroxide decomposition and superoxide scavenging." *Biomaterials* 34, no. 3 (2013): 765–773.

Iravani, Siavash. "Bio-based synthesis of magnetic nanoparticles and their applications." In *Magnetic Nanostructures*, pp. 13–31. Springer, Cham, 2019.

Kasote, Deepak M., Surendra S. Katyare, Mahabaleshwar V. Hegde, and Hanhong Bae. "Significance of antioxidant potential of plants and its relevance to therapeutic applications." *International Journal of Biological Sciences* 11, no. 8 (2015): 982.

Kumar, R. Senthil, B. Rajkapoor, and P. Perumal. "Antioxidant activities of *Indigo feracassioides* Rottl. Ex. DC. Using various in vitro assay models." *Asian Pacific Journal of Tropical Biomedicine* 2, no. 4 (2012): 256–261.

Lee, Sang Kook, Z. H. Mbwambo, H. Chung, L. Luyengi, E. J. Gamez, R. G. Mehta, A. D. Kinghorn, and J. M. Pezzuto. "Evaluation of the antioxidant potential of natural products." *Combinatorial Chemistry & High Throughput Screening* 1, no. 1 (1998): 35–46.

Matkowski, A., W. Jamiolkowska Kozlowska, and I. Nawrot. "Chinese medicinal herbs as source of antioxidant compounds—Where tradition meets the future." *Current Medicinal Chemistry* 20, no. 8 (2013): 984–1004.

Rea, P. A., J. Crowe, Y. A. B. D. Wickramasinghe, and P. Rolfe. "Non-invasive optical methods for the study of cerebral metabolism in the human newborn: A technique for the future?" *Journal of Medical Engineering & Technology* 9, no. 4 (1985): 160–166.

Rees, D. D., R. M. Palmer, and S. Moncada. "Role of endothelium-derived nitric oxide in the regulation of blood pressure." *Proceedings of the National Academy of Sciences* 86, no. 9 (1989): 3375–3378.

Regoli, Francesco, Marco Nigro, Stefano Bompadre, and Gary W. Winston. "Total oxidant scavenging capacity (TOSC) of microsomal and cytosolic fractions from Antarctic, Arctic and Mediterranean scallops: Differentiation between three potent oxidants." *Aquatic Toxicology* 49, nos. 1–2 (2000): 13–25.

Rho, Taewoong, Hyun Woo Jeong, Yong Deog Hong, Keejung Yoon, Jae Youl Cho, and Kee Dong Yoon. "Identification of a novel triterpene saponin from *Panax ginseng* seeds, pseudo ginsenoside RT8, and its anti-inflammatory activity." *Journal of Ginseng Research* 44, no. 1 (2020): 145–153.

Sangeetha, Gunalan, Sivaraj Rajeshwari, and Rajendran Venkatesh. "Green synthesis of zinc oxide nanoparticles by *Aloe barbadensis* miller leaf extract: Structure and optical properties." *Materials Research Bulletin* 46, no. 12 (2011): 2560–2566.

Sanguansri, Peerasak, and Mary Ann Augustin. "Nanoscale materials development—A food industry perspective." *Trends in Food Science & Technology* 17, no. 10 (2006): 547–556.

Schimmer, B. P., and K. L. Parker "Adrenocorticotropic hormone; adrenocortical steroids and their synthetic analogues; inhibitors of the synthesis and actions of adrenocortical hormones." In J. G. Hardman and L.E. Limbird (eds.), *Goodman & Gilman's the Pharmacological Basis of Therapeutics*, 11th ed. United Kingdom: McGraw-Hill (2005): 2314–2342.

Shahidi, Fereidoon, and Priyatharini Ambigaipalan. "Phenolics and polyphenolics in foods, beverages and spices: Antioxidant activity and health effects–A review." *Journal of Functional Foods* 18 (2015): 820–897.

Shanmugam, Saravanan, Isla Alcântara Gomes, Marina Denadai, Bruno dos Santos Lima, Adriano Antunes de Souza Araújo, Narendra Narain, Maria Terezinha Santos Leite Neta, Mairim Russo Serafini, Lucindo José Quintans-Júnior, and Parimelazhagan Thangaraj. "UHPLC-QqQ-MS/MS identification, quantification of polyphenols from *Passiflora subpeltata* fruit pulp and determination of nutritional, antioxidant, α-amylase and α-glucosidase key enzymes inhibition properties." *Food Research International* 108 (2018): 611–620.

Shinde, U. A., A. S. Phadke, A. M. Nair, A. A. Mungantiwar, V. J. Dikshit, and M. N. Saraf. "Membrane stabilizing activity—a possible mechanism of action for the anti-inflammatory activity of *Cedrus deodara* wood oil." *Fitoterapia* 70, no. 3 (1999): 251–257.

Siddhuraju, Perumal, and Klaus Becker. "Studies on antioxidant activities of mucuna seed (*Mucuna pruriens* var utilis) extract and various non-protein amino/imino acids through *in vitro* models." *Journal of the Science of Food and Agriculture* 83, no. 14 (2003): 1517–1524.

Singh, Priyanka, Abhroop Garg, Santosh Pandit, V. R. S. S. Mokkapati, and Ivan Mijakovic. "Antimicrobial effects of biogenic nanoparticles." *Nanomaterials* 8, no. 12 (2018): 1009.

Valgimigli, Luca, Andrea Baschieri, and Riccardo Amorati. "Antioxidant activity of nanomaterials." *Journal of Materials Chemistry B* 6, no. 14 (2018): 2036–2051.

Wellen, Kathryn E., and Gökhan S. Hotamisligil. "Inflammation, stress, and diabetes." *Journal of Clinical Investigation* 115, no. 5 (2005): 1111–1119.

Winston, Gary W., Francesco Regoli, Alton J. Dugas Jr, Jessica H. Fong, and Kristie A. Blanchard. "A rapid gas chromatographic assay for determining oxyradical scavenging capacity of antioxidants and biological fluids." *Free Radical Biology and Medicine* 24, no. 3 (1998): 480–493.

Yang, Eun-Jeong, Seungjae Kim, Jong Soo Kim, and In-Hong Choi. "Inflammasome formation and IL-1β release by human blood monocytes in response to silver nanoparticles." *Biomaterials* 33, no. 28 (2012): 6858–6867.

Zhishen, Jia, Tang Mengcheng, and Wu Jianming. "The determination of flavonoid contents in mulberry and their scavenging effects on superoxide radicals." *Food Chemistry* 64, no. 4 (1999): 555–559.

Index

A

B

C

D

E

F

U

V

W

X

Z

For Product Safety Concerns and Information please contact our EU
representative GPSR@taylorandfrancis.com
Taylor & Francis Verlag GmbH, Kaufingerstraße 24, 80331 München, Germany

www.ingramcontent.com/pod-product-compliance
Ingram Content Group UK Ltd.
Pitfield, Milton Keynes, MK11 3LW, UK
UKHW052216190926
475702UK00006B/1036
9781032137285